Public Private Partnerships in the Water Sector

Public Private Partnerships in the Water Sector

Innovation and Financial Sustainability

X. Cledan Mandri-Perrott and David Stiggers

Publishing
London • New York

Published by **IWA Publishing**
Alliance House
12 Caxton Street
London SW1H 0QS, UK
Telephone: +44 (0)20 7654 5500
Fax: +44 (0)20 7654 5555
Email: publications@iwap.co.uk
Web: www.iwapublishing.com

First published 2013

British Library Cataloguing in Publication Data
A CIP catalogue record for this book is available from the British Library

ISBN 9781843393207 (Paperback)
ISBN 9781780401058 (eBook)

Contents

About the Authors

The Authors have collaborated on a wide variety of projects and schemes related to major infrastructure development and use of Public-Private Partnerships (PPP) over the last 20 years. These have ranged from development and improvement of utility operations, establishing high level national and regional infrastructure policy and development of associated schemes, including developing associated public and private funding mechanisms. They have authored a variety of papers and books on various aspects of PPP in Infrastructure. They both have a wide experience of water sector issues and have demonstrated and implemented successful and effective operational and capital investment solutions to infrastructure challenges. This book is a distillation of this knowledge and experience, linking practical use of Private Sector Participation to more effective financing, development and operation of Water and Waste Water sector and related schemes and utilities.

X. Cledan Mandri-Perrott is a Lead Infrastructure and PPP Specialist in the Singapore Global Infrastructure Finance Centre of the World Bank. Cledan is a member of the World Bank's Global Expert Team on Public-Private Partnerships and has been involved over nearly 20 years in numerous infrastructure PPP projects. He has recently been recognized by the World Bank through the Award of Excellence for his particular work on an innovative PPP arrangement for Irrigation, as well as some cutting edge financing structures in Brazil.

Cledan combines an engineering, financial and legal background with sound commercial and legal expertise. Prior to joining the World Bank, he has advised both Governments and Private companies on issues related to public private partnerships. He has a strong regulatory background and has designed and implemented regulatory policy frameworks.

He has practical operational experience having managed a utility in Trinidad and Tobago. Furthermore, he has assisted a number of utilities in evaluating the potential benefits of privatization, with associated capital investment ranging from US$20 million–US$1.2 billion. These have been in a broad range of sectors including water and waste water, light rail transit, road sector and energy sector, and in a diverse spectrum of developed and developing economies. He has been instrumental in promoting the development of the appropriate PPP solutions and their implementation. He has gained a reputation as a pragmatic problem solver helping resolve regulatory disputes and PPP contracts in distress, with his solutions providing a strong framework for a practical and continuing approach for future successful service provision. He has authored operating and service agreements for the provision of basic infrastructure services. Cledan has written extensively on issues of PPP and infrastructure and is a regular presenter and trainer on PPP related topics. He is regularly consulted on this area of expertise within the World Bank and by their Clients.

Cledan holds a PhD in Project Finance, an LLM in International Law, MSc in Finance and a Bachelors of Civil Engineering with honours.

David Stiggers is an Independent Private Sector Participation and Infrastructure Specialist He has over 35 years experience in the international infrastructure, finance and utility business. He has been involved in most aspects of the infrastructure sector, making public sector projects happen more effectively. Initially he worked as technical & economic consultant on large regional schemes, then as a project developer with a major international contractor and followed this by working for a major private operator on international utility business improvement and development projects. For a period at the World Bank he was Task Team Leader dealing with development projects in all stages from project concept through to loan and project management, all involving some element of Public-Private Partnership (PPP). He now acts as independent adviser in this sector.

His involvement in PPP projects has spanned the last 20 years, including experience in project feasibility, development and funding; working as Transaction Adviser on a variety of Management Contract, Lease and Concession Contracts; and reviewing potential PPP reform, including commercial, operation, legal, financial and regulatory issues. His work with an international private operator included work establishing and then, as Director, managing one of the first major World Bank sponsored Management Contracts, in the Trinidad water and sanitation sector. He brings considerable, wide ranging and practical international experience including work in Middle East, South Asia, Ethiopia, Pakistan, East and West Africa, Eastern Europe, European accession countries, Latin America and the Caribbean.

His current work involves provision of specialist high level advisory services to governments and regional utilities, particularly in implementing effective PPP and infrastructure strategy options. Recent projects have ranged from provision of national sector reviews of PPP options (e.g. Armenia, Cambodia, ECA transport sector), project development & transaction advisory work for PPP utility projects (e.g. Jordan, Uganda, Albania), water utility performance improvement, managing workshops in PPP infrastructure development for ministers and decision makers (e.g. Bosnia, ECA finance ministers in Washington, Honduras, China, Pakistan, Dubai), and editor and contributor to a World Bank book on PPP in the transport sector. Recently he was adviser to the Ethiopian Government on development and practical implementation of an innovative PPP arrangement to bring irrigated agriculture to a population of over 100,000.

David holds a BA (Systems and Organization) and is a Chartered Engineer (Civil Engineering and also in Water & Environmental Management).

Acronyms & Abbreviations

Acronym or Abbreviation	Expanded Definition
AP	Availability Payment
BAFO	Best and Final Offer
BOO	Build Own Operate
BOOT	Build Own Operate and Transfer
BOT	Build Operate Transfer
BTO	Build Transfer Operate
CIPX	Consumer Price Index - excluding mortgage interest payments
Commission	Means the European Commission
CP	Conditions Precedent of a PPP agreement
CPI	Consumer Price Index
CSFB	Credit Suisse First Boston
DBFM	Design Build Finance Maintain
DBFO	Design Build Finance Operate
DBOM	Design Build Operate Maintain
DCMF	Design Construct Manage Finance
DFID	Department for International Development (UK)
DLR Ltd.	Docklands Light Railway Limited (DLR's contracting authority)
DSCR	Debt Service Coverage Ratio
EC	European Commission
EU	European Union
Ex-Im	Export-Import

GNI	Gross National Income
GPOBA	Global Partnership on Output Based Aid
GSP	Generalized System of Preferences
ICSID	The International Centre for Settlement of Investment Disputes
ICT	Information and Communication Technologies
IFC	International Finance Corporation
IFI	International Funding Institution
IPP	Independent Power Producer
IPPPs	Institutionalised Public Private Partnerships
IPR	Intellectual Property Right
IRR	Internal Rate of Return
KPI	Key Performance Indicator
LIBOR	London Interbank Offered Rate
M&E	Monitoring and Evaluation
NPV	Net Present Value
O&M	Operation and Maintenance
OBA	Output Based Aid
OECD	Organization for Economic Co-operation and Development
PFI	Private Finance Initiative
PPIAF	Public Private Infrastructure Advisory Facility, a multi donor trust fund in support of PPP
PPP	Public-private partnership
PQQ	Pre-Qualification Questionnaire
PSP	Private Sector Participation
RfP	Request for Proposals
RPI	Retail Price Index (UK)
RPIX	Retail Price Index (UK - excluding mortgage interest payments)
UK	United Kingdom
ULBs	Urban Local Bodies
UNCITRAL	The United Nations Commission on International Trade Law
UNDP	United Nations Development Programme (UNDP)
US	United States
USAID	United States Agency for International Development
VfM	Value for Money
WB	World Bank
WSPF	Water and Sanitation Pooled Fund
WSS	Water and Sanitation Services
WTO	World Trade Organisation

List of Boxes

List of Tables

List of Figures

Chapter 1
Introduction

The objective of this book is to determine and make recommendations for optimizing the use of Public Private Partnership (PPP) in developing infrastructure whilst ensuring the sustainable long term provision of Water and Waste Water Services. The outcome is focused on providing detailed recommendations on contractual issues and contract structures to achieve this objective. These recommendations are primarily aimed for use in developing countries but may of course be applied in other circumstances.

The use of the term Developed or Developing country is hotly debated and there is no precise definition[1]. Different classifications are used by different institutions such as for example, the World Bank which uses a strictly numerical classification, considering all low- and middle-income countries as 'developing'. These categories (low, middle, high income) are based on the Bank's operational lending categories (World Bank, 2009). In its most recent classification, economies are divided using 2008 Gross National Income (GNI) per capita, and divided further into four income groups. Low income countries have GNI per capita of US$975 or less. Lower middle income countries have GNI per capita of US$976–$3855. Upper middle income countries have GNI per capita between US$3856–$11,905. High income countries have GNI above $11,906. So, while the World Bank classifies all low- and middle-income countries as developing it notes, 'The use of the term is convenient; it is not intended to imply that all economies in the group are experiencing similar development or that other economies have reached a preferred or final stage of development. Classification by income does not necessarily reflect development status'. The International Monetary Fund, on the other hand, uses a flexible classification system that considers "(1) per capita income level, (2) export diversification—so oil exporters that have high per capita GNI would not make the advanced classification because around 70% of its exports are oil, and (3) degree of integration into the global financial system." According to the classification from IMF before April 2004, all the countries of Eastern Europe (including Central European countries which still belonged to 'Eastern Europe Group' in the UN institutions) as well as the former Soviet Union (U.S.S.R.) countries in Central Asia (Kazakhstan, Uzbekistan, Kyrgyzstan, Tajikistan and Turkmenistan) and Mongolia, were not included under either developed or developing regions, but rather were referred to as 'countries in transition'. However, they are now widely regarded (in international reports) as 'developing countries'. In the 21st century, the original East Asian Tiger countries (Hong Kong, Singapore, South Korea, and the Taiwan) are considered 'developed' regions, along with Cyprus, Israel, Malta, and Slovenia which are considered 'newly developed countries'. Moreover, it is interesting to note that the United Nations states that there is no established convention for the designation of 'developed' and 'developing' countries and that it is a classification system intended for statistical convenience not necessarily to express a stage reached by a particular country or area in the development process. Nevertheless, Japan in Asia, Canada and the United States in northern America, Australia and New Zealand in Oceania, and Europe are considered by the UN to be 'developed' regions or areas. In international trade statistics, the Southern African Customs Union is also treated as a developed region and Israel as a developed country, whereas countries emerging from the former Yugoslavia are treated as developing countries and Eastern European countries and the Commonwealth of Independent States in Europe are not included under either developed or developing regions. The World Bank classification of developing countries is used in this context.

It became apparent that one risk issue, the problem of imperfect data, is not adequately addressed in the existing contractual approach to PPPs and has resulted in failure or ineffectiveness of many PPP schemes, and so a specific recommendation on a contractual remedy for this has been included.

This Chapter presents an Introduction which starts with a background note on the major problems resulting from the increasing demand for more and better Water and Waste Water Services and the associated and massive demands for investment and effective management. The public sector funding and resources are often inadequate to meet these demands, and a growing case history shows increasing involvement by the private sector in provision of infrastructure and services through PPP Arrangements. The aim of this book is to identify best practice in legal and contractual arrangements, and identify ways that these can be best structured and procured to ensure long term viability and sustainability of effective provision of Water and Waste Water Services.

In order to establish some key determinants of a PPP Arrangement, some of the key PPP and financing issues affecting both public and private sectors have been looked at. These include the roles of the public and private sector partners, allocation of risks, project financing issues, the broad types of PPP contractual arrangements, as well as the issue of ownership of all or part of the assets.

The methodology is developed from international case studies, literature and contractual document reviews, expert opinion and fact finding mission reviews. The approach is based on:

- Review of Water and Waste Water Sector issues, and general review of use of PPPs, taking into account levels of service and cost implications.
- Establishing key issues in effective PPP development.
- Review and recommendation of detailed contractual issues for optimal and sustainable long term provision of Water and Wastewater Service.

Finally:

- Recommendations to improve critical contract provisions including maintaining the economic and financial equilibrium of the PPP Agreement.
- Recommendation of an approach for dealing with limited or unreliable data, to ensure long term contractual sustainability.
- Suggestions for improved contractual structures.

In any reform process in the Water and Wastewater Sector there are a number of terms used to describe the main players and their relationships. For consistency, throughout this book, the public sector or government entity buying the service to be developed under any specific form of PPP is referred to as the Grantor, Contracting Authority or Government. In those instances of general policy setting, the reference is made to Government. However, the entity responsible for establishing the PPP scheme can either be Government or a Grantor, or a delegated entity responsible for awarding and managing the Agreement. The public sector counterpart is referred to as the Developer. The Developer is a private party responsible for all or part of the development, construction and operation of the PPP scheme. The contractual arrangement detailing the roles and responsibilities of the Grantor and the Developer is referred to as the PPP Agreement.

1.1 PUBLIC PRIVATE PARTNERSHIPS – SOLUTIONS TO MEET INFRASTRUCTURE INVESTMENT AND MANAGEMENT IMPROVEMENT NEEDS

The infrastructure challenge

It could be said that the provision of physical infrastructure to supply public water and wastewater services has been at the core of the development of many great civilisations. Massive water irrigation works date back to the times of the Egyptians, Greeks, Aztecs and Incas. In modern times the provision of Water and Wastewater Services has been a necessary pre-requisite for the effective economic and social development of modern urban centres. Yet, over the past 50 years, governments have neglected investment in these basic infrastructure services. Not only has this resulted in an inability to meet a growing demand for services and for higher service standards, but the existing water and wastewater structures and networks are deteriorating as they age and lack investment in essential structural maintenance. As a result, they require even more investment to restore them to existing standards. The effect is a huge backlog of investment needs. A recent report by the European Commission estimates that for the ten countries of central and Eastern Europe alone, the cost of all investments needed to comply with European environmental requirements would exceed euro 120 billion in the next five years. In the United States alone investments in basic water and waste water infrastructure are estimated to be over US dollars 325 billion today, and the rest of the world's investment needs (Figure 1.1) estimated at US dollars 1.8 trillion (World Bank, PPIAF PPI Database, 2009).

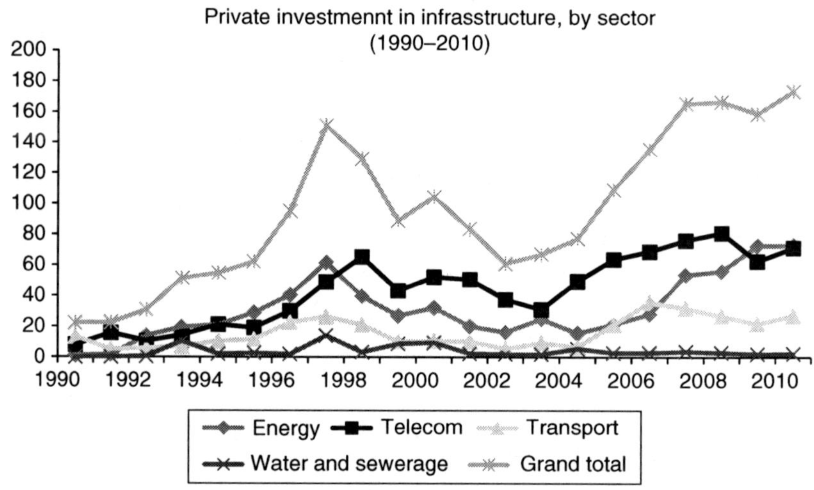

Figure 1.1 Diagrammatic representation of the research methodology. *Source*: World Bank and PPIAF, PPI Project Database.

The lack of investment in public water and wastewater services is not just limited to renewal or expansion of existing systems, but often includes limitations on public funding for the management, operation and maintenance of the systems. Several studies (Johnstone, 2007; Whittington, 1994; Vives, 1999) also suggest that, under public management, water utilities are sometimes exposed to short-term political agendas that restrict available funding. This situation, coupled with governments' inability to fund massive investments required for upgrading infrastructure, has made it apparent that new and innovative structures need to be devised to 'bridge' these gaps in funding and management. A principal option for national and local governments to help close those gaps is to turn to the private sector for help. This has been seen in developing countries which increased from US$25 billion up to a staggering US$297 billion (World Bank, PPIAF PPI

Database, 2011). Of this amount, water was a very small percentage of private investment averaging only 4% over the last two decades.

The provision of a public service requires a careful balance between the potentially divergent interests of the public and private sectors[2]. On the one hand, the public sector has a responsibility and an obligation to provide services to its inhabitants and ratepayers. At the same time the private sector is entitled to receive adequate compensation for the risk of managing the provision of that infrastructure service, as well as being compensated for putting its Private Capital at risk. Since 1989 we have seen a surge in the number of PPPs. The considerable case history that has been amassed shows that these PPP arrangements can be an effective tool for Water and Waste Water Scheme development and long term management and operation. However, it is also clear that there is a considerable range in the effectiveness of these PPP arrangements, including some outright failures.

We should be clear that the PPP approach is not the only potential solution for effective reform and provision of Water and Waste Water Services. Policymakers and governments need to find the most adequate solution and mechanism to allow for public reform. In this respect, and in an effort to improve the provision of public services, a number of public sector reform strategies are being developed. For example, instruments for public reform might include the creation of autonomous public service organizations, the development of industry service agreements to ensure accountability, and the growing use of financial incentives for managers and staff.

Similar reform strategies are necessary to implement PPP Arrangements for private provision of public services and increasingly, the additional benefits of including the private sector as part of the reforms for provision of these basic infrastructure services, are seen as offering a practical and effective alternative to public sector management.

In the absence of a uniform legal framework for PPPs and in view of the diversity of national laws, the challenge is to ensure transparency and effective competition between market participants, while at the same time ensuring the most effective provision of the Water and Waste Water Services.

Public private partnerships – a possible solution

Countries in the developing and developed world are increasingly interested in creating arrangements that will enable them to better meet the challenges of developing basic infrastructure. Public interest to cooperate with the private sector is driven by budgetary constraints as well as the search for increased efficiency, in terms service (such as quantity and quality of water and wastewater) and the allocation of finite financial resources.

It has been noted previously that in the last two decades there has been a general surge to use PPP arrangements for provision of public services. The water and sanitation sector has gained considerable PPP experience during this period, but many other infrastructure sectors linked to public service provision have also made use of the PPP approach. There is considerable variation of capital investment, materials and technology transfer, management and operational resources involved in each sector. PPP Arrangements have been used in a variety of forms in infrastructure sectors including heavy capital investment in projects related to Ports, Airports, Roads, Bridges and Power Generation and Distribution projects and have also included more highly service oriented PPP Arrangements such as those in Education and Health services projects.

PPPs can take many forms. In its simplest form it refers to a cooperation between a public authority or authorities and the private sector (Kerf *et al.* 1998). Its primary aim is to fund, construct, renovate, manage and/or, maintain a given piece of infrastructure or providing a service. A PPP is a cooperative partnership between the private sector and the government, which can contribute to the economic growth and quality of life in urban centres. Involving the private sector in the delivery of water services can also serve to eliminate the problems that typically blight public sector run utilities, such as patronage and short-termism. Efficient, cost-effective and sustainable water delivery systems require long term planning strategies and systems with government support. Crucially, private sector involvement in water services broadens the range of policy options available to the government.

PPPs are usually relatively long term arrangements, involving cooperation between the public partner and the private partner on different aspects of a planned project. Funding generally involves a combination of sources, both public and private, which in some instances means complex arrangements between the various players. Notably, the role of the private sector is as a Developer, who participates at different stages in the project for example design, completion, implementation and long term management, operation and funding (Klein & Hadjimichael, 2003). In its recent interpretative communication the European Union (2008) defines PPPs as a form of cooperation generally geared to the longer term, where the role of the private partner (the Developer), who is involved in the various phases of the project (planning, implementation and operation), is intended to bear risks that are traditionally borne by the public sector and who often contributes to financing the project. It should be made clear that the term Developer is used throughout to describe an 'economic operator' able to perform all these functions and not solely responsible for the technical operation of plant and equipment. In some of the literature reviewed the word 'operator' is used without this distinction being made. This distinction is important as it reflects the key contractual responsibilities under PPP Arrangements. The public partner or Grantor focuses primarily on defining the objectives to be attained in terms of public interest, quality of services provided and pricing policy, and it takes responsibility for monitoring the compliance of these objectives.

For the PPP Arrangement to be effective, the design of the structure for private sector involvement involves a clear definition of the roles to be played by the government and the private sector, the type of structure to be adopted for private sector participation, the applicable regulatory regime and its effects, and also the influence of the specific current economic and political conditions. The type of PPP structure chosen will have a direct influence on the provision of services and it is essential that an adequate legislative and regulatory framework be in place to ensure the long term sustainability of the chosen PPP Arrangement.

The PPP design should seek to strike a balance between establishing incentives for the private sector to invest and operate efficiently, and protecting the interests and responsibilities of the public sector (Camdessus & Winpenny, 2003). Additionally, the types of risks involved in the provision of water and wastewater services needs to be adequately quantified, qualified and, most importantly, mitigated, as these are elements that will determine the success of a PPP project. All risks must be properly understood and allocated in an equitable manner between all parties (Cowan, 1997a). This is a key issue and will be looked at in some detail further on.

The allocation of risk between the parties is generally based on the specific risks to be taken by the party best able to manage them, and within the overall context to suit the conditions relevant to the specific project, sector and economy. Projects which involve PPPs need to be 'bankable' (Stiggers, 2004). A 'bankable' project is one that is considered to combine key characteristics in a way that gives an equitable balance of risks and rewards both for the public and the private sector partners and that the overall result will include overall human, social, economic and environmental sustainability. Typically, projects considered bankable include characteristics such as clear contractual structures, a stable economic and financial basis and a clear definition of roles and responsibilities (Hodges & Dellacha, 2007; UNIDROIT, 1995; Salacuse, 2000; Shirley & Patrick, 2000).

Generally PPP projects are based on project finance[3] in which a private operator is granted a right and, in return for that right, has the obligation to provide a service (Seidenstat & Hakim, 2002). Under a project finance scheme, the right granted by the public sector includes the right to expropriate, use, treat and sell water, in exchange for charging tariffs to consumers for such services. The types of PPP structures are defined in Chapter 3.

Risks, asset responsibility and types of PPP structures

The array of options open to the private sector wishing to be involved is characterised by a variety of different kinds of contractual arrangements. These contractual forms differ in the degree of responsibility and risk transferred from the public to the private sector. Choosing the appropriate option depends on what the government in question requires from the private sector, that is the objectives to be satisfied by entering into such contractual agreements, and, as a result, how much control of the service it is prepared to relinquish (Smith, 1997a). The different forms of PPP examined here are based on the degree of private involvement in the development and ownership of the assets.

The idea is to assess the ways in which the PPP contractual arrangements may be developed to optimize the long term project viability. The PPP structures reflect not only the allocation of risk, but also the manner in which asset ownership is dealt with. There are two principal ways:

- The first group, where ownership of the assets remains with the government or public sector, usually includes the following type of contracts: Service and Management contracts, Lease arrangements and Concessions. Each type of contract will be reviewed according to the amount of private involvement.
- The second group, where the assets are all or partially in private ownership, include: BOOT (Build-Own-Operate-Transfer), variations to BOOT contracts such as Design-Build-Finance-Operate-Maintain (DBFOM), BOT (Build-Operate-Transfer), BOO (Build-Own-Operate) and reverse BOOT, joint ownership or mixed companies, and outright sale, or partial divestiture.

1.2 CHARACTERISTICS OF URBAN AND PERI-URBAN WATER AND WASTEWATER SERVICES

The use of PPP in the water and wastewater sector has been chosen specifically to illustrate these kinds of contracts. Although this sector exhibits most of the characteristics of PPP in other sectors, the water and wastewater sector is unique in that the service provision is seen to be a public good providing a service that is essential to life and health. In every country of the world the provision of water and wastewater services is deeply ingrained as a public service obligation. As will be discussed later, major issues in the sector are poor existing infrastructure, massive demand growth and high capital investment needs. Over the past two decades, the use of PPP in this sector has been adopted as a possible solution to bridge the gap in resources and funding. Whatever form of PPP Arrangement is developed (and even if there is no PPP) these public service obligations cannot be divested to the private sector one hundred percent, and it is this that differentiates the approach needed to ensure viable long term PPP Arrangements that will reflect the long term public service objectives in the water and wastewater sector. Ultimately, even under a PPP Arrangement, Governments cannot countenance an ultimate failure in the sector arrangement. Therefore, it is to be understood that Water and Wastewater PPPs are more about delegated responsibilities rather than divesting them, unlike the focus of other PPP infrastructure arrangements. As a consequence, the characteristics of this sector have a profound influence on the structures developed.

The current state of play in the water sector

Investment in private water projects has changed significantly in the last decade. As mentioned above water investment has been in the range of 2–3US$ billion since 2001, well below the peak of 1997 but within the US$2–3 billion range of the previous three years. One explanation for this has been the reduction of the size of water PPPs mainly in the form of concessions and divestitures. In the 1990′s, there were several large tenders for water utilities (Figure 1.2). The peak in 1997, was due to the signing of two large water concessions for Manilla's water utility. The concessions were divided into Manila East (US$2.9 billion) and Manilla West (US$4.5 billion). Manila West has since been canceled. In the previous decade, there were also several large water concessions in Latin America. Chile partially divested Aguas Andinas in 1999 to Aguas de Barcelona, with a share purchase of over US$1 billion and investment commitments reaching US$2 billion, the total project cost amounted to US$3.1 billion (World Bank, PPIAF PPI Database, 2011).

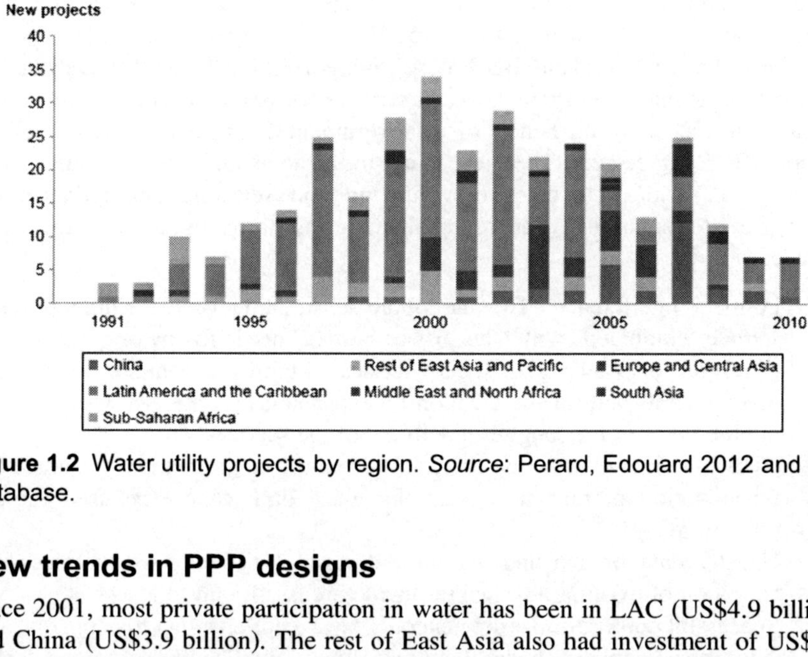

Figure 1.2 Water utility projects by region. *Source*: Perard, Edouard 2012 and PPI Database.

New trends in PPP designs

Since 2001, most private participation in water has been in LAC (US$4.9 billion) and China (US$3.9 billion). The rest of East Asia also had investment of US$3.9 billion (Perard, 2012) As the chart indicates, in the last three years, the number of projects has also been declining across all regions with the exception of Latin America. There have been several new developments, some of which are not yet reflected in the data. In 2011, through the federal water commission Conagua, has raised the level of private participation by creating a PPP scheme that supports large scale water projects. Conagua provides a subsidy for 49% of the project cost in the form of a capital grant and the remaining financing is provided by the private sector. With this PPP modality, Mexico has financed the largest wastewater treatment plant in the world, Atotonilco with investment commitments of US$793 million and a capacity of 1,990,000 cubic meters per day. In 2011, the US$214.8 million El Realito Acueduct also reached financial closure, with investors ICA of Mexico and FCC, Spain. In El Realito, the subsidy was funded by municipal trusts that earmarked 50% of the payroll tax to recover the debt and equity provided by the sponsor. Operational costs, fixed and variable are covered by Interapas, an inter-municipal authority covering the state of San Luis Potosi. The project received Latin America Water Deal of the year by Project Finance Magazine.

In 2011, Peru financed the Taboada Wastewater Treatment plant with a unique instrument that uses payment certificates backed by the national water authority to issue bonds, called a RPICAO. In February 2011, ACS closed a S./942 million ($340 million) local currency bond financing for the project. The issuance used RPICAOs, certificates with payment obligations for completed construction milestones up to 2013, as collateral for the issuance. The bonds stretched the ACS payment obligations by maturing in 2025. The financing was led by BNP Paribas, and received a AAA local and BBB global rating from Fitch (Jett, PPI Database).

Decisions to pursue a PPP in the water services sector require considerable initial preparation and study on behalf of a Government acting as the grantor of the proposed PPP Agreement. There are many useful texts and aids to guide readers through the various processes of analyzing and selecting appropriate water and wastewater solutions[4]. The key challenges and problems faced by the sector include:

- Political interference: The monopolistic situation of the single provider service combined with the basic human need for water, makes the provision of water and waste services highly susceptible to political involvement, both in the creation or expansion of water services, but also in all aspects of the long term delivery of the services.
- High investment costs for basic infrastructure (e.g. creation, expansion or rehabilitation of pipe networks) that once built cannot be used for any other purpose.
- Usually data on existing services, demand satisfaction for services and condition of existing physical infrastructure is difficult to assess.
- Successful public utilities are generally the exception rather than the rule and, since most people are under the jurisdiction of public utilities, a good deal of the world's population remains inadequately served by their water and wastewater providers (Baietti, Kingdom & van Ginneken, 2006a).
- Many public utilities suffer from weak performance incentives, low willingness of customers to pay cost recovery tariffs, and insufficient funding for maintenance, leading to a deterioration of assets and the squandering of financial resources (Baietti, Kingdom & van Ginneken, 2006a).
- The monopolistic nature of the sector along with weak and misdirected policies exacerbates the situation.

- Water, by its very nature is a politically sensitive topic and policymakers have tried and, more often than not, failed to adequately balance the trade-offs between affordability and expansion of coverage to poorer sections of society against the utility's need for financial viability.
- Water and wastewater resources are more generally suited to local management and in many cases, responsibility for these services is decentralized to local government levels.
- Policymaking in the sector is characterized by short-term political interests.
- Governments face strong pressure to keep prices below costs, and to have to subsidize the difference between prices and costs. The biggest challenge for governments, either with public or private operation, is to address these problems in a way that will encourage investment to improve services.
- Corruption is also a persistent challenge within the sector deriving from collusion between contractors and kickbacks in contracting (van Ginneken & Kingdom, 2008).

Successful incorporation of private sector participation into this sector depends largely on the chosen PPP Arrangement's ability to handle the above mentioned challenges. Experience shows that efforts to reform the Water and Wastewater Sector often face considerable hurdles in the form of opposition from those most likely to be affected by the impact of reforms. Reforms are also often triggered by crises that have created a consensus of opinion that the current situation is untenable. Such events may be droughts, drops in service levels or a financial crisis. Van Ginneken (2008) argues that external pressures may also trigger reforms, such as donor pressure or political forces as in elections. The last ten years has seen a renewed effort by governments to reform their Water and Wastewater Sectors but the vast majority have met with limited success. In some cases, systems have been 'overbuilt' and do not reflect the true needs of consumers nor what they are willing to pay for. Inefficiencies and poor management often mean that the only viable reform is complete restructuring – a step that most governments are unwilling to countenance. Moreover, many reform packages solely focus on reforming the utility and do not attempt to reform the business and social environment or the means of interaction with other actors in the sector. Various sectoral pieces of research (World Bank, 2006a) recommend that governments first realize that their *ownership interest in a utility is distinct from their policy interest*. Therefore, the financial impact of their policy and ownership decisions should be reconsidered to emphasize the financial incentives required to induce a better performance from poorly-performing utilities (Baietti, Kingdom & Van Ginneken, 2006a).

Wastewater service issues

Improved access to wastewater services in developing countries is a critical factor within the development process, particularly in the urban sphere. A World Bank assessment (World Bank Water Toolkit, 2001) shows that wastewater services are often neglected as compared to the attention granted to the water sector in general. This can be illustrated by recent comparisons of investment and coverage levels. The wastewater sector differs from water in the following manner:

- Wastewater services are less likely to be a networked or utility service and by nature are generally based on extremely localized services geographically.
- Wastewater services are expensive and most professionals believe that a high level of government subsidy is required to maintain the sector.
- It is difficult to achieve economic viability in the sector.
- Wastewater services are largely the preserve of municipal government and any reforms in service delivery would necessarily be linked to a wider process of municipal reform.

The technical challenges and institutional complexities of the wastewater sector have posed a challenge to the private sector. However, once the sector is physically connected, issues of design transaction are common to both the waste and wastewater sectors. Wastewater requires more challenging policies and legal issues and may require institutional reform vis a vis establishing appropriate roles for local governments and private operators or other entities responsible for contract enforcement. Many governments decide not to pursue a PPP in the wastewater sector because they perceive it as unsuitable for such a reform, preferring local and decentralized technical solutions, and believing that the private sector will simply not be interested. Clearly the sector requires deeper thought and the solutions may appear less obvious. However, arguments making the case for water sector reform – such as improving levels of service, expanding coverage, adopting technical standards – are equally appropriate for the wastewater sector.

1.3 CAN PPPs BE AN EFFECTIVE SOLUTION FOR THE CHALLENGES OF THE WATER AND WASTEWATER SECTOR?

The past few decades have seen a growing awareness of the planet's fragile ecological balance. Water resources are being stretched to their limits and are putting pressure on urban centres worldwide. Nevertheless, there is a growing consciousness of the need to achieve public policies that will assure the protection of the precious resources and the environment within their economic framework. Water as a resource needs to be protected and adequate strategies need to be devised to dispose of wastewater. We are concerned here with the Water and Wastewater Services sector.

We propose to undertake a detailed examination of the factors leading to the decision to develop a public-private partnership to ensure an effective and appropriate contractual arrangement. The rationale behind this is that public-private partnerships have become an increasingly important mechanism in the provision of long term water and waste water services, and that they are 'here to stay'. However, PPPs are not a panacea for all situations. Experiences over the past two decades demonstrate that there has been a considerable range of effectiveness of these long term PPP Arrangements. At best they have provided all the required investment and services efficiently and effectively, in a sustainable manner, but there have also been some less effective cases, and even some catastrophic failures. However, given the current reality experienced by the world's infrastructure sectors, public funds will be insufficient to meet the growing demands of the world's population. The public purse cannot fund all investment needs, and the demand for greater use of private sector involvement is increasing.

Given the varied effectiveness of the many and diverse PPP examples over the last decade, there is a clear need to provide practical guidance on how best to structure PPPs to obtain optimal performance. Successful contracts require a careful selection of appropriate and robust contractual mechanisms, but at the same time allow for sufficient flexibility to deal with changing circumstances and conditions.

Key questions

We have asked the following questions related to the effective design, development and implementation of PPP arrangements for the Water and Wastewater Services Sector:

(1) How best do we capture the benefits of the private sector and blend them with the strengths of the public sector?
(2) What PPP arrangements work and what do not?
(3) Are there any common lessons to be learnt?
(4) What special legal and financial instruments are needed to deal with PPPs in infrastructure, particularly when applied to the socially sensitive sector of water and sanitation services?

The use of PPPs alone does not guarantee a resolution of inherent problems in the sector. Key issues considered by Governments when seeking improvements in water and wastewater services include a reform strategy that takes into account key management and financial issues, an understanding of the state of the existing water infrastructure and systems, and the system of rules and incentives governing the sector. Typically, developing a strategy to deal with the problems that beset the water and wastewater sector will require Governments to answer the following questions (World Bank Water Toolkit, 2007):

- How can competent managers and management systems be developed and handled?
- How can the services provided by the water and wastewater company/utility be made efficient?

- How can the utility's solvency be ensured in order to cover all its costs (tariffs and subsidies)?
- How can the utility guarantee that new capital expenditure is directed to the most pressing areas?
- How can service expansion be financed in the absence of sufficient internally generated funds?

Here we will establish the approach in answering these questions regarding reform strategies involving PPP.

Furthermore, we will test how the use of a PPP scheme can bring tangible accountability and decision-making benefits to the water and wastewater utility. In particular, whether the private sector can bring:

- More focused service and commercial performance.
- Easier access to finance.
- A more disciplined approach to tariff and subsidy levels to ensure long term sustainable service delivery.

Achieving value for money in PPP arrangements

A widely used measure of economic effectiveness for governments implementing water and wastewater PPP projects is the Value for Money (VfM) obtained from the project. VfM is defined by the United Kingdom's HM Treasury (2006) as 'the optimum combination of complete-life costs and quality (or fitness for purpose) of the good or service to meet the user's requirement'. VfM is not focused solely on the lowest financial cost. Accordingly, throughout this paper focus has been placed on answering questions related to ways that the drivers of VfM can best be achieved in any PPP arrangement. These include:

- What should be the optimum allocation of risks between the various parties?
- Does the PPP focus on complete life cycle costs of the asset rather than just the upfront costs involved?
- Is there use of an outputs specification approach to describe the Grantor's requirements, thereby, amongst other things, allowing potential bidders to develop innovative approaches to satisfy the service needs of the Grantor?
- Does the PPP transfer rigorously the risks to parties responsible for them, ensuring the enforced allocation of risks and the associated costs are borne by the parties in the manner allocated and agreed upon?
- Are there sufficient incentives within the procurement structure to ensure that assets and services are developed and delivered in a timely, efficient and effective manner?
- What is the most suitable term of the PPP Agreement?
- Are there sufficient skills and expertise in both public and private sectors, and how can these be utilized effectively during the procurement process and subsequent delivery of the project? and

- What are the most appropriate strategies to manage the scale and complexity of the procurement to ensure that costs are not disproportionate to the project(s)?

1.4 OUTLINE OF THE BOOK

This book is presented in three main parts:

PART 1 – Background to the sector and PPP

The first part, Chapters 2–4, presents background information on the urban and peri-urban water and sanitation sector to give an understanding of some of the sector specific issues. Following this, there is a review of the use of PPP in the sector, the issues to be considered when developing a bankable project, and some of the key requirements for success. The potential relationships between the Public and Private institutions are looked at and the staged approach to PPP development and Implementation is described. The key subject of Levels of Service, the associated costing and funding issues, and ways of measuring the PPP's performance in the longer term are reviewed.

PART 2 – Main factors influencing effective development and operation of PPP schemes

The second part, Chapters 5–9, looks at the main factors that can influence the effective development and operation of water and waste water service reform when using a PPP arrangement.

We start, in Chapter 5, with the Legal, Regulatory and performance monitoring issues, looking at how the planned service provision can be ensured in the longer term within a specific legal regime. The issue of economic regulation, and how it can best be effected through policy and contractual provisions, is considered, and the effects of the regulatory framework and institutions for setting and revising economic and service standards, monitoring and controlling, are described.

Chapter 6 looks at the issue of Understanding and Managing Risk, which is at the core of the development of a successful and sustainable PPP project. Through review of methodologies for analyzing responsibilities and risk the authors look at common risks and responsibilities, as well as risk associated with political and macroeconomic factors, water sector specific issues and risks associated with project development and implementation.

Funding and Finance (Chapter 7) are two of the key drivers for choice of PPP for Water and Waste Water Sector schemes. This section revolves around issues related to bankability of PPP projects: sources of funds (public and private funds, multilateral and bilateral institutions); structuring finance, as well as the potential (and possibly major) effects of contingent liabilities for the Grantor. We look at some financial issues that may not occur in all schemes, but may require consideration (e.g. hedging and refinancing).

The section on General Procurement Issues of PPP's (Chapter 8), covers issues that are a major concern of the Public Sector when implementing large infrastructure projects. Generally existing public procurement methods have to be adapted to ensure most effective design, implementation and regulation of PPP projects. We look at issues that can arise over the whole of the procurement process, and look at best practice options for dealing with PPP procurement. This process includes: choosing the right method for Developer selection; management of the bidding process, setting selection criteria and bid evaluation, establishing pre-qualified bidders, selecting the appropriate bidding process and developing the request for proposal; submission of bids, and how they are evaluated and finally the selection of the successful Developer. The process of negotiating with bidders and a number of 'non-standard' issues that can occur with these schemes are looked at. Also given the importance of the European Union and how it has developed specific procurement issues related to transparency and competition, briefly reviewing the EUs position on PPPs and how national and EU legislation deals with the monopolistic nature of the water and wastewater sector.

In the final section, recognizing the fact that large numbers of water and waste water PPP arrangements have had to be renegotiated within a few years of contract award (Guasch, Laffont & Straub, 2005), we look at some of the practical provisions for establishing contractual and legal mechanisms to allow for potential future renegotiation of new contracts, as well as ways of dealing with bidding for existing PPPs in Chapter 9 (Dealing with Existing PPPs and Renegotiation).

PART 3 – Developing sustainable PPP arrangements

In this third part (Chapters 10–14) specific recommendations for improving the design and sustainability of PPP arrangements are presented, dealing with key issues through detailed contract drafting, in order to ensure most effective procurement and implementation of the chosen PPP scheme, to achieve optimal results. Sustainability in the context of water and wastewater services should be considered in its broadest of interpretations that is it is not only about financial sustainability but should also include overall human, social, economic and environmental sustainability. In the context of a sustainable PPP arrangement, it is important to integrate and reconcile the human, social, economic and environmental aspects within a holistic and balanced sustainable development framework (Munasinghe, 2008).

The water sector by its very nature involves human, social, economic and environmental aspects. A good – and therefore bankable – PPP arrangement should be inclusive of these. Human sustainability means maintaining human capital. Access to basic services such as water constitutes human capital.

Investments in education, health, and nutrition of individuals have become accepted as part of economic development (Goodland, 2002) and as such access to water and wastewater services has a demonstrable effect on human and economic sustainability. Similarly, social sustainability means maintaining social capital (Goodland, 2002), and social capital is investments and services that create the basic framework for society (El Serafy, 1989). Water and wastewater services are key to society's basic framework. Last but not least, environmental sustainability itself seeks to improve human welfare by protecting natural resources (a definition of environmental sustainability has been given by Daly (1973, 1974, 1992, 1996, 1999) and Daly & Cobb (1989)). In this respect, an adequate PPP arrangement will seek to ensure that the water resource as well as water ways are protected as part of the contractual obligations.

The contractual arrangements for the PPP scheme have to embody ways of dealing with the long term nature of the provision of water and wastewater services (determined in the previous section), as well as the main factors influencing effective and sustainable development as described above. The long term success of the PPP arrangement will in a very large part be determined by the way that these factors are dealt with contractually.

Chapter 10 (Contracts: Key Contract Provisions) contains the bulk of the work in this section, and works on the basis of an outline contract form developed for the most complex types of PPP arrangements (i.e. the Concession contract). This is developed as a practical checklist covering all the typical contract elements presented in the order typically expected. I note but do not cover in detail standard or project specific clauses. However I do give suggested ways of dealing with all clauses that are considered to potentially influence contractual and scheme performance.

The importance of ensuring the continuing success of PPP schemes during their long term implementation and operation phases is recognized by a separate Chapter 11 (Maintaining the Economic and Financial Equilibrium of the PPP arrangement). Over long periods it is inevitable that changes will happen and that not all can be anticipated at the contract design stage and which may materially affect the financial and economic equilibrium of the arrangement. Accordingly, adequate flexibility needs to be allowed for in the design and implementation of the PPP agreement. At the same time, it is important that this flexibility does not open the door for full re-negotiation of the basic terms of the contract. Adequate balance needs to be struck between price, long-term flexibility and certainty of whole-life costs. There are certain costs associated with the financing structure of the PPP agreement that need to be included so as to provide for changes that cannot reasonably be foreseen by the Developer either at the time of bidding or during the implementation of the PPP agreement. All these issues require a pragmatic solution that implies including in the PPP agreement specific mechanisms that allow for these changes. These issues are explored and specific recommendations made.

In Chapter 12, we review all the contract recommendation made in the previous sections, and building on some of the contractual and structural limitations of the 'traditional' contract forms, propose new contractual models namely Output Based PPP agreements and long term PPP agreements incorporating the use of Trust structures to more effectively leverage financial resources.

Many PPP schemes are designed around information on conditions of existing and future water and waste water systems using available data and estimates that are less than perfect. In particular this applies to the considerable element of the underground assets (e.g. pipe work, water mains and sewers) associated with provision of water and waste water services. Chapter 13 (Dealing with Imperfect Data) looks at effective ways in which risks of underground asset condition and other imponderables can be transferred to the Developer in a systematic and sequential manner. This can have a major impact on the execution of the agreement, and this section looks at ways that bidding and contractual mechanisms can be incorporated that will handle this and any future improvements in data. This is an important subject as case history shows how lack of these mechanisms has affected PPP performance adversely, and in some cases has led to outright contract failures.

Arising from the specific issue of Imperfect Data and the particular characteristics of water and wastewater PPP schemes, there is a particular need to improve public bidding and procurement procedures to deal with this in the bid process (Chapter 14). Given that there will be gaps in the information on which Developers will make their proposals; the procedures can be strengthened by use of an approach that recognizes the need for data improvement and validation in a structured way through an improved bidding and procurement approach. We give a recommendation for a structured approach that runs through the pre-qualification process, the development of the bidding process to incorporate revisions and improvements in data and assumptions, and finally continues to deal with improved data in the initial period of the PPP contract. In regard to this last point we include a proposal for dealing with the adjustment of the Developers' risks and payments as the improved information becomes available during these initial years of the contract.

Summary conclusions and recommendations

In this final part we present some thoughts on optimizing the PPP approach for long term provision of water and waste water services. We summarise some of the key recommendations made primarily in Part III which include contractual clauses and provisions, specific strategies to ensure the long term financial and economic equilibrium within the PPP arrangement, overall PPP structural improvements in the form of new structures (OBA and Trust), how to deal with imperfect data and how procurement processes can be re-engineered to allow for progressive and systematic approach to improved data. Some areas for future research are explored as well.

ENDNOTES

[1]Other institutions such as the World Trade Organisation (WTO) leaves the definition of 'developed' and 'developing' countries to its members to decide. Members announce for themselves whether they are 'developed' or 'developing' countries. However, other members can challenge the decision of a member to make use of provisions available to developing countries. Developing country status in the WTO entails certain rights. There are, for example, *provisions in some WTO Agreements which provide developing countries with longer transition periods* before they are required to implement the agreement and developing countries can also receive technical assistance. That a WTO member considers itself as a developing country does not mean that it will benefit automatically from the unilateral preference schemes of some of the developed country members, such as the Generalized System of Preferences (GSP). In practice, it is the preference giving country which decides the list of developing countries to benefit from the preferences (UNCTAD, 2009).

[2]The main challenge facing policymakers in meeting ever-increasing demands for good public services is how to balance the level of service provided with the massive costs required to provide the infrastructure to support this. If partnering between the public and private sectors is to be effective, the main consideration is to ensure that these 'partnerships' are sustainable. Involving the private sector implies cost. How can policymakers balance their responsibility to provide basic infrastructure with their limited public funds? How can public funds be best utilised, and how can these funds leverage private funds? How best to ensure that governments meet their social obligations and keep control of the provision of basic services? Sustainability of the proposed public-private partnership is at the heart of this response.

[3]Project Finance is a method of funding by which the lender looks primarily at the revenues generated by a single project both as the source of repayment and as security for exposure.

[4]World Bank offers resources specifically for this purpose available at: www.worldbank.org/water

Part I

Background to the Sector & PPP

Chapter 2

Sector issues in the water & waste water service

2.1 INTRODUCTION

Although there has been considerable use of PPP arrangements in the Water and Waste Water Service sector over the last two decades, there have been some intrinsic challenges in applying the PPP approach in this sector. Essentially the issues that, in combination, differentiate the sector from other public service sectors are:

- Water is a basic human need for survival. Arguably, unlike other public services (e.g. electricity, telecoms or airports) individuals have no alternative but to use this service to survive. This generally brings with it an obligation to provide the service to all (designated) customers, whether they can afford to pay or not. Because it is seen as a basic essential, there has long been a discussion about whether profit should be made in the provision of such services. Generally this manifests itself in the low profitability of this service sector, compared with, say, the Telecoms or Power sectors. In the context of PPPs this poses particular funding challenges.
- Water and Waste Water services are usually provided under a tendency for a natural monopoly condition because of the high costs of the of the infrastructure installations needed to treat, distribute and collect water and waste water. Generally it is not economically feasible for more than one company to provide these services. This monopolistic situation, combined with the basic human need for water, makes the provision of water and waste services highly susceptible to political involvement, both in the creation or expansion of water services, and also in all aspects of the long termdelivery of the services. In the PPP context the monopolistic situation has the advantage of an already defined market for services, but implies a

high expection of political involvement to arrive at the form and type of scheme and services provided. Usually in the longer term monitoring and service control there is a well defined function of public Regulation. Any well defined regulation, whilst potentially constraining or limiting the Developer in key business areas (e.g. related to investment, tariffs or levels of service), will also bring a certain stability since the business elements and the way that they are monitored and controlled will be defined in the contract.

- The initial investment costs in the necessary infrastructure (e.g. creation, expansion or rehabilitation of pipe networks) are typically high, yet these are one-off purpose built costs and are irreversible.
- Generally the data on the existing services, service demand satisfaction and the condition of existing physical infrastructure is difficult to assess. Often much of this information is not fully available prior to the contract award. This affects the design and pricing of any new schemes, and where there is uncertainty this generally increases prices paid to cover this perceived risk. Since much of water related infrastructure is underground, this complicates data collection on the condition of the asset, leading to major cost liabilities involving rehabilitation, renewal or replacement of existing systems. To be effective, a method to deal with discrepancies in data has to be built into in the PPP design and implementation process, to ensure long term scheme viability. Contractual mechanisms are being developed that allow for data collection and greater accuracy during the early years of the contract, and potential structured re-negotiation or adjustment of associated PPP elements.

This Chapter serves as an introduction and background to these and other Water and Waste Water Service Sector issues that are integral to development of effective and sustainable PPP arrangements in the sector. These issues may not all apply to specific PPP schemes considered, but will, in general, influence the specific types and forms of PPPs which are considered here.

2.2 STRATEGIC PLANNING, POLICY CONSIDERATIONS & THE COUNTRY CONTEXT

Water and Waste Water Sector issues generally form an important part of government public service policy. Given the importance of the sector, and the long term planning horizons associated with any sector investment, design of any new PPP scheme is best carried out within the context of a strategic plan for the sector.

A good starting point for strategic planning is to obtain a full diagnosis of the problems associated with the existing water and wastewater systems. With this information in hand, the government can begin developing solutions that best address the issues identified in the diagnosis and best suited to the country context. Some common issues prevalent within the sector are:

- *Poor tariff setting capability*: Utilities are often beholden to politics and political calendars. The ability to set tariffs at, or above cost recovery levels may be out of their hands with consequences resulting from under-funding and poor incentives. A potential solution could be to grant the tariff-setting authority to an independent body/watchdog with the prerogative to override political decisions.
- *Poor accountability*: Often the reason given for poor accountability for the provision of services is that there is insufficient performance information and systems. of analysis. A potential solution would be to create an agency whose sole purpose is to collect data on the utility's performance and loaded with monitoring and evaluation capabilities. The agency would serve the dual purpose of improving the information available to the government (and thereby improve decision-making) as well as improving the information available to the public.
- *Insufficient technical capability*: Managing water and wastewater resources and services is a difficult and technically demanding task. Involving the private sector in the management brings even more demands. Governments need to consider whether they possess the technical skills necessary to undertake these challenges within their existing institutions. If not, in some cases it may be prudent to set up an independent regulator/body to oversee the sector and control the activities of the service providers. This solution is not appropriate for all country contexts, instead, it might be more appropriate to set up an independent panel of experts that advises the government on sensible policies or assists it in analyzing the state of the existing system when called upon. The findings of the panel can also be made available to the public, increasing the levels of transparency and accountability.

The water and wastewater sectors are natural monopolies and any form of government intervention to improve operational efficiencies must be considered with this in mind. In competitive markets, competition between suppliers generally keeps the prices charged in line with the costs, and the market itself acts as a regulatory force. Such competitive forces do not normally exist in the water and wastewater sectors as the infrastructure does not allow customers to choose between competing suppliers, eliminating the competitive pressure that would normally force a service provider to deliver the services their customers require, at an affordable price. Water and wastewater services are worth more than it costs to provide them and, with few other options, customers are often willing to pay a premium for the services rather than go without. Without sufficient oversight or regulation Operators/Developers are prone to charge high prices with little incentive to provide high quality services, or conversely, they may charge low prices and deliver a low quality of service even in the face of evidence that customers would pay more in exchange for better services.

Regulation for publicly owned utilities is traditionally focused on ensuring that prices do not rise too high, and it is often observed that publicly owned providers charge too little. By charging below cost the utility is too reliant on government subsidies and generally will not have sufficient resources to spend on service upgrades, expansion and maintenance. Subsidies are rarely sufficient and can often work to undermine the customer-focus that should be the key driving factor behind a utility service provider. Although regulation to maintain low tariffs also maintains more friends than enemies, the challenge is for the government and the private Developer to earn the public's confidence and support to create understanding and acceptance of tariff increases (Groom, Halpern, & Erhardt, 2006).

Vital government roles in the water sector that complement regulation include:

- *Policy Making*: sector objectives, principles and allocation of risks and responsibilities, the general strategy for controlling tariffs and service standards.
- *Ownership, Service Provision and Governance*: Water sector performance is largely controlled by four factors:
  - Who owns the assets?
  - Who owns the service provider?
  - Who is responsible for delivering services?
  - How is the utility's management controlled?

  The challenge for governments is to provide competent solutions to the challenges created by these factors.
- *Coordination*: Governments must coordinate the water sector to ensure policy decisions and implementation plans are consistent and effectively implemented.

Level of decentralization

Decentralization in the water sector refers to the level of responsibility for water and wastewater services that is allocated to municipalities by the central government. The French model of water sector management deals with greater decentralization by adjustments in the contract between the utility and the local government. Any tariff adjustments or service standards are treated through adjustments in the contract and any disagreements are settled through dialogue between the parties or by courts/arbitration. In comparison the 'English' model is based on a regional system held accountable to a national water watchdog. The watchdog is also empowered by the government to make binding decisions on the utility. Greater decentralization is often associated with more responsiveness to local needs and issues. At the same time, municipal governments generally lack the regulatory capacity of central governments, leading to conditions that may be conducive to potentially higher levels of corruption. Many countries have attempted to develop hybrid systems, with mixed success. These issues are dealt with in more detail in the chapters focused on Regulation (Chapter 3) and Contracts (Chapter 7).

Managing water resources

Water resource management is a complex task that must be considered when defining water and wastewater sector policies that in turn will condition any water sector PPP arrangement. Water Rights Laws may differ across states and countries and therefore it will be imperative that governments use existing frameworks or create a new framework to handle this issue. The key points to consider are:

- Understand the concepts of prior appropriation – while no single person may own the water in a stream, all persons, corporations, and municipalities have the right to use the water for beneficial purposes.
- Understand the differences between groundwater rights and surface water rights.
- Co-ordinate with the agency or official that is responsible for the administration of all water rights within the region or the country.
- What is the availability of raw water? What are the prior-use rights established by other users (e.g. hydropower companies)?
- What is the current state of water treatment by utilities? (This affects the quality of the water available to users downstream).
- Is the government capable of assuring Developers about future access to raw water supplies and of the quality of the supply? Any private sector Developer will require such guarantees before engaging in a partnership in order to protect itself from the likelihood of circumstances such as depletion of groundwater or rivers or pollution of surface water or groundwater or rivers.

Generally, governments entering long term contracts with private Developers may be required to provide substantial guarantees of water resource availability. For governments lacking the resources and capabilities to manage the complexities of water rights and the issues raised above, it may be more appropriate to first enter into short term PPP arrangements (such as a management contract), since this would grant the government a window of opportunity to create a better system for water rights management before entering a longer term commitment.

Integrated water resources management

Integrated water resource management refers to the integration of water quantity and quality management for both groundwater and surface water. Ideally the management of these resources is conducted with a full understanding of how natural resources and populations within the service area are impacted by developments or policies that affect the resource itself. Major stakeholder groups are often involved in the management of the water resources which may have a crucial and definitive influence on policy decisions. Often the management of water resources depends on river basins and situations where river basins overlap into different regions or countries in which the river may fall under the remit of different legal and political institutions with no existing mechanisms for co-ordination. Such situations emphasize the importance of a sound, over-arching 'integrated river basin management system' (World Bank, 2006a). The key elements behind such a system would include:

- Establishing a basin-wide institutional framework that ensures all the main government institutions operating within the basin are involved. The framework should be detailed, strong and clear and backed by legislation, statutes or decrees along with a clear financing and budgeting system.
- Solid and reliable data on the status and condition of the natural resources of the basin (not just water quality).
- Development of an integrated strategy that ensures representation and participation by all parties in all decisions and elements of resource management.
- Creating a monitoring and evaluation (M&E) system that can reliably and effectively determine whether the goals and objectives of the integrated are being carried out, that the health of the river basin is being maintained and supported by systems of accountability among the key staff and institutions involved in the management of the system.

Whilst PPP arrangements are generally not directly used for water resource management, either nationally or for river basins, the influence and importance of the water resource management situation can have a direct influence on the provision of PPP arrangements for water and waste water services. Furthermore, PPP can be used for small scale providers, as a recent study by the Water and Sanitation Program (2002c) shows. The principles are similar although the scale will clearly be different. In some instances the tariff levels may be insufficient and may need to be complemented through subsidies (Veevers-Carter, 2008).

2.3 SOCIAL ASPECTS

Small private operators

Governments in developing countries will have to develop strategies to deal with the plethora of small scale providers that may exist in their countries. Many small providers are associated with high prices, lack of environmental and health standards and inferior services. However, particularly for low-income and unserved populations in urban and peri-urban areas, small providers have become established service deliverers and are formally recognized by their governments (Mann, 1993). Mehta (2003) shows that these small providers have inherent weaknesses – lacking access to finance, insufficient technical and managerial expertise, no legal status or tenure and unregulated price and service provisions. Despite these problems, small private providers continue to supply unmet demand and government strategies can incorporate their services in the short and medium term (and in some instances they may remain the only feasible long term approach).

Government approaches to small service providers are determined by the country context: however the following general points are noted:

- As part of the long term goal of improving access and quality of services, governments may wish to recognize and legitimize small providers.
- Governments can consider introducing a limited form of regulation for quality, safety and environmental standards for these providers.
- Over the long term, as a public utility increases its coverage, it may decide to collaborate with the small providers under a more formalized contractual arrangement.

Here we are concerned with the use of PPPs for larger public utilities, and not specifically on these small scale private operators. However, any PPP arrangement may have to accommodate these providers in some form or another, particularly if they are currently providing a service not already provided by the utility, or can mobilize resources for investment or operation more economically than the utility.

Ensuring access for all

Unsafe and inadequate access to water and wastewater services can lead to unbearable living conditions, high rates of waterborne illnesses, loss of livelihood and loss of human dignity. Evidence suggests that improvements in water supply and wastewater services can have major positive impact on health, efficiency and productivity. Poor households are disproportionately affected by inadequacies in both sectors and improved provision of services to this segment of society is a development priority. Poor households need access to water and wastewater services and in many countries, they have shown their desire to be considered legitimate customers of services, buying water and wastewater services like other residents. There is a clear and undeniable willingness to pay for a better quality of service. As poorer households comprise the majority of potential new customers in most cities in developing countries, it is imperative that utilities develop the appropriate expertise to cater to this demand and design services considering the requirements and constraints of low-income customers. The main concerns include (World Bank, 2001):

- *Price* of the service being delivered.
- *Expansion* of the system in order to add new connections in previously unconnected neighbourhoods.

- *Service levels,* such as water quality, hours of service, speed of response to service calls, and administrative services such as billing and connection applications.

One of the strategic reasons for considering use of PPP is often for expansion of services. In many cases the unserved populations are in the poorer categories. In some cases the need is to expand services into informal communities, with no formal land tenure, besides having limited incomes (e.g. favelas, barrios). To incorporate this sort of expansion into a PPP arrangement would involve clear resolutions in the associated legal issues, and provisions of clear and adequate financial mechanisms (e.g. special grants or aid provisions related to measurable units of expansion achieved, such as increased output or numbers of connections).

The customer approach

Water and wastewater provision is driven by the need to provide good and effective services. Therefore, focus should be on 'customers', and I use this term throughout, instead of 'users'. Simply put, customers (unlike users) have rights, including the right to safety, to be informed, and to have redress, choose, satisfaction of basic needs and the right to a healthy environment (United Nations, 2003). In the planning and implementation of water and wastewater systems, governments and Developers alike need to ensure that the customers' rights (as well as those of other stakeholders) are protected.

2.4 SUMMARY AND CONCLUSIONS

There are four main characteristics that differentiate the Water and Waste Water Sector from other forms of infrastructure sectors for public service provision. These are that Water and Wastewater Services are a basic human need, with the associated social and political implications, the fact they are provided through a natural monopoly with high capital investment costs and low rates of return achievable, and that data on current and future service conditions and demands is scarce, including condition of underground assets, causing a major impact on the levels of investment required.

These characteristics influence the use of PPP arrangements for long term service provision in this sector. Bearing in mind these key characteristics, the authors have reviewed some of the sector's specific issues in need of accommodation in any successful PPP arrangement. These issues are described from the existing public sector aspect, but with specific comments on issues needed to be dealt with more effectively under the PPP arrangement.

The focus is on issues of urban and peri-urban utilities, in which a number of sector challenges have been identified and listed, including: current poor sector service performance, strategic planning issues, political involvement, pricing and tariff issues and finally, how to deal with possible corruption issues.

Similar challenges apply in the waste water sector, but there are additional issues related to the localised nature of these services, and the perceived higher subsidies required for this sub-sector. Successful incorporation of private sector participation into the sector depends largely on the chosen PPP arrangement's ability to handle these challenges.

Since Water and Waste water sector issues generally form an important part of government public service policy and given the importance of the sector and the long term planning horizons associated with any sector investment, the design of any new PPP scheme is best carried out within the context of a Strategic Plan for the sector. Therefore, a number of the strategic planning issues that may be relevant have been considered, including those related to the on-going roles and responsibilities of Government and Developer:

- Tariff Setting and Regulation.
- Accountability issues.
- Technical capability required.
- Government Responsibilities and Policy Making issues.
- Ownership, Service Provision and Governance issues.
- Coordination of sector operation and development.

Additionally specific issues of decentralization in the sector have been considered and how water resource management and integrated water management influence the way that any PPP arrangement is applied.

Finally, some of the social issues related to this sector are considered. In many developing countries public water supplies only meet a portion of the actual water demand, with the balance provided by small and informal private water vendors. Even though we are not so concerned here with the smaller water service providers, their role in provision of services needs to be accounted for in any PPP arrangement. There are two other socially oriented issues to be looked at: the need for projects to focus on all aspects of Customer Service provision and the need to 'Ensure access to all'. This last issue is linked with the policy issues of service provision expansion, and a key challenge for expansion of services into poor and disadvantaged urban and peri-urban areas is how to deal with the associated special issues (such as land tenure, low possibility of realistic payment and special aid and funding requirements) under a PPP arrangement.

Chapter 3

Incorporating private sector participation

3.1 INTRODUCTION

The use of the private sector as an ally for the development of infrastructure, particularly that related to basic services such as roads, electricity, telecommunications and water is not a new phenomenon. Throughout history governments have made partnerships with the private sector to help them meet their social obligations. It is also clear that poor infrastructure impedes a nation's economic growth and international competitiveness (World Bank, 2005; World Bank, 2006c). Calderón and Servén (2004) argue that inadequate and or insufficient infrastructure lead to a loss of quality of life, illness and ultimately death. Furthermore, research by the international financial institutions such as the Interamerican Development Bank, European Bank for Reconstruction and Development, African and Asian development banks, World Bank and the OECD, confirm that infrastructure projects have high social rates of return. Research by the World Bank (2008) indicate that the growth generated by infrastructure investment is generally pro-poor, with the income levels of the poor rising more than proportionately to the overall income increases. The public sector provides financing for the vast majority of infrastructure services, but private investments have not matched demand.

Private participation in infrastructure is one of the tools policy-makers have to help increase investment in infrastructure services and improve its efficiency. There is evidence that, if appropriately and properly procured, PPPs can provide significant improvements over the results that can be achieved from traditional forms of procurement (Engel & Galetovic, 2000; Estache *et al.* 2000; Estache & Rossi, 2002; Frangano *et al.* 2001; United Kingdom's H.M. Treasury, 2003; Kerf *et al.* 1998; Kikeri, 1998; Klein, 1998a; Klein, 1998b; Klein, 1998c; Levy, 1996; Littlechild, 2002; Llorens & Soler-Couteaux, 2001; Lobina & Hall, 2003; Lorrain 1997; Mandri-Perrott, 2003; Mehta, 2003; Pickering, 1998a; Vives, 2000).

Notwithstanding the use of PPP as a strategy to complement single source public provision of water services, clearly, the use of the private sector is no panacea and there are numerous options, structures, solutions and strategies for its use.

PPPs are often complex and generally take longer to procure than traditional projects. Transaction costs, for both the public and private sector, also tend to be higher (Platts, 2001; Plummer & Nhemachena, 2001; Ward, 2002; Vickers & Yarrow, 1988). There is generally a poor level of understanding of PPPs since they are a new means of procurement and public sector officials have no experience in implementing such projects. Therefore, they need to develop new skills and capabilities in order to undertake PPP procurements effectively. This requires an investment on behalf of the public sector. There is a need for better sharing of knowledge and experience between different parts of the public sector. This will be conducive to greater efficiency in the development of PPP approaches and processes and consistency in procurement (Tynan, 2003; Torres de Mästle & Izaguirre, 2008; Vives, 2000).

Decisions to pursue a PPP in the water services sector cannot be taken without considerable study and initial preparation on behalf of the Government acting as the grantor of the proposed PPP agreement. There are many texts and other useful aids that can guide readers through the various processes of analyzing and selecting appropriate PPP contract solutions (World Bank, 2006). An obvious first step, but one often overlooked, is for the government to identify its objectives and the main challenges faced in achieving them. From this starting point, the decision on the mode of PPP to pursue can be determined.

The range of options for engaging the private sector depends upon the level of control the government wishes to retain or cede. Responsibility for system operation and maintenance (O&M) and financing and commercial risk can either be completely devolved or partly shared between the government and a private Developer. All of these arrangements will take place within a system of rules and regulations created by the Government. The challenge for the government is how to structure this framework to prevent Developers from acting in a monopolistic manner and to make sure that operational standards and services are maintained, environmental standards are respected and some form of services are extended to the poor.

Public Private Partnerships (PPPs) may take many forms. In its simplest form, the term PPP refers to a structure of cooperation between a public authority or authorities and the private sector. Its primary aim is to fund, construct, renovate, manage and/or, maintain a given piece of infrastructure or the provision of a service. At its core, a PPP is an arrangement whereby the private sector delivers infrastructure services to the public sector or assists the public sector in its task of delivering infrastructure services to the public. In today's world, infrastructure is primarily a public sector issue, in which public sector annual investment in infrastructure vastly exceeds that invested by the private sector. Delmon (2009) argues that even for the most public of infrastructure service providers, private involvement forms an essential part of successful service delivery, whether through construction contracts, service agreements or delivery of goods or joint ventures.

PPPs are typically arrangements of relatively long duration (generally varying between 3 and 30 years) involving cooperation between the public partner and the private partner in different aspects of a planned project. Funding often involves a combination of sources, both public and private, which in some instances mean complex arrangements between the various players. Notably, the role of the private sector is in the form of an 'economic operator', who participates at different stages in the project for example, design, completion, implementation, funding. The public partner focuses primarily on defining the objectives to be attained in terms of public interest, quality of services provided and pricing policy, and it takes responsibility for monitoring compliance with these objectives.

However, if the private sector is to be involved effectively there is a need to balance the interests of the public and private sectors and allocate the risks between them fairly. This will permit the most suitable contractual form to be developed. It is to be noted that PPPs are not a miracle solution, and so for each project there is the need to assess whether the suggested partnership can in fact add value, compared to other more traditional contractual forms.

This chapter looks at the conditions which lead to considering the use of a PPP structure in the water and wastewater context, the various potential PPP forms as well as the conditions necessary to develop an effective PPP scheme.

3.2 LITERATURE REVIEW AND CONSIDERATIONS FOR PPP IN THE PROVISION OF WATER AND WASTEWATER SERVICES

Much has been written on PPP structures in the water and wastewater sector (see Gómez-Ibáñez, 2003; World Bank, 1998, 2004; Delmon, 2009 amongst others), but what is evident from the literature reviewed[1] is that PPPs involve a complex process requiring input from a number of different parties, each playing an essential role in managing project risk. A variety of factors motivate the public sector to seek PPPs, including:

- Underperformance of public sector utilities, often linked with opaque funding structures and inefficient or corrupt procurement methods.
- The need for improved overall performance or meeting of specific standards for quality and quantity of the utility service being provided.
- Inadequate technical and management resources in the public sector.
- Investment demands exceeding public resources, in particular given the large up-front capital costs associated with infrastructure investment and the 'lumpy' cost implications of periodic major maintenance.

Gassner *et al.* (2007) and Andres *et al.* (2009) have undertaken a comprehensive review of the effects of private sector participation in infrastructure using a number of examples in different sectors such as electricity, telecoms and water over the past decade. Interestingly, they argue that PPPs are generally perceived to provide enhanced efficiency by delegating specific functions to the private sector, enhancing transparency and accountability for the provision of public services, improving technology, innovation and know-how as well as developing a long term view to asset management. Furthermore, the literature review and in particular a recent database developed by the World Bank on public private investment in infrastructure (World Bank & PPIAF, 2009), show clearly that PPPs can encourage the mobilization of new or additional sources of finance for infrastructure development, and provide new opportunities for local financial markets. Given the importance of the effects of PPPs, each are dealt with in more detail below.

Efficiency – It is often considered that the private sector provides higher levels of efficiency in running infrastructure projects than the public sector. This increased efficiency results from many factors, which may include, amongst others, improved financial engineering leading to an overall reduction in the cost of capital, a more cost conscious effort towards problem solving through more effective use of labour and materials, and improved governance leading to more cost effective driven decision making.

Enhanced Transparency – PPPs are often seen as an instrument of good governance by providing transparency, equal treatment and open competition. Delmon (2009) argues, for example, that lack of good governance makes investors and lenders worry (increasing the cost of money), reduces competitive pressure on bidders (increasing costs and reducing quality of solutions proposed) and increases the likelihood of rent-seeking/bribery and other forms of corruption (which add cost and delay to project implementation)[2]. In this regard, Bakovic *et al.* (2003) suggest there are a number of actions that come hand in hand with a PPP structure which include, among others, the use of financial and fiduciary managements, in particular, ring fencing revenue and subsidy flows from the government, to demonstrate the viability of the project and attract investment. Improving public access to information about the project and the procurement process, enhanced project procurement approach with greater control to package contracts to increase competition, transparency and control is also seen as an effective by product of PPPs (Beato, 1997; Brook & Locussol, 2001). Furthermore, the over-riding monitoring function of the lenders (who stand to lose money if corrupt practices are encountered in the project) to, for example, spot check works on a regular basis with external, specialized engineering consultants, add a further level of oversight and transparency (Project Finance, 2005).

Technology, innovation and know-how – the grantor may look to the private sector to provide technology, innovation and know-how. This will include access to skills and technologies otherwise unavailable to the government, or developed for the project specifically that would not be developed through other procurement mechanisms, for example, given the alignment of incentives created by a PPI project.

Whole asset life solution – depending on the type of PPP contract, a contractual arrangement granted to a project company or developer can last for 25 years or more. Some of the literature reviewed (Raymundie, 1996; Hodges, 2003; Harris *et al.* 2003; Irwin, 2003; Kerf, 2000; Saghir *et al.* 1999; Saltiel, 2003; River, 1996; Ringskog, 2000; Roche *et al.* 2001; amongst others) suggest that PPPs may have the effect of forcing the project company to adopt a more appropriate long term commercial approach to problem-solving and asset management. Public funding of infrastructure maintenance often falls short of requirements, particularly in developing countries. PPPs may help manage this funding shortfall, by designing sufficient financial resources and technical solutions into the project from the start.

Sources of financing – PPPs can encourage the mobilization of new or additional sources of finance for infrastructure development, and provide new opportunities for local financial markets. Harris (2003) argues that private participation in infrastructure has in some instances helped mobilise local financial markets not accustomed to providing financing to infrastructure projects directly, but which are looking for long-term, stable investment opportunities. Vives (2000) further argues that PPPs have had a catalytic effect for local pension funds in different countries in Latin America. Irwin *et al.* (1997) imply that PPPs may help a country maximise fiscal space by using capacity of private balance sheets and sharing risk in a manner able to extend the amount of investment globally. Other positive side effects identified by them include reducing fiscal strain on public companies by sharing the financing burden, and enhancing access to foreign financial markets and capital.

PPP arrangements are ultimately flexible, limited only by the creativity of those involved and the access to funding. Moreover, it must be emphasised that a careful *balance between the interests of the public and private sectors* is needed. On the one hand, the public sector has a responsibility and obligation to provide services to its inhabitants and ratepayers and on the other hand the private sector is entitled to receive adequate compensation for the risk of investing in, and managing, the utility as well as being recompensed for risking its private capital (Bradbury, 1992).

Accordingly, the design of the structure for private sector involves a clear definition of the roles to be played by the government and the private sector, the type of structure to be adopted for private sector participation, the applicable regulatory regime and its effects, as well as the specific current economic and political conditions (Cook & Kirkpatrick, 1998). Clearly, the type of structure chosen will leave an indelible mark on the provision of services and it is essential that an adequate legislative and regulatory framework be in place to ensure the sustainability of such a process. Regulation has to strike a balance between giving the private sector incentives to invest and operate efficiently, but at the same time protecting the interests and ultimate responsibilities of the public sector (Rees, 1998; Trémolet, 2002c).

The types of risks involved in the provision of water and wastewater services need to be adequately quantified, qualified and, most importantly, mitigated, as these are elements that will determine the success of a PPP project (Heming & Mansoor, 1988)[3]. It was noted previously of the need to balance the interests and roles of the public and private sectors in the PPP design, and that the types of risks involved in the provision of water and wastewater services need to be adequately dealt with. It is to be noted that all risks must be properly understood and allocated in an equitable manner to ensure an effective PPP arrangement, and what measures should be taken to mitigate the risks wherever possible. As an example, according to Rivera (1996), typical risks to the private sector in a PPP arrangement might include a variety of technical risks (e.g. related to lack of sufficient technical knowledge of the installations and the associated operations and the extent of necessary capital works). He notes commercial problems related to non-payment or inability to recover costs, tariff rates that do not allow adequate cost recovery with a reasonable level of profit and financial risks (including currency devaluations and conversion of local and foreign currency). Other risks include legal uncertainty related to the manner of resolving disputes, and political risks due to a change in government policy or the public sector's reluctance to increase tariffs. On the other hand, risks to the public sector might include services not in accordance with the agreed pre-defined standards and a possible political backlash.

As a basic premise, the allocation of risk between the parties should suit the relevant conditions in question, as well as the generally accepted precept that risk should be borne by the party most able to bear them. In stable economies with reliable financial institutions and institutional frameworks, the private sector might reasonably be expected to take commercial risks and compete for credits in the financial markets. Conversely, it is unlikely that the private sector would be willing to bear full financial risk in countries with developing economies, rapid population growth, low coverage of water and wastewater services, and where financial institutions and existing legislative and regulatory framework are still in their infancy. Given the importance of this issue Understanding and Managing Risk issues are considered in more detail in Chapter 5.

Frameworks conducive to PPP in the water and wastewater sector

Successful PPPs require a framework that includes a combination of factors to support a process conducive to attracting and maintaining the private sector Developer and at the same time ensuring that the Grantor is able to meet its development objectives. This is true for all forms of PPP, but in the water sector is especially important due to the public service nature of the service and the fact that the commodity is essential for human survival. The diagram in Figure 3.1, represents the four fundamental factors generally considered in developing an effective framework to supports the PPP process and the PPP agreement:

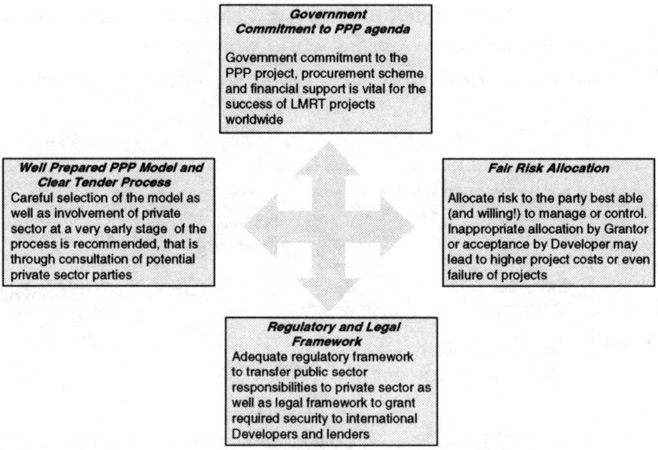

Figure 3.1 Frameworks conducive to PPP.

The basis for the entire development of these proposals is the assumption that both the Grantor and/or the Government have full commitment to the PPP agenda, otherwise the whole issue of the scheme development and implementation would have a basic structural flaw. As described, underpinning any PPP program is the concept of a fair risk allocation (Chapter 5) ensuring that risk is allocated to the party that is best able to undertake, manage and accept such risk. The overall legal framework and, to a lesser extent, the regulatory environment[4] will be important, as any PPP agreement that is developed will be enshrined within it (Chapter 6). The less these are developed, the more the PPP agreement will rely on the conditions of the agreement itself. However, this is likely to increase the perceived risk of any such transaction and will require within the PPP agreement itself, giving protection to the Developer and the lenders. This is discussed later on in Section 3.6, and in the Tender Process in Chapter 8. Consideration of all these issues should lead the Grantor to the selection of the most suitable type of PPP and the PPP Contract Structure (some generic structures are described in the following section).

3.3 OVERVIEW OF POSSIBLE PPP CONTRACT STRUCTURES

In the Introduction a variety of typical PPP contract structures that can be used were noted. The selection of the actual contract structure will be to meet specific project requirements, but in this section a brief overview is given of the key characteristics of the main contract types, starting from the general types of infrastructure, PPP contracts from the European Union view on PPPs and working down to the specifics for the water and wastewater sector. The final contract version will be chosen and tailored to suit the exact project needs, with the overall aim of ensuring long term sustainability of the water and waste water services.

General infrastructure PPP structures are represented by Delmon (2009) in Figure 3.2:

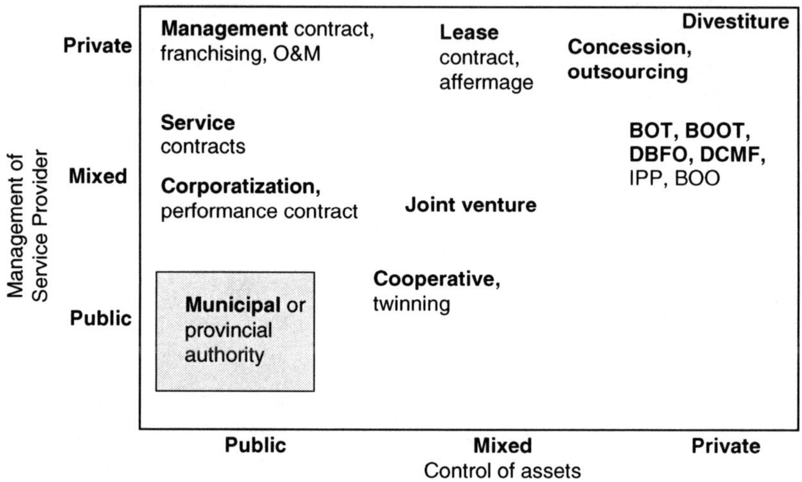

Figure 3.2 Characterization of general infrastructure PPP structures (Delmon, 2009).

Corporatization – involves a utility that is in public ownership being run in a manner similar to that of a private sector entity, using incentive mechanisms for staff and management similar to those used in the private sector. A corporatized utility may be structured as a limited liability company, with its share capital controlled by the public, while publishing the equivalent of an annual report containing a profit and loss account, balance sheet and cash flow data, giving a clear picture of the utility's finances and where any inefficiency may be found.

Performance contracts, service contracts, management contracts, operation and maintenance (O&M) – are all structures whereby a private company provides services to a utility/grantor, for example management services, selecting contractors, setting prices, billing and collection, overseeing construction and/or other services for a fee, generally based on performance or total cost.

Franchising, Cooperative, Twinning – is about using an experienced operator or utility's expertise and reputation through a local company's man-power and resources.

Build-Operate-Transfer (BOT), Build-Own-Operate-Transfer (BOOT), Design-Build-Finance-Operate (DBFO), Design-Construct-Manage-Finance (DCMF), Independent Power Producer (IPP), Build-Own-Operate (BOO) are similar in nature, looking to the project company to build (or refurbish if you replace the 'B' with an 'R') a facility – be it power generation, water treatment, a road, a hospital or otherwise.

Lease contract, affermage, concession – these generally involve the project company delivering services directly to the consumer, and differ primarily in whether the project company is responsible for new asset investment (in an affermage, generally it is not) and whether the project company owns the assets (in a concession, generally it does).

PPPs and the European Union

This section describes specifically the approach that the European Union has taken with respect to PPPs. EU law on public contracts and concessions does not prescribe whether or not a PPP arrangement should be used. A European Union Member Country, and thus by implication an Accession Country, is free to provide a public service itself or to entrust it to a third party. Moreover, EU law allows a pragmatic approach to PPP whereby it recognises that under PPP arrangements, public partners are primarily national, regional or local authorities. There may also be public law bodies created for the specific purpose to fulfil certain specific interests under State control which may enter into PPP arrangements. For the purposes of EU interpretation, the term 'contracting body' or 'contracting authorities' is used to designate all entities[5] which may enter into PPP agreements. In this way, EU law may be a useful guide for developing countries seeking to create an enabling PPP environment and associated legislation.

The EU is not prescriptive about the structure of PPPs but does seek to ensure that in selecting the partner of a PPP, the general principles of the EC Treaty apply, that is, directives on public contracts, procurement, equal treatment, proportionality and mutual recognition. Furthermore, all agreements in which a public authority awards a contract are subject to the general principles on freedom of establishment and freedom to provide services in Articles 43 to 49[6]. In addition, PPP agreements that count as public contracts are subject to the application of the public procurement directives. This is further explored in Chapter 8 related to PPP procurement issues.

The EU's definition of 'PPP' is a blanket term for agreements between public authorities and private industry. Such agreements may concern public infrastructure projects or public services and are categorised as:

(a) Purely Contractual PPP which refers to a partnership based solely on contractual links between the different players. It covers a variety of set-ups where one or more tasks are assigned to the private partner, and which can include the design, funding, execution, renovation or exploitation of a work or service.
(b) Institutionalized PPPs refer to the establishment of an entity held jointly by the public partner and the private partner. The joint entity has the mandate to deliver work or a service for the benefit of the public. IPPPs may be created

by the formation of a mixed-economy enterprise or company, but may also arise when a private company assumes control of what was a public concern, that is, by changing the share ownership.

(c) Cooperation agreements: these are usually between local authorities[7].

Recently significant steps have been taken by the Commission to develop an approach to PPP. To this end the Commission adopted the Green Paper on Public-Private Partnerships and Community Law on Public Contracts and Concessions[8] as well as subsequent communications[9]. This paper was the result of a consultative process[10] and was one of the actions resulting from the European Initiative for Growth[11]. The aim of this Green Paper has been to determine the suitability and applicability of Community law on public contracts and concessions involving PPP. It sets out the rules that should be applied when taking a decision to entrust a given activity to a third party (usually the private sector) under a PPP. It must be noted however, that these rules are downstream of the economic and organisational choice made by a local or national authority. The Commission makes no value judgment regarding the decision to use PPP and thus externalize the management of public services or not. This decision remains squarely within the competence of public authorities.

Specific PPP forms in the water sector

Specifically, in the water sector there is an array of options for involving the private sector characterised by a variety of contractual forms. These contractual forms differ in the degree of responsibility and risk transferred from the public to the private sector (Lee, 1996)[12]. The typical PPP structures that are used can be looked at based on how they deal with the following key differentiators:

- Dealing with Management and Operational Risk.
- Dealing with Commercial Risks (e.g. revenue and collection risks).
- Dealing with Capital Investment Risk.
- Dealing with Asset Ownership.

Table 3.1 illustrates, after Stiggers (2007) and Idelovitch and Ringskog (1995), the various levels of Developer risk in several key areas of responsibility for specific PPP contract forms for the water sector. Each checkmark represents an increased level of risk taken by the Developer. Broadly the table shows that there are two main groups of PPP contracts based on the level of risk the developer is prepared to take. In the first group, ownership of the assets remains with the government or the public sector, whereas in the second group partial or full ownership is transferred to the private sector. The first group, where the assets remain in public ownership, include the following type of contracts: Service and Management contracts, Lease arrangements and Concessions. The second group, where the assets are partially in private ownership, include: BOOT (Build-Own-Operate-Transfer), variations to BOOT contracts such as BOT (Build-Operate-Transfer), Design Build Finance Operate

Maintain (DBFOM), BOO (Build-Own-Operate) and reverse BOOT, joint ownership or mixed companies, and outright sale, or partial divestiture. These contractual forms will be enumerated in Table 3.1 and the manner in which each contract allocates roles and responsibilities will be described.

Table 3.1 Main PPP structures and developer risks.

Developer Risks: / Contract Form:	Operational	Management	Commercial	Financial	Capital Investment	Asset Ownership
Service Contract	☑☑					
Management Contract	☑☑☑	☑☑	☑			
Affermage - Lease	☑☑☑	☑☑☑	☑☑☑	☑		
Concession	☑☑☑	☑☑☑	☑☑☑☑	☑☑☑	☑☑☑	☑
BOOT	☑☑☑	☑☑☑	☑☑	☑☑☑	☑☑☑	☑☑
Joint Ownership	☑☑☑	☑☑☑	☑☑☑☑	☑☑☑	☑☑☑	☑☑☑
Sale/Divestiture	☑☑☑	☑☑☑	☑☑☑☑	☑☑☑	☑☑☑	☑☑☑

Source: Adapted from Stiggers (2007).

Choosing the appropriate option depends on what the government in question requires from the private sector, that is, the objectives to be satisfied by entering into such contractual agreements, and, as a result, how much control of the service it is prepared to relinquish. The following section examines the different forms of PPP based on the degree of private involvement in the development and ownership of the assets. Throughout this section reference will be made to various examples where these forms of contract have been used. These examples draw upon the author's personal experience unless otherwise stated.

Service contracts

In a Service contract the public authority retains overall responsibility for the operation and maintenance of the system but contracts out specific, limited-scope services. It is important to note that in these types of contracts the public authority bears all the commercial risk and must finance fixed assets as well as working capital. The private sector's responsibility is limited to managing its own personnel and services efficiently (Haarmeyer, 1992).

The types of activities that may be covered by Service contracts include: maintenance, emergency repairs, meter reading, billing and collection, upgrading of existing facilities or construction of new ones, and rental of equipment. These contracts are for a short period of time, usually not exceeding two years, and are generally renewable. The basic premise of these contracts is that they require little or no fixed investment on the part of the private operator. Efficiencies are effectively limited to the short contract period.

Service Contracts are used most effectively as a mechanism to bring about specific improvements[13]. These have been used successfully by many public entities to improve, for example, billing and collections. In these schemes, such as with the Guatemala City Municipal Water Company, the utility entered a three year service contract with a private entity specialised in the issuing and collection of bills. Using a simple contractual structure, this was an effective mechanism to improve the financial standing of the entity.

One of the main advantages of these contracts is that in large urban areas, for example, different firms can be contracted in separate geographical areas to deliver the same services. Multiple contracts ensure adequate competition and enable the water authority to compare costs and performance on an ongoing basis. However, it must be noted that if Government is to use this type of contract they will need to be prepared to manage the private sector, that is, their role is to shift from that of execution to supervision (Idelovitch & Ringskog, 1995).

Management contracts

In a Management contract the public sector transfers responsibility for management of the entire operation and maintenance of a system including associated workforce, to a private company (Figure 3.3). This gives the private company freedom to make day-to-day management decisions without assuming any commercial risks. What is important to note in this type of contract is that the private sector acts in the capacity of 'agent' to the public sector, that is, it acts at all times on behalf of the public authority. However, payment to the private sector is in the form of a fee, generally linked to the performance of the utility. Financial responsibility continues to be vested in the government or public authority and it has the responsibility to provide working capital and investment funds (Stiggers, 1997).

Payments to a management contractor are usually proportional to certain physical parameters, such as improved efficiency, volume of water produced, improved collection rates (i.e. some very limited commercial risk), or reduction of unaccounted-for water.

The duration of these contracts is usually three to five years. Therefore management contracts do not entail large investments with long payback periods to be made by the private sector. Often the intention is to act as a kind of 'interim' management prior to finding a more permanent and robust form that brings capital investment. Thus Management Contracts may feature an option to convert to more intensive, longer-term, PPP arrangements.

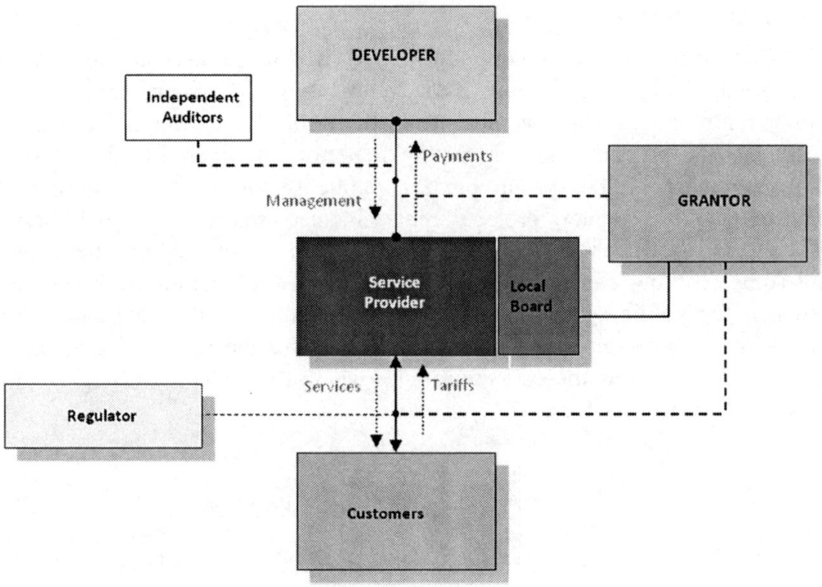

Figure 3.3 Typical management contract.

Management contracts have not been as effective as originally hoped. From my personal experience, an example of a management contract that worked well, for the most part, was that of Trinidad and Tobago, where I was Director of a management contract that had a sound contractual structure and some very clear private operator performance indicators. The idea behind the management contract was to achieve real improvements through private management of the water company (the Water and Sewerage Authority of Trinidad and Tobago – WASA) enhanced by a targeted investment programme. However, at the time the management contract was designed and subsequently let, a loan from the World Bank was envisaged, specifically designated for the implementation of major capital investment involving major system rehabilitation. In reality, what happened was that the management contract was implemented for a period of three years but there was no capital programme as the World Bank loan did not materialise. The pipework and infrastructure required major rehabilitation of old and leaking pipework and structures, and this was not possible without the planned investment Therefore, the ability of the management contractor to bring about significant operational and financial improvement was limited in aspects directly related to capital investment, such as continuity of water supply or reduction of leakage. This lack of linkage between the Grantor's obligation to provide investment and the Developer's obligations under the Management Contract is a potential issue that can hinder performance improvement. This issue can be dealt with contractually by ensuring the Grantor's funding and capital investment obligations are clearly set out (Suleiman, 2002).

Lease contracts

Lease Contracts have been commonly used in France and are also known as 'Affermage' contracts (Figure 3.4). In this type of arrangement a private operator rents or leases the facilities from the public sector for a given period. In essence, the private sector is renting a permit for operation, maintenance, and management of the system but the public sector is still responsible for capital expenditure for new projects, replacement of major works, debt service, tariffs and cost-recovery policies. Occasionally the Developer may be asked to bring some working capital to support his day to day operations. It should be noted that the public sector maintains the ownership of all assets and there is no transfer of ownership rights. The Developer takes the risk of operations and management as well as the collection risk – where the collected revenue is used to pay the Developer's fees.

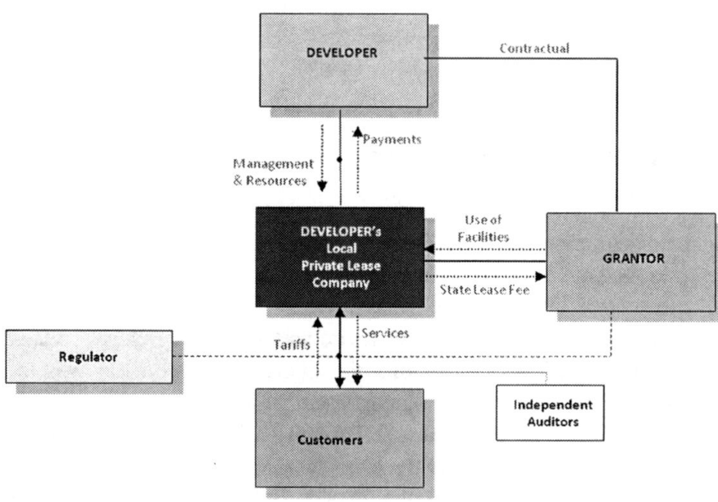

Figure 3.4 Typical lease contract.

Lease contracts can be medium or long-term duration. They usually last five to ten years but can be extended for as long as twenty years (Camdessus, 2003). Lease contracts have a number of benefits as they allow for:

- Automatic periodic revisions of the rate being paid to the private sector through the use of price index formulas.
- Periodic re-negotiations of the rates and cost index formulas on the basis of results achieved by the private sector. In this way, savings that the leaseholder has been able to achieve during the previous years can be passed on to consumers.
- Specific penalties that will apply in the case of poor performance.

Given the above factors, the risks involved in a lease arrangement tend to be limited (i.e. Operation, Management and Commercial collection risks) and the arrangements generally allow the private sector to become more comfortable with the system and associated risks. Accordingly, Lease Contracts can be used effectively for more extensive potential involvement in the future.

However, Lease Contracts run the risk of placing undue burden on the Government and its ability to raise funding. In the case of Dar es Salaam – Tanzania, where one of the authors acted as senior advisor to the Government with funding from the World Bank, the main issue was to ensure that the lease payments made were sufficient to cover some of the associated costs of the loan programme that was designed for capital improvements. It is interesting to note that, in this contract, it was a requirement for bidders to design a capital improvement programme to be financed with proceeds of a World Bank loan against which the Private Operator would have to perform.

As in the Management Contract, there are potential problems caused in a Lease agreement if there is a lack of linkage between the Grantor's obligation to provide investment and the Developer's obligations under the Lease agreement, as any lack of planned investment will generally hinder performance improvement. This issue can be dealt with contractually by ensuring the Grantor's funding and capital investment obligations are clearly defined.

Concession contracts

In a Concession, the Developer, or concessionaire, has overall responsibility for the services (Figure 3.5). The distinctive feature of this type of contract is that the private sector is not only responsible for operation, maintenance and management, and all commercial risks, but it also has full responsibility to source funds for the capital investments needed for the expansion of services, either directly or indirectly (i.e. through banking or financial institutions). The fixed assets, however, remain the property of the government or public authority, but they are entrusted to the concessionaire for the duration of the concession contract and must be returned in the same condition at the end of the concession period. There may be some temporary ownership of newly constructed assets, but with eventual asset transfer to the Grantor.

The duration of Concession contracts is usually twenty to thirty years, depending on the level of investment and the payback period needed for the concessionaire to recover investment costs. If some capital expenditures have not been fully amortised by the end of the contract, there is usually a provision in the contract for the contractor to be compensated accordingly (Dumol, 2000). The concessionaire retains exclusive rights for the duration of the contract. When the contract expires, all works and equipment are turned over to the public sector.

Under a Concession contract, the Developer is paid directly by the customer, based on the defined user fee, generally related to consumption. The private sector assumes 'collection risk'. The private sector retains the balance of revenues after paying back any taxes and charges levied on customers by the public authority. If expenses (i.e. cost of capital works program in addition to operating expenses) exceed revenues, the private company suffers losses. Depending on the size of the capital works program these losses could be quite heavy. Penalties in Concession Contracts may be levied for failures to meet pre-defined targets for investment as well as service related performance measures (Dinar, 2000).

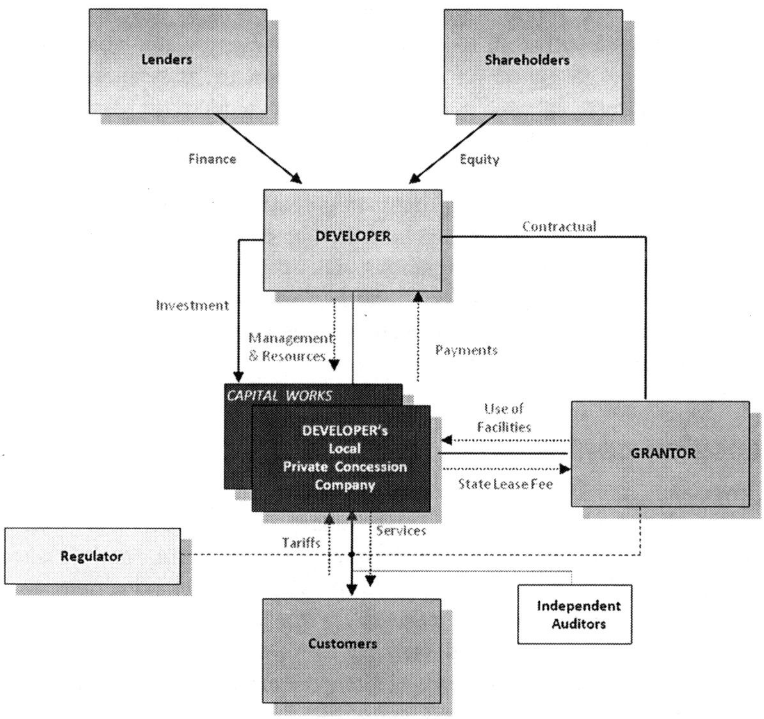

Figure 3.5 Typical concession structure.

Concession contracts by their nature tend to be for long periods of time in excess of twenty years. Conditions change over such long periods and assurances made at the time of the signature of the contract might not necessarily hold true after a given period. According to a study by Estache *et al.* (2002) there is an interesting example in what happened with the concessions in Argentina, where the Government defaulted when the foreign currency protection mechanism was exceeded due to the devaluation of the national currency (peso). This is an example of the need to ensure that concession contracts are robust enough to allow the private sector to raise capital against these contractual structures, but that at the same time the contracts are sufficiently flexible to deal with unforeseen changes in circumstances.

Concessions have some of the most complex contractual PPP structures, covering many issues found in other models, and for this reason are used as the starting point to develop the Recommendations in this research for development of Key Contract Provisions for Optimizing PPP arrangements (Chapter 9).

BOOT contracts

Build Own Operate Transfer (BOOT) contracts encompass the private sector financing, building, owning, and operating a specific new facility or system.

The BOOT contracts are made on a similar basis to the Concession structure (see all comments above), but are generally used for a new facility or system, rather than operation and further development of an existing system (Delmon, 2009). After a predetermined period of time, ownership of the facility is transferred to the public sector. The public sector is usually responsible for determining the demand for the service being contracted, and therefore for the size of the facility. Demand is often guaranteed by Grantor. In this regard the Developer may consider the commercial risks to be less than in, say, a Lease contract. One point to be noted is that problems may arise if there are differences between real and estimated demand, with associated financial risks to the Grantor, as well as potential operational problems for the Developer.

The duration of BOOT contracts is related to the time needed to cover the debt incurred, as well as the necessary time to provide adequate return to equity investors. At the end of the pre-defined period, the private sector transfers the facility back to the public sector. As to the risk borne by each one of the parties, it is important to note that BOOTs may represent a substantial risk for the private sector, particularly if there are no assurances that the investment output will be paid by the public sector, or that the quality of services that can be achieved will be uniform and according to design standards.

The traditional BOOT schemes may have a number of variations depending on the level of risk allocated to the parties. For example, one slight variation of the BOOT system is known as BOT. In this case ownership is transferred to the public sector as soon as the facility is completed, and the function of the private firm is only to construct and operate it (Franceys, 1997; Frangano, 2001).

One more variation is BOO, whereby ownership is not transferred to the public sector but remains with the private firm that constructs and operates the facility (Gray, 1997).

Another variation is used in countries where economic or political risks are high and where the private sector may be requesting very high risk premiums in return for their participation. In this case the public sector finances and builds the plant itself and then contracts a private firm to operate it over a long period of time. It is then possible for the private firm to acquire the plant gradually by paying the public sector an annual fee which usually covers the full debt service of the entire investment cost (Gray, 1997; Kerf *et al.* 1998).

Joint ownership

Joint ownership is a good solution for situations where private sector participation is desirable but it is not possible to pass on all the risk of investment and operations to the private sector (Figure 3.6). Under this scheme, a private sector firm and the public sector incorporate a firm under the country's existing commercial code. The basic principle here is that, although the parties start with a fairly equal number of shares, the public sector has the option to sell part, or all, of its participation to the private sector in the future.

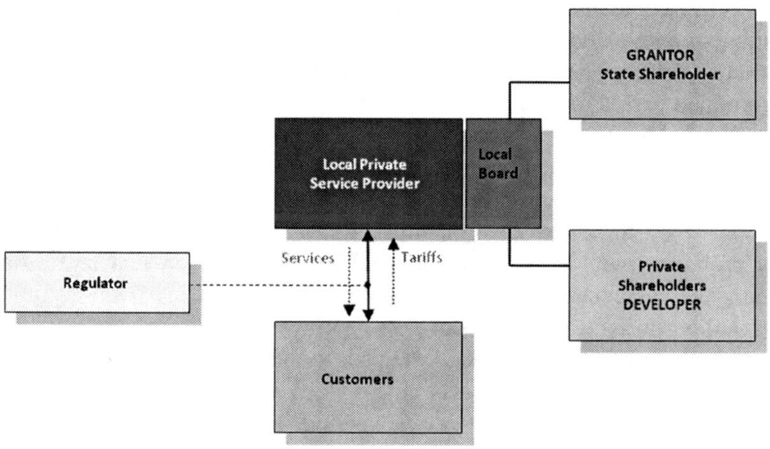

Figure 3.6 Typical joint ownership structure.

What becomes crucial in this type of agreement is the contractual agreement that regulates the relationship of the shareholders, that is, the private and public sector. The roles and responsibilities are detailed in a shareholders agreement. Particularly relevant to this agreement is not only the mechanics for distribution of profits but the level of profits to be distributed between the partners.

Idelovich and Ringskog (1995) argue that successful jointly owned companies can be expected to establish creditworthiness and to raise capital by floating bonds or issuing notes. This has the advantage of limiting public sector debt, an important sector objective. From my experience, raising capital through bond flotation is unlikely due to the typical immaturity of the financial markets where these types of agreements tend to occur. In addition, supporters of joint ownership argue that the shares of the company would eventually be put on the local stock market and that pressure to operate efficiently will be exerted through stock market operations (Mandri-Perrott, 2005). I consider this to be definitely feasible in more advanced economies such as the UK where the public owns shares in the water and wastewater companies but, judging by this research, there is also a growing use of these in developing countries.

However, it must be noted that in countries with a weak regulatory tradition, the benefit of joint ownership is conducive to satisfying regulatory requirements since the public sector will be represented by a board of directors who will have a broad insight into the firm's operations.

Outright sale/divestiture

Some governments have opted for an outright sale of the water assets and wastewater services (Figure 3.7). The sale may be prompted by the public sector's desire to completely separate ownership from operations and maintenance. It is also a way for the public sector to raise revenues. This has been a strategy used effectively in the sale of other utilities such as telecommunications and energy providers. However, it must be noted that in the water and wastewater sector its 'attractiveness' to the private sector will depend mainly on the rates or tariff they will be allowed to charge. It must be clearly understood that the assets themselves have virtually no alternative value and that the business is modelled solely on its capacity to generate revenue and not on the value of the assets per se (Hudson Institute, 1999).

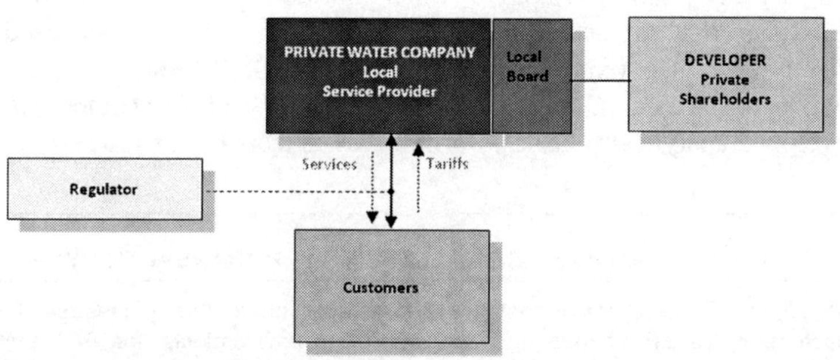

Figure 3.7 Private water company structure.

The most notable example of this is the sale of water companies in England and Wales (Ofwat, 2001). However, the particular political and commercial situation of the UK market at the time, with a general desire for full privatization of public utilities, together with a well established capital market, may make this a difficult model to replicate elsewhere.

3.4 KEY REQUIREMENTS FOR A SUCCESSFUL PPP

When coming to the decision on whether to involve private sector participation, governments have to assure themselves that this creates value in excess of associated costs (Whittington & Venkateswarlu, 1994). Compared to public entities, private firms usually have higher costs of capital[14] as well as profitability requirements which significantly impact the cost of infrastructure initiatives.

Accordingly, structuring a PPP is to define a legal and financial optimal structure in order to maximize a given objectives such as (a) efficiency in the management issues, (b) efficiency in the costs (such as ensuring a cost efficient tender process), and (c) categorization of the level of finance from private and public sources. A well designed PPP arrangement should in principle enhance 'value for money' through a combination of factors including: financing, operational efficiencies, superior risk management, greater implementing capacity and enhanced service quality (Whittington, 1992; World Bank, 2001; World Bank, 2004; Gómez-Lobo & Hinojosa, 2000).

Designing, constructing, operating and maintaining water and wastewater systems is complex and requires substantial resources. Public institutions may lack the institutional capacity required to implement traditional public procurement of water sector components. The value added to the design and contractual integration of water and wastewater systems may justify the additional costs of private sector participation.

Achieving value for money

As previously discussed in Chapter 1, when considering PPP options, the Grantor needs to ensure that the scarce public resources available to it are used in the most effective way. Governments with experience with PPP (for instance Canada, Australia, South Africa (Box 3.1), the Netherlands and the United Kingdom) have developed analytical tools to inform policy makers in this process[15].

BOX 3.1 SOUTH AFRICA – ASSESSING PPP TRANSACTIONS

South Africa's Treasury Unit contains a dedicated group that oversees all PPP transactions in the entire country. Before approving new projects, the PPP unit will verify the following:

Affordability – Any public institutions involved must be able to meet their financial obligations under the PPP agreement throughout the life of the project. Current and anticipated budgets must be able to support any PPP related expenses including the costs of public oversight.

Value for Money – Incorporating private sector participation must offer additional value to the public sector through risk transfer, increased quality or some combination thereof. In the absence of Value for Money, traditional public procurement would be the preferred method for realizing new infrastructure services.

Substantial Risk Transfer – Private parties must assume meaningful risks as part of their participation in a proposed PPP arrangement. Value for money often results when private partners can manage project risks better than public counterparts.

South Africa's Treasury unit employs a series of reviews and approvals throughout the project development process. In some cases, funds from a special Project Development Facility can help local and regional planners engage additional resources to aid in planning projects that meet Treasury standards for PPPs. To learn more about South Africa's PPP unit, please visit: www.ppp.gov.za.

Quantitative analysis is needed to assess the value for money proposition of a water PPP agreement. This usually involves a comparison between net costs associated with public service provision compared to those under the PPP scenario. This is often captured within a Public Sector Comparator (PSC) or Value for Money model against a PPP reference model. Together with qualitative assessments, this quantitative analysis can help public institutions make rational choices regarding infrastructure procurement. However, using the quantitative models, the Grantor should consider a number of issues including:

- *Completeness and accuracy*: both PSC and PPP reference models may omit or incorrectly value critical risks. In the case of water PPPs these mistakes may involve:
  - Contingent liabilities.
  - Contract negotiation risks.
  - Explicit but variable liabilities such as operating subsidies and minimum revenue guarantees.
- *Model inputs*: selecting appropriate inputs for project models is always a difficult task. The subjectivity of inputs can provide an avenue for manipulation and undesirable biases. Many developed nations now use sophisticated databases as a check on modeller discretion. However, this is often not possible in developing countries where data from previous projects may be unavailable or lacking in quality. Discount rates for the opportunity costs of funds are particularly influential model inputs and may be especially contentious. Small alterations in discount rates can inordinately affect analyses as values are typically compared on an NPV basis[16].
- *Development costs*: conducting thorough quantitative analysis of a project's value can require time and considerable costs – especially where institutional capacity is lacking and public authorities rely on external advisors.
- *Timing*: determining the initial value for money of a PPP based procurement approach requires some indication of final project costs. However, these costs are not entirely 'knowable' until later procurement stages and financial close. Even where PPP reference models are well crafted, their results still may differ substantially from actual procurement costs.
- *Reasonableness*: as suggested by Leigland and Shugart (2006) constructing public sector comparator (PSC) models may not make sense when public authorities cannot conceivably implement a project for lack of funds or institutional capacity. Traditional financial and economic analysis assuming private sector participation may be the only quantitative analysis available for assessing projects when this is the case.

In the United Kingdom the concept of Value for Money is defined as the optimum combination of whole-of-life costs and quality (or fitness for purpose) of the goods or services to meet the Grantor's requirements. Value for Money is *not* the choice of goods or service based on the lowest cost.

A widely referred to study in the literature on PPP contracts carried out for the UK Government by Arthur Andersen and Enterprise LSE in 2000, identified six key factors that drive VfM, namely:

- *Risk transfer* – optimally allocating risks to the party best able to manage them over the contract period.
- *Long term contracts* – these enable the private sector to recover the initial investment, develop alternative approaches to service delivery and focus on life cycle costing (*not applicable to Management and O&M/Service Contracts*).
- *Use of performance-based specifications* – specifying services provided under PPP project contracts as outputs enables payments to the private sector to be linked to the quality and delivery of those outputs.
- *Competition* – high levels of competition between interested private sector Developers can result in improvements in pricing and service delivery innovations.
- *Performance measurement and incentives* – contractual incentives act as a means of ensuring that the standards and specifications set in place through the original deal or contract will be implemented.
- *Private sector management skills* – the private sector's ability to effectively manage the delivery and operations of a project is critical to the success of the PPP.

Importantly, it must be noted that the PSC represents a hypothetical methodology based on a series of assumptions rather than actual cost to government (Figure 3.8). This should be remembered throughout the process. Importantly, PSC are not an exact metric and should also be undertaken after the contract has been awarded.

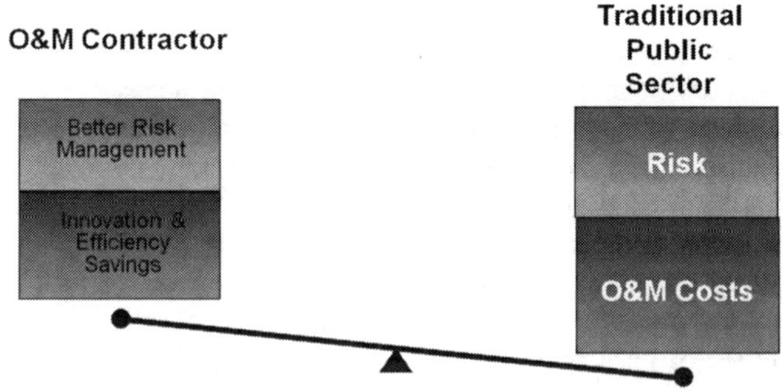

Figure 3.8 Value Added of O&M Contractor Approach. *Source*: Halcrow.

Under the PSC/VFM methodology, the 'value' created by the Grantor through transferring risks to the Developer is added to the 'public' cost estimate. The principle is that these risks are 'included' in the price of the Developer cost estimate through the risk premium ('Profit Margin') charged by the Developer/contractor, and therefore, should be *added* to the public cost estimate to allow for an accurate comparison of the 'true cost' of public versus private service provision (Figure 3.9).

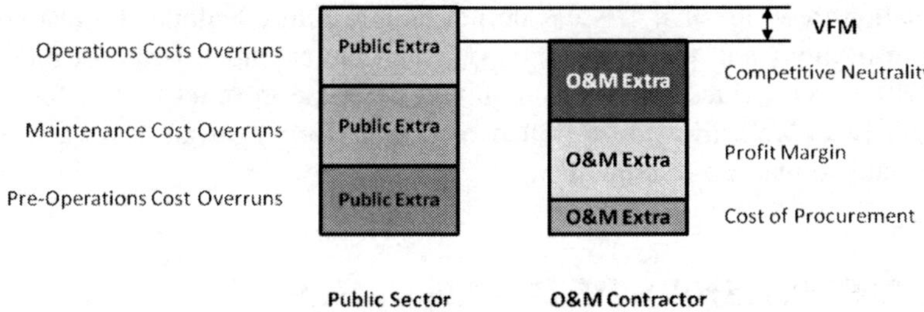

Figure 3.9 PSC/VFM Methodology. *Source*: John Leber (2012).

A different approach to calcaulating Public Sector Comparator using an approach developed by the Government of Victoria, Australia for its PPP project analysis. The PSC estimates the hypothetical risk-adjusted cost if a project were to be financed, owned and implemented by government. The PSC is developed in accordance with the required output specification, the proposed risk allocation reflected in the contract, and is based on the most efficient form and means of government delivery.

Contingent liabilities

Contingent liabilities represent commitments to future expenditure (e.g. by the Government) if certain events occur (Leigland & Shugart, 2006). Many of the risks associated with private sector participation in infrastructure create sizeable contingent liabilities for public institutions. Because such liabilities are uncertain and do not correspond to definite cash flow events, simply relying on cash based budgetary analysis will fail to consider their potential impacts on affordability. For example, a public guarantee on a water project's debt will not result in a direct cash outflow, but nevertheless, the value of such a guarantee has *potentially* enormous future budgetary implications that should be considered. The challenge for planners is valuing and accounting for such contingent liabilities appropriately. Typical methods used to accomplish this include:

- Actuarial or Statistical Techniques.
- Econometric Models.
- Contingent Claims Analysis.

Contingent liabilities and the opportunity cost of cash reserves established to cover their potential impacts should form an important component of affordability analysis (Lewis & Mody, 1997). Later sections of this book will discuss the key issues related to valuing risks and contingent liabilities (Chapter 6).

Budgetary implications

The budgetary implications of water and wastewater initiatives have a major impact on public sector finances. The massive nature of water and wastewater expenditures has to be justified against the opportunity cost of scarce public funds. Planners should also closely consider contingent liabilities associated with private sector participation to ensure that services do not require future bailouts at great cost to public institutions and taxpayers. Indeed, given the essential nature of the water and waste water service provision, it should be borne in mind that governments are unlikely to walk away from failed or non performing water or waste water service, and so this is a true liability.

3.5 PARTNERSHIPS BETWEEN PUBLIC AND PRIVATE INSTITUTIONS

Refining and improving the ways that water systems can incorporate private sector capital and expertise makes good sense given the size and scope of such projects. There is a clear need for increasing the long term sustainability of private sector participation in the water sector considering the large amount of money (both public and private) at risk as well as the lengthy duration of contractual relationships. Partnering with the private sector can deliver substantial value by augmenting limited government capacity (both operational and financial) to the benefit of customers. However, the mechanics of actually formulating complementary relationships between public and private entities is never simple. Failing to 'get it right' can have severe consequences for tax payers, investors and customers alike.

Preserving private incentives for performance, while at the same time using contributions of substantial public funds for investment in new services, requires delicate management. Achieving the possible benefits of a PPP arrangement for water schemes requires getting two big things right:

- Giving the Developer the ability and incentives to make good operating and investment decisions. This means giving the Developer enough freedom to make decisions while exposing it to the related business risks—so that it gains when getting decisions right and loses when getting them wrong.
- Protecting the Developer from the risk of losing through the government changing the rules of the game rather than from bad operating and investment decisions. For example, this means protecting the Developer from the risk that the government will opportunistically cut prices after the Developer has invested (or take similar actions that undermine the investor's profitability).

Neither goal is easy, but protecting the Developer from policy risk is especially hard. The challenge is to embody the proposed policy objectives in a functional PPP contractual arrangement. A good PPP arrangement will be reflected in a PPP agreement (contractual terms) that match risks and rewards and establishes adequate 'controls' for both the Grantor and the Developer. This whole subject is covered in more detail in Chapter 7 and in the contracts section, Chapter 9.

A private Developer's profit incentive may cause it to operate more efficiently than its public counterpart. For example, it may provide the same level and reliability of services with improved water quality and quantity fewer staff. However, on the other hand, the PPP arrangement must be designed to ensure that this profit incentive is harnessed to the benefit of both the public and private partners.

Often a key element in private operation is the know-how, technology and other managerial techniques employed. It is important that the benefit of the private Developers' experience in provision, management and operation of water and wastewater systems is taken advantage of. This can be gained at the initial design and development stages, through market consultation, or at the bid stage through contractual arrangements that include appropriate degrees of responsibility for the plant and equipment design and operation (Contracts, Chapter 9).

Finally, although typically private Developers can bring skills to improve services through more effective management and operation of services it is important to realize that there is a limit to what can be achieved in operational improvements without the necessary capital investment for refurbishment or new infrastructure. A key part of any PPP arrangement is to establish how this capital investment will be funded, who will have responsibility for carrying this, and how this will be linked to sustaining or improving levels of service.

3.6 MANAGING THE STAGES OF DEVELOPMENT AND IMPLEMENTATION OF PPPS

This section describes the review of current best practice for the preparation, implementation and management of PPP arrangements. Usually it involves three stages that may overlap each other in time as detailed in Figure 3.10.

The time required to complete the preparatory stages varies by country and by the arrangement being pursued. Countries with laws supportive of private participation in water and sanitation services and with good-quality information on the system may proceed relatively rapidly. In addition, a management contract usually takes less time to prepare and implement than a concession. With strong political commitment, a management contract could typically be designed and implemented in under 12 months, while a concession and its associated financing could easily require two or more years.

Governments may, with benefit, choose to allow more time initially for consideration of the issues and setting in place the upstream sector strategy and policies, as well as management of social and political concerns, to ensure the successful development of the water and wastewater scheme. One outcome is that generally this will allow a more effective design and management of a PPP arrangement.

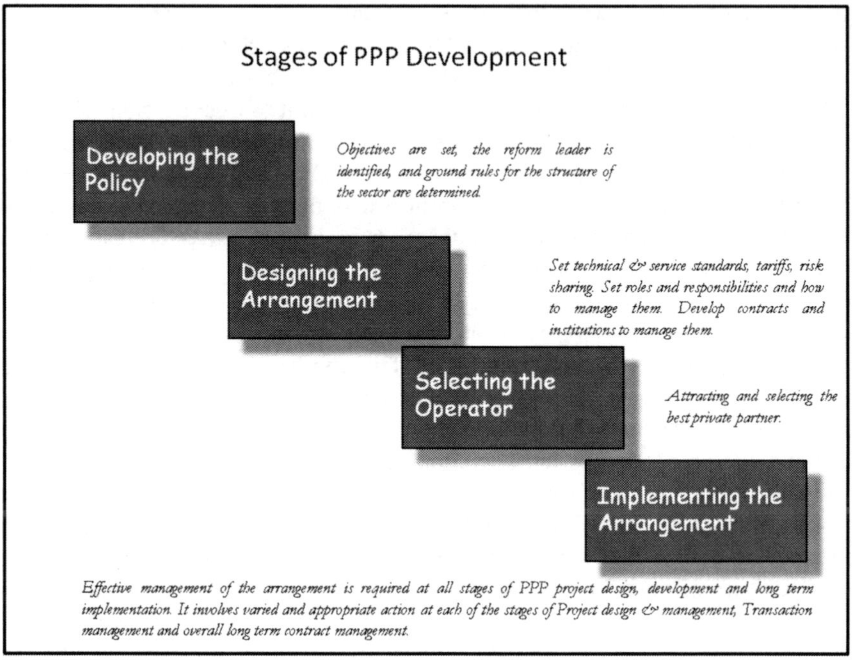

Figure 3.10 Stages of development & implementation. *Source*: Modified from World Bank Water Toolkit.

Financing of water and wastewater PPP projects is always an important, complex and time consuming process, and more details of the issues and possible approaches involved are given in Chapter 7.

Developing PPP policy

The first step in developing policy includes deciding on whether to consider private participation as a tool for achieving the government's objectives. As a first step the government will need a decision on what the scheme is intended to achieve and its link into the overall plan for the sector. This starts with setting objectives and recognizing the trade-off between different objectives.

Clearly articulating and agreeing objectives at the start of the process allows everyone to work toward a similar end and provides a solid framework for choosing between options and resolving disputes during the design process. Some key issues that will need to be considered at this stage include:

External constraints, that may impact the type of PPP arrangement and for which the Grantor needs to take into account any over-riding legal and physical restrictions on the operating environment. Some of these could include the legal and regulatory environment. The choice of contract type will be determined (and in some cases limited) by the legal environment under which the water and wastewater scheme operates. The implications of the legal regimes are further discussed in Chapter 5. The business environment will also affect the monitoring regime that is developed.

Budgetary constraints may result given the fact that water and wastewater schemes are capital intensive for both the public and private sector. The public sector through the Grantor (and depending on the chosen form of PPP) may have significant financial exposures to the proposed scheme. Working within its budgetary limits, the Grantor has to assess the level of subsidy and or public contribution it is able and prepared to pay.

Detailed service planning issues, which although typically done by government in defining the sort of public water services required, can be augmented by the Grantor by for example, detailing tariffs, service levels and subsidies. The PPP contract has to clearly set out each party's role. In some instances, the service planning function is completely carried out by the public authority, even down to tariff setting and level of subsidy.

Designing the details of the arrangement

Once the objectives, vision, and structure for the sector are set, the details necessary to make it work need to be developed. In particular, the following should be considered as integral activities in designing the PPP arrangement and how it gets reflected in the PPP contractual agreement (Kerf *et al.* 1998; Klein & Hadjimichael, 2003; Littlechild, 2002):

- Stakeholder consultation and communication are important – knowing what they want and their contribution to the reform to make it sustainable. Identify early on issues that are likely to be politically sensitive or require policy decisions.
- The heart of the arrangement is the level of service to be provided and the tariffs and any subsidy that will have to be paid. This involves technical and financial work, and a large degree of consultation.
- The technical and physical determination of the water and wastewater infrastructure requirements has to be made, and the costs and the resulting Investment Plan have to be established.
- The choice of who will bear risks and costs of the chosen Arrangement has to be made (both private and public sector) and these decisions may influence the type of contract chosen.
- Institutions need to be developed to manage the development, and management of the PPP agreement, as well as supervision and adjustment of service standards and tariffs.
- Legal frameworks and contracts have to be drafted to ensure an effective legal arrangement (PPP agreement) to reflect all the design points and ensure stability of the long term arrangement.

Designing and implementing an arrangement requires economic, financial, technical, and legal expertise, and the coordination of that expertise. Detailed work is needed to refine the option to be implemented and the legal measures to support it, and to prepare complex documents, such as laws, bidding documents, and draft contracts. Governments or municipalities usually lack the full range of expertise within the civil service to carry out these tasks and so will need advisers[17] to provide some of these skills and specialized expertise. Management of transactions of major water and wastewater schemes with private sector involvement needs a specialist management expertise and it is likely that transaction advisory support from specialist advisers will be needed to help lead the whole process. There will be times when more or less work is needed, and the appropriate combination of advisors will always depend on the particular circumstances. The costs of advice always need to be weighed against its benefits.

Selecting the private service operator

Selecting the Developer/Operator involves both ensuring that the opportunity is attractive to potential private service Operators and using good processes to determine which Operator will be the best partner (Lorrain, 1997; Saltiel, 2003). Research by Shirley and Walsh, (2000) suggests that involving private service Operators early in the process is usually a good idea because it increases the likelihood Operators will be interested in the arrangement. This is further confirmed by Brocklehurst and Janssens (2004), Bradbury (1992), Budds and McGranahan (2003), Caplan and Jones (2001a & 2001b), Chapman and Cuthbertson (1999), Cook and Stevens (2005), Engel *et al.* (2000), and Guasch *et al.* (2005) The government will also need to consider the kind of Developer it wants and can expect to attract. In the various PPP contracts evaluated under this research the possibilities for Developers range from large international investment firms with experience in private participation in water and wastewater provision schemes, to small local firms, alternative providers, and individual entrepreneurs. After having decided on an arrangement a government needs a suitable private partner. Processes to select and reach agreement with a private sector Developer can be divided into three broad types[18] (a) competitive tendering, (b) competitive negotiation and or (c) direct negotiation.

Managing the arrangements

Early on, the government may wish to choose a 'reform leader'—that is, a government entity that has appropriate skills, capacity, and responsibilities and that can champion and coordinate the overall process. Choosing the right entity is sometimes difficult. For example, if local water and wastewater services are a municipal responsibility, should the reform leader be the municipal administration (which has the appropriate responsibility) or a central government agency (which has more power and capacity)? According to Sommer (2001), it is an advantage to have this 'Reform Leader' established at the earliest stage, whilst policy is still being developed. Other tasks under 'Developing the Policy' which should be addressed include allocating responsibilities to different tiers of government, deciding on the market structure including who is responsible for setting and or regulating tariffs.

To manage the process, the government needs to be clear about which level of government is responsible for managing the process in the long term, and set up a streamlined management structure with strong analytic capacities and a reporting structure that brings powerful decision makers into the process in an effective way (Clarke *et al.* 2002). For reform to proceed smoothly, the authority leading the transaction (e.g. the Grantor) may need to establish a project team. The team could consult with interested stakeholders or representative forums, while viewing the process from a broad social perspective that focuses on achieving the government's objectives (Box 3.2).

BOX 3.2 PROJECT TEAM MEMBERS

Project Team Members might include:

- A high-ranking official from the contracting authority.
- A legal official with applicable policy and procurement experience.
- A technical officer with appropriate engineering knowledge and experience.
- A representative with expertise in communication skills.
- A representative of the local water and wastewater authority.
- A finance officer with experience in the financial management and funding of public utilities and, if appropriate, the negotiation of financing arrangements with private investors and public-private lenders.
- Political representatives (such as municipal councillors).

Source: Modified from Clarke *et al.* 2002.

A project team's skills are crucial. From the research it appears that successful PPP transactions generally involve a team which includes senior individuals drawn from the Grantor and other agencies with a particular interest or area of responsibility related to the project. The project team will typically have an advisory role with the Grantor approving all key decisions. The project team's delegated powers should allow the reform process to proceed in a timely, efficient, and transparent manner with appropriate checks and balances. The project team needs to report to a suitable political decision making group, generally a steering group. The steering group may be a cabinet subcommittee, a committee of municipal leaders, or a combined local and national-level committee. The steering group should make decisions, with recommendations from the project team.

There are many ways of arranging the project management and decision making structure. My research seems to suggest that there are a number of characteristics of a successful structure and which are likely to include a competent, dynamic, and focused project team, with a mandate to develop options and proposals, and to describe them clearly. According to the World Bank (2006a), a decision making group including enough influential people to ensure that the group's decisions are not undermined, while being focused enough to provide clear and rapid responses to proposals from the project team, as well as a reform champion—a senior individual who is committed to moving the process along and overcoming inertia, are key success factors[19].

Once the arrangements have been designed, the emphasis shifts to marketing the transaction. This requires a Transaction Manager who knows the potential Developers well. Technical, financial, and economic specialists will all be involved in putting together a request for proposals and information memorandum. Lawyers will help ensure that the legal aspects of the procedure are in order and that the transaction is not challenged on procedural grounds. Communication with different stakeholders[20] during this process is vital, and communications and human relations specialists may be brought in to assist.

Prequalification of potential Developers typically involves assessing their financial and technical strength and experience, and specialists with good judgment in these areas will be needed. Similarly, once bids are received, it may be necessary to assess them from a technical and financial perspective, depending on the bidding procedures used. The Grantor and its team will lead the negotiations with the Developer. Lawyers and financial specialists will be involved in negotiations to ensure that all the necessary documentation is executed to make the arrangement legally effective and binding.

No matter how well designed the rules and institutions governing the arrangements are thought to be, it will take time and experience for the parties to truly understand each other and work together—something that should be taken into account when the arrangements are designed.

After the Developer is selected, the hard work of managing the relationship starts. If the design stage was done well, the rules and institutions created should keep the relationship on track and serving the public interest. During most arrangements, there are likely to be tariff reviews and other adjustments. Once the Developer starts work, performance under the contract needs to be monitored to ensure agreed standards are met, and also that contractual and financial obligations are met by all parties. Tariffs and service standards may eventually need to be adjusted and disputes may arise that need to be resolved.

Financial close (i.e. when the lender is able to start drawing-down on its loans) marks the beginning of the long term relationship between the Grantor and the Developer. Good institutions and rules for maintaining and governing the relationship will be needed. The institutions set up to manage the arrangement may already have the capacity necessary to carry out these tasks. Some of the experiences reviewed in my research, for example in the case of Tallinn, Estonia show that some Grantor's may need assistance, especially early on, to help or augment skills required to manage and monitor the Developer. Technical specialists can help in monitoring service performance and assessing operating efficiency. Financial analysts and economists will be needed for tariff resets, and lawyers for enforcement and dispute resolution.

According to Vives (1997), to manage the process Governments need to be clear about which level of government is responsible for managing the process in the long term, and set up a streamlined management structure with strong analytic capacities and a reporting structure that brings powerful decision makers into the process in an effective way.

3.7 SUMMARY AND CONCLUSIONS

This section looks at the wider issues of incorporating private sector in design, development and implementation of PPP arrangements to provide long term water and waste water services. At the core of a PPP arrangement is the sharing of risks between the Grantor and the Developer. In order to ensure the optimal performance of the arrangement the design and implementation has to be carried out within a supportive framework that includes government commitment to a PPP agenda, fair allocation of risks between the parties to the PPP agreement, an adequate legal and regulatory regime, and a well prepared PPP model and clear tender process. This research includes discussion on all these points, and particularly leads to recommendations for optimising the last point.

The allocation of responsibility and risks for any PPP scheme can help us to analyze the needs of the scheme. A typical PPP scheme will need to deal with management and operational risks, commercial risks (e.g. revenue and collection risks), capital investment risk and allocation of asset ownership.

Using these differentiators, I have described a variety of PPP contractual models, where the Developers share of risks varied, starting from the limited risks of a Service Contract, through the increasing risk taking in Management Contracts, Lease schemes, Concessions, BOOT's, Joint Shareholding with the public sector partner, through to totally privately owned water and waste water service companies. Each of these structures involved contractual linkages, and institutions, that reflected the risks and obligations taken, the responsibilities for investment and capital works programs, as well as the means of monitoring and evaluation the long term performance (e.g. regulation and auditing).

Concessions have some of the most complex contractual PPP structures, covering many issues found in other models, and for this reason are used as the starting point to develop the recommendations in this research for development of Key Contract Provisions for Optimizing PPP arrangements (Chapter 9).

I note some overriding economic and financial drivers and issues that need to be considered by the Grantor in developing and implementing PPP schemes:

- *Value for Money (VfM)* – a key driver and measure for assessing the level of benefit of a PPP scheme.
- *Contingent Liabilities* – 'hidden' potential commitments to future expenditure that can have major implications for the Grantor or Government.
- *Budgetary Implications for government* – for instance use of scarce public capital.

At the heart of this is the realization that however much governments want to place risk with the private sector, if things go wrong with the PPP arrangement, then for social and political reasons it is unlikely that the government will be able to walk away from taking back these obligations, which might be very costly. Hence this is another reason for the importance of ensuring PPP effectiveness and sustainability.

The arrangement not only has to look at the skills, resources and responsibilities that it expects from the Developer but also has to consider the manner in which the profit motive of the private Developer can be harnessed, through the arrangement structure, contract and associated incentives.

To obtain an understanding of some of the procedural issue of design, development and implementation we looked at the four stages of PPP development:

- Policy Development, where objectives are set, the reform leader is identified, and ground rules for the structure of the sector are determined.
- PPP Arrangement Design, where the technical and service standards, tariffs, risk sharing are developed, and the roles and responsibilities defined and how to manage them. During this stage the contracts are developed and institutions to manage them are identified or created as may be appropriate.
- Developer Selection, where the rules for attracting and selecting the best private partner are developed.
- Managing the Arrangement, refers to all actions required for all the stages from project concept through to the end of the contract period. This involves Project design and management, Transaction management and overall long term contract management. This occurs over the long term, from project concept, and throughout the long term operation.

Each stage requires an appreciation of the key issues to ensure sustainability and effectiveness in the PPP arrangement, ideally all eventually embodied in the contractual arrangement. The whole process is over a long duration, varying according to the arrangement structure chosen, but possibly varying between 2 years through to 30 years for a Concession. The Management of the Arrangement has to span this whole period, but with a changing role according to the changing needs of each phase (e.g. starting with design management, and culminating in the long term management during the long operational period). The message from this is that care taken in the management of the design and development of the Arrangement at the earlier stages will result in reduced need for management in the later stages. This section concludes with the recommendation that to manage the process the government needs to:

- Be clear about which level of government is responsible for managing the process in the long term.
- Set up a streamlined management structure with strong analytic capacities and a reporting structure that brings powerful decision makers into the process in an effective way.

The point to be made from this is that, however robust, comprehensive and clear are the contractual arrangements, there is still a need for effective management by the Grantor at all stages over the whole life of the PPP Arrangement.

ENDNOTES

[1]Evidence and arguments about the effects of private participation, including case studies reviewed here include Abdala, 1996, Barlow and Clarke, 2002, Bitran and Valenzuela, 2003; Brocklehurst and Janssens, 2004, Clarke and others, 2004, Crampes and Estache, 1996, Estache and Rossi, 2002, Gray, 2001, Harris, 2003, Lobina and Hall, 2003, Megginson and Netter, 2001, Nickson and Vargas, 2002, Plummer, 2002, Public Citizen, 2003, Rivera, 1996, Saghir and others, 1996, Shirley, 2002, Shirley and Walsh, 2000.

[2]See generally www.transparency.org and United Nations Economic and Social Council, Governance in Public Private Partnerships for Infrastructure Development (2005). Furthermore, Delmon (2009) suggests that PPI provides an opportunity to implement good governance into every aspect of project implementation, and thereby reduce the opportunities for corrupt practices.

[3]Hemming, R., and Mansoor, A.M., *Privatization and Public Enterprises,* (1988), Internal Paper No 56, International Monetary Fund, Washington DC, at 2: Careful analysis of the risks involved is required at an early stage of the process, and the sharing of risks between the private and public sector should follow the guiding principle that whoever can control the risk best should assume it and should receive adequate compensation for doing so.

[4]The regulatory environment is important as it sets out the 'rules of the game'. However some countries may not have a fully developed regulatory environment specifically for water schemes. Having a fully developed regulatory framework is not a pre-condition to PPP although clearly it is beneficial.

[5]The term 'contracting body' or 'contracting authorities' is considered the same and falls within the meaning of Directives 92/50/EEC, 93/36/EEC, 93/37/EEC and 2004/18/EC and the contracting entities of the type 'public authorities' and 'public undertakings' within the meaning of Directives 93/38/EEC and 2004/17/EC.

[6]The rules on the internal market, including the rules and principles governing public contracts and concessions, apply to any economic activity, that is, any activity which consists in providing services, goods, or carrying out works in a market, even if these services, goods or works are intended to provide a 'public service', as defined by a Member State.

[7]See ruling of 13 January 2005 in Case C-84/03 (Commission v Spain), This ruling has caused significant debate amongst local authorities in Europe as to whether, and in what circumstances, forms of cooperation between local authorities are subject to procurement law. The ruling says that cooperating local authorities cannot as a rule be excluded from procurement law by a national provision. The law on contracts always applies when two legally distinct entities conclude a contract for pecuniary interest. The status of the entities is not at issue.

[8]Green Paper on Public-Private Partnerships and Community Law on Public Contracts and Concessions, COM(2004) 327 final Brussels, 30.4.2004

[9]Green Paper on Public-Private Partnerships and Community Law on Public Contracts and Concessions, COM(2005) 569 final, Brussels, 15.11.2005

[10]See Communication from the Commission to the Council and to the Parliament 'Public finances in EMU 2003', published in the European Economy No 3/2003 (COM (2003) 283 final).

[11]Communication from the Commission 'A European initiative for growth: Investing in networks and knowledge for growth and jobs', COM (2003) 690 final, 11 November 2003. This report was approved by the Brussels European Council on 12 December 2003.

[12]According to Lee, (1996) at 14: The many options for private sector participation oscillate between one extreme of almost total public sector responsibility (e.g. leases) to totally private responsibility (divestiture) at the other.

[13]Examples of service contracts include the Mexico City contracts where the city was divided into four sections and contracts negotiated with four international operators.

[14]Typically private capital is a combination of debt and equity finance. This issue is discussed further in Chapter 7.

[15]For links to many government websites please visit the World Bank Institute's Global Public-Private Partnerships in Infrastructure web portal at: http://info.worldbank.org/etools/PPPI-Portal/links.htm

[16]Net Present Value (NPV) is the sum of future cash flows discounted to present time by some opportunity costs of capital.

[17]A useful work addressing the appointment and use of advisers in PPP and Infrastructure development is *A Guide for Hiring and Managing Advisors for Private Participation in Infrastructure* by the Public Private Infrastructure Advisory Facility.

This Toolkit provides a comprehensive and thorough guide to the issues surrounding the hiring of advisors to support government officials involved in increasing Private Participation in Infrastructure (PPI). The Toolkit focuses on the hiring of advisors to increase private sector participation in the following sectors: electricity generation, transmission and distribution; natural gas transmission and distribution; water and sewerage; solid waste; telecommunications; railways; ports; airports; and roads. http://www.ppiaf.org/content/view/236/485/.

[18]Often the best result can be achieved by competitive tendering, with prospective Operators competing in a formal, structured process. However, sometimes other approaches are suitable, for example when there is limited bidder interest, or innovative solutions are needed which are hard to define in advance. I deal with this issue in some of the recommendations section of my research.

[19]The composition of the project team and steering group may change during the process. For example, different structures may be suitable for the phases of policy development and the transaction implementation. Following selection of a preferred bidder the Grantor may form a separate negotiation team, and, following contract implementation, a long term contract management team.

[20]Adequate stakeholder consultation is essential as changes in water sector provision generally have high perceived social and political impact. Adverse stakeholder reaction can severely impede successful implementation of PPP schemes (e.g. Cochabamba, Bolivia, where offices were burned and the private sector contract was cancelled).

Chapter 4

Balance of costs and service standards

4.1 INTRODUCTION

The prime aim of a public water and waste water entity is the provision of a sufficient, effective and reliable service to its customers. The size and complexity of a scheme will be derived from the chosen level of service to customers.

In this chapter some of the key steps in the iterative process of designing, establishing and maintaining a sustainable, bankable scheme for water and wastewater service provision will be considered. Whether the scheme is funded by public or private funds the process to be followed would be in a similar manner.

The process starts with the Grantor defining the level of service that it believes will be politically and socially acceptable to the community. Size, complexity and funding of PPP schemes all revolve around what level of service is chosen and the standards to be met. This directly affects the overall cost of the project. It will be the task of the Grantor (and as appropriate, its transaction advisers) to review the technical and financial requirements to provide this level of service and determine the funding requirements from tariff and other sources necessary to reach this level of service (e.g. expansion of service area to new customers, increase reliability of supply to 24 hours per day, additional sewerage connections and so on. This process is graphically represented in Figure 4.1).

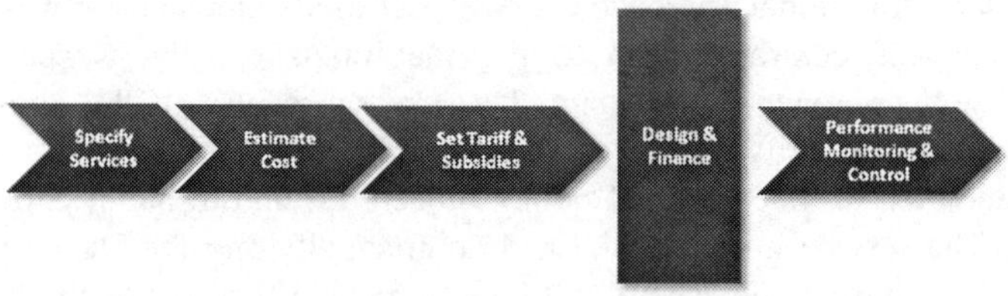

Figure 4.1 Balancing service standards, tariffs & subsidies.

Once the final balance between services, costs and subsidies has been established then the Grantor proceeds to establish the final design, financial approach and the contractual basis of the PPP arrangement.

In the long term, there is a need to monitor and control the services to be provided by the Developer, and link this to some form of performance payment mechanism.

It is to be argued that by developing a standard set of more customer oriented performance measures, acceptable to a broad range of stakeholders, will allow transaction advisors/policy makers/Grantor to introduce more appropriate performance monitoring systems into the design of PPP schemes, in a more cost effective manner and ultimately lead to more stable long term PPP arrangements. This leads on to the premise that for long term effectiveness and sustainability it is necessary to formalize customer and stakeholder consultation throughout the life of the scheme, by means of contractual arrangements.

4.2 SETTING SERVICE STANDARDS AND LINKING TO COSTS

In this Chapter we are interested in establishing the first step of the project design, where the Grantor sets service standard targets to be achieved, and the parameters to be used to achieve this. Regardless of whether there is public or private funding, the Grantor needs to set out its service goals in terms of coverage (number of people served or to be served) and quantity and quality (physical and bacteriological characteristics of water, reliability, continuity and pressure, customer interface and service, effluent treatment and coverage and so on). Clearly, there must be a balance in terms of what government wants as regards *coverage* and *quality* and what customers are willing (and able to pay). Once initial objectives have been set, the Grantor should estimate the cost of providing the service. However, much confusion still surrounds the definitions of 'cost of service' and 'cost recovery'. The average cost of investment and the income required would be the starting point for determining the level of tariffs, as well as some issues about the role of tariffs in water PPP schemes[1].

In relation to *coverage* and *quality* specifications, the trend with modern water PPP arrangements is to design them on an *output basis*, rather than an *input basis*[2]. Under a traditional approach every aspect of a water scheme was designed in detail, and the contractor had to provide inputs exactly as specified. An 'Output Based' approach is well suited to make best use of the ability of the private developer to provide the most economic and practical solutions within defined standards, to meet defined policy objectives, and to attain defined levels of service. The service goals are defined contractually and the Developer has to meet or exceed these goals. There will be a number of key issues that will be set as service goals that have to be met throughout the life of the PPP arrangement. These have to be defined early in the design process, and chosen to ensure that in achieving these levels of service the long term goals of the PPP arrangement will be achieved.

A full list of service level goals can cover most aspects of the water scheme, but for successful contractual implementation it is more practical to consider the smallest number of 'overarching' levels of service goals that will ensure that the project meets the overall scheme objectives. Further details on this are provided in a later section.

Cost recovery is considered to be at the heart of sustainability of the water and wastewater sector. Regardless of whether a utility is public or private, if it cannot cover its costs the service provided to customers will suffer. Full cost recovery implies that the tariffs are high enough to cover the costs of service. In this regard, the cost of service has three commonly accepted elements (adapted from World Bank, PPIAF, 2006):

- Operating and maintenance expenses – These are the day-to-day expenses involved in providing services and keeping the system functioning. They include labour costs, electricity, materials, repairs to equipment, and the like.
- Depreciation – Depreciation is the reduction in value of system assets over time. This is roughly equivalent to the amount of money needed to replace assets as they wear out.
- Rate of return on investment – Sponsors and lenders will require a return on the debt and the equity which they raise for the project. The return is the amount of money gained or lost on an investment relative to the amount of money invested and includes money earned (a return) on both debt and equity. Typically the cost of debt is the interest lenders charge, whereas the cost of equity is the return to the sponsors of the project less the return committed to debt servicing. The so-called weighted average cost of (debt and equity) capital is usually considered an appropriate measure of the return on capital.

The capital costs of building the infrastructure for a water system are significant. These costs depend on many factors, a dominant one of which is the underground asset condition and the validity of the data presented on the condition of the assets, its material (e.g. led, concrete, PVC and so on), age and other factors. Furthermore, there will be a number of financial factors that also need to be considered and include:

- Financing costs which may be affected by the willingness/or not of the private sector and other lenders as well capital markets to service the debt on the project[3].

- The interest rates and maturities available for such debt.
- The debt service covers ratios required by the lenders and other associated conditions imposed by the lenders, including amount of debt and equity of the Developer.
- The availability of bilateral, export credit, government and other funding sources for the project.
- Cost of exchange and currency rates between the local currency and the currency in which the debt has been negotiated[4].
- Land costs.
- Labour costs.
- The tax and accounting regime which will influence issues such as the level of asset depreciation allowable, tax treatment on financing and interest payments and so on.

Full recovery of these costs will only be achievable if the associated level of the tariff proves affordable to the customers (Ehrhardt, 2003; Baietti & Raymond, 2005).

In many developing countries, the low level of the domestic economy makes it likely that only a portion of the tariff allowing full cost recovery will be affordable by many sections of the community. Although the international lending institutions (e.g. World Bank, Inter American Development Bank) would ideally aim for full cost recovery, there is an understanding that in most cases in developing countries this would result in tariff levels not currently affordable.

Frequently the aim of new schemes is to operate at a minimum tariff that will at least cover Operating and Maintenance costs. As the scheme becomes established, and the customers begin to benefit from the improved levels of service, it is often planned to increase the tariffs further, in a gradual manner over time, until the maximum economically affordable tariff rate is reached.

4.3 FINANCIAL SUSTAINABILITY, SETTING TARIFFS AND SUBSIDY REQUIREMENTS

Having estimated the cost of providing the required level of water service, the next step is to check how much of this amount could be recovered through tariff income, and to what extent the funding arrangement for the water project can recover the balance of capital and operating costs (Figure 4.2).

Basically a water and wastewater service provider, whether a Utility or a simple water user association, has to recover its operations and maintenance costs and the cost associated with the development of its infrastructure. In the context of a PPP Agreement, the Grantor needs to consider not just the full cost of service as described previously, but also the annual cash requirements of the

Developer, and the financial ratios required by lenders. Notwithstanding ownership structure, two approaches of accounting are possible[5]:

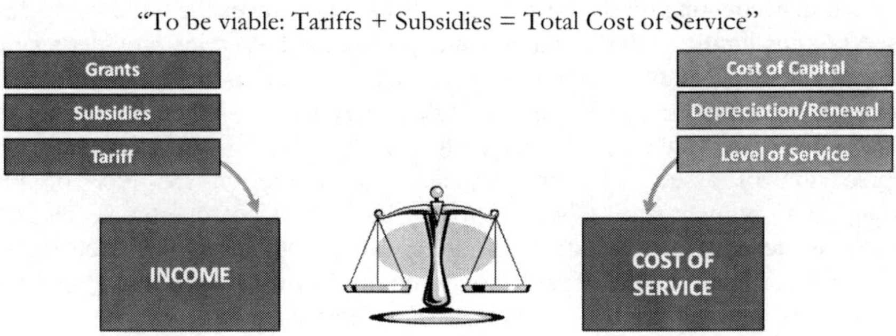

Figure 4.2 Balancing income with cost of service. *Source:* Stiggers, 2006.

(1) *Cash accounting*: Revenues generated has to be sufficient for covering the service provider's 'cash needs', that is, its operations and maintenance costs and the repayment of the principal and interest on its loans. This implies that the entity's revenues are directly affected by the capital structure of the service provider. In this way the manner in which capital investments are financed ie either by commercial loans, grants or a combination thereof play a critical role on the revenue base of the service provider. This method of estimating revenues, based on cash accounting, is mostly consistent with the vision of service providers using cash-based budgets, such as government departments or water user associations[6].

(2) *Accrual accounting*: Revenues generated have to be sufficient for meeting all the service provider's costs, that is, its operating and maintenance costs, the depreciation of its fixed assets and a return on assets sufficient to service the debt and remunerate the equity invested by the Developer. In this case, revenues are not directly affected by the capital structure of the service provider. Here the principle is that investments have to be recovered with an excess or profit to cover the cost of financing, whether debt or equity. This method of estimating revenues, based on accrual accounting, is consistent with the vision of Utilities managed as a corporate entity, whether public or private owned.

In practice what cash accounting means is that given costs are covered this approach is financially sustainable in the short-term. However, one important caveat of this approach is that it perpetuates the service provider's dependency on debt. In many instances this form of debt comes either from government of is grant money, and does not encourage building up cash reserves to protect the future operation. Furthermore, what this approach means in practice is it translates in the need for steep tariff increases each time a major investment is to be financed.

Accrual accounting on the other hand actively encourages the building up cash reserves, thus limiting reliance on debt and protecting the service provider against external shocks. Noteworthy however is that in order to be meaningful, this approach requires that fixed assets are adequately valued. When this is not the case, assets may have to be revalued needing also to set clear rules for depreciation of assets. In some instances depending on the type of PPP arrangement, a pragmatic approach to this issue may be to estimate required Developer revenues to be sufficient to cover O&M costs, the debt service (principal and interest), the depreciation of equity financed assets and a return on equity, allowing cash needs to be met and cash reserves to be built.

However, whether the provision of services is public or private when setting tariffs the following should be considered:

- *Willingness to pay*: Given the social nature of the water and wastewater services and the fact that the main revenue stream can only come from its customers[7], willingness to pay is crucial in assessing the overall sustainability of the services to be provided. Research suggests that willingness to pay surveys show that most consumers (especially unconnected consumers) are willing to pay substantial amounts for better water services (World Bank, PPIAF, 2006). An assessment of willingness to pay for various levels of service, together with customer demand forecasts will be useful preparatory work for water system design. The willingness to pay is greatly affected by water availability and service levels.
- *Social acceptability*: It must be noted that willingness to pay should be coupled with improved service levels. There are notable examples of where tariffs were increased without the necessary service improvements[8].
- *External benefits*: As mentioned earlier, water and wastewater services are at the core of the development of society. Accordingly, governments and policy makers may wish to subsidise the sector in various forms for the sake of external benefits such as health, hygiene and the reduction of water borne diseases and so on.

Decisions on whether to raise tariffs or decisions on levels of service provision have a direct impact on the operational performance of a utility (regardless of whether it is a publicly run utility or whether it is a private undertaking managed by a Developer under a PPP scheme) and its customers (Estache & Rossi, 2002). It must be noted that recovering costs does not systematically require increasing tariffs. Progressive steps can be taken such as reducing operations and maintenance costs by enforcing operating performance objectives or optimising capital costs by ensuring that least cost solutions are implemented, or a combination of both. Tariff can also be set with various objectives in mind such as pre-defining minimum levels of efficiency (such as the case of the 'empresa modelo' in Chile, see Box 4.1), consideration for the long run marginal cost (LRMC) of water, and so on.

BOX 4.1 CHILE – AN EXAMPLE OF A TARIFF SETTING MECHANISM DESIGNED TO ACHIEVE SEVERAL OBJECTIVES

The economic regulation in Chile clarified tariff setting principles and established an effective and independent Regulator. Currently, eighteen PPP Agreements are in place which finance their development exclusively through cash generated from operations and funds raised on local capital markets without financial assistance from the government (except for low income families for which there is a specific subsidy scheme in place). The principles introduced by the 1988 Tariff Law can be summarized as follows:

(a) *Dynamic efficiency*: Tariff is set by reference to a model 'efficient' company to against which Developer performance is benchmarked.
(b) *Economic efficiency*: The tariff is set relative to the long run marginal cost of water (LRMC). LRMC calculations should include the cost of externalities, such as the collection and treatment of wastewater. LRMCs are estimated by the Regulator for 320 typical water supply systems and 270 typical waste water systems, by dividing incremental capital and operation expenses by incremental sales.
(c) *Intelligibility*: The tariff structure guides producers and consumers in production and consumption decisions talking into account water production, water distribution, waste water collection, and waste water treatment and disposal.
(d) *Equity*: The tariff should not allow for discrimination between individual, although there is a recognition of low income households which benefit from a discount on their water bills financed by a Central Government program.
(e) *Self-financing*: If LRMC yields too low a tariff the law allows a minimum tariff to meet short term financial obligations. Interestingly, the average debt equity ratio of concessions is approximately 0.99 and the average return on equity invested is 18 percent.

Source: Chilean Water Regulator website and discussion with a leading private water operator in Chile (2012).

Even if one were to ponder whether subsidies are justifiable in the context of PPP schemes, subsidies may play an important role. Having considered the cost of service, willingness to pay, social acceptability, externalities, and subsidy options, the government will face a choice of what tariff level should be applied. In some instances the Grantor may decide to set tariffs below costs and provide a subsidy to make up the difference. Making these decisions may involve consultation with customers and other stakeholders. It may be necessary to develop a range of options, and iterate toward a solution, until finally an acceptable tariff level trade-off is reached. It must be noted that there is a trend to disburse subsidies on an output basis. This can be a particularly useful technique to augment the bankability of a PPP scheme. See Box 4.2 for further details.

BOX 4.2 APPLICATIONS OF OBA – SUBSIDY DESIGN MECHANISMS

One-off subsidies are the most common application of OBA approaches and usually involve capital subsidies for access to a given service. An example is when a large portion of the subsidy is paid *after* the targeted beneficiaries are connected to a network and connections are verified. Given that in OBA approaches the emphasis is on service delivery rather than on physical connections, even in the case of one-off subsidies a portion of the subsidy may be withheld and paid only after verification of a certain number of months of satisfactory service delivery (hence a mixture of 'outputs' and 'intermediate outcomes' against which one-off subsidies are disbursed to provide a measure of improved access).

Transitional subsidies can be used to support tariff reforms, where a subsidy is used to fill the gap between what the user is deemed able and/or willing to pay and the cost-recovery level (e.g. long-run marginal cost) of the tariff. The subsidy is fazed out after a specified period of time (e.g. months or years) as the user contribution increases (and possibly as tariff levels required for cost recovery decrease with efficiency gains). In these cases, the output against which the subsidy is paid is the service delivered and billed by the provider. The review identified only a handful of transitional OBA schemes and very few of those are still in place, often because of differences in actual demand and willingness-to-pay as compared to estimates made ex ante.

Ongoing subsidies may be required in cases where there is a continuous gap between affordability and cost recovery – including for consumption costs. As in the case of transitional schemes, ongoing subsidies should be paid out against pre-determined targeted outputs in order to be considered OBA. On-going schemes are largely used in the road and health sectors, although several ongoing OBA schemes have been identified for the utility sectors outside the WBG.

Source: Global Partnership on Output Based Aid, 2009.

4.4 SETTING TARIFFS FOR ACHIEVING ECONOMIC, EFFICIENCY AND EQUITY OBJECTIVES

Much has been written on the mechanisms for effective tariff setting. Below we offer some basic principles for consideration when setting tariffs (Figure 4.3).

Setting Realistic Objectives: As mentioned above it is important to strike a balance between costs of running and maintaining an efficient water operator. In this regard it is important to develop, through the PPP agreement a series of realistic objectives with realistic metrics for measurement. The International Benchmarking Network (IBNET) administered by the World Bank for example, could help by helping set realistic objectives to be achieved for key efficiency indicators such as metering, collection and staffing ratios, energy consumption or non-revenue water (NRW).

Setting standards needs to be done carefully (Box 4.3). For example, as regards NRW there is a lot of debate about what should be the metric as well as the actual value which should be used. Expressing NRW as a percentage of water production is often misleading when per capita consumption decreases as a result of a successful social connections programs or demand management through pricing. It is often more representative to express NRW as losses in m^3 per connection or per km of distribution network[9].

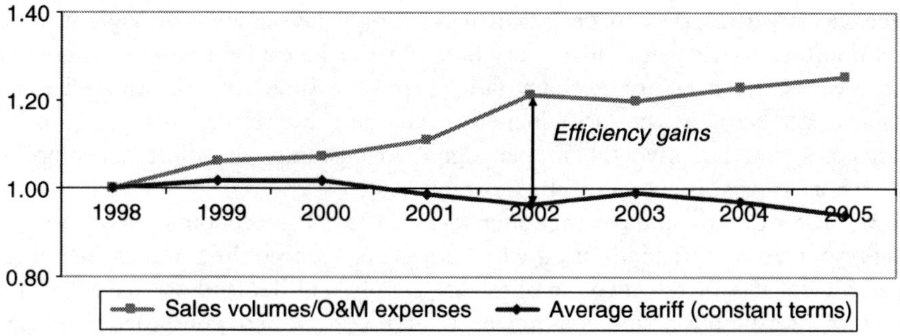

Figure 4.3 Senegal – Efficiency Gains achieved through Contractual Obligations.

BOX 4.3 THE ARMENIAN WATER SECTOR – BALANCING REALITY WITH A STORY OF SUCCESS MANAGING EXPECTATIONS WITH REGARDS TO NRW REDUCTION

The Armenian water sector has experimented with PPP since the mid-1990s with all urban centres of the country served by three professional private operators. Prior to the introduction of the private sector, water was provided for only a few hours per day despite very high per capita water production of 885 lcd in Yerevan and 670 lcd in secondary cities served by the Armenian Water and Sewerage Company. The PPP arrangements had a clear focus on increasing the number of hours of service and complying with bacteriological quality standards. Service improvements have been noticeable, but NRW has remained extremely high at 83 percent of the production in Yerevan and AWSC service area. Around half of NRW is believed to be the result of physical leaks on poor quality distribution pipes and connections. The economic justification of replacing the distribution system has to be carefully assessed especially given the fact that Armenia as a country does not have a water resource problem. Furthermore, private operators have significantly reduced energy consumption and costs by refurbishing pumping plants and switching production to gravity sources whenever feasible. About half of NRW is believed to be of commercial origin. Despite a significant programme of meter replacement and a change in ownership of the meters from customers to operators (previously the operator was unable to read a meter as this was not his property), the absence of adequate legislation means that billings and collections are still inaccurate. Achieving significant reductions in NRW continues to be a challenge under these circumstances.

Source: Conversation of Authors with Ms Zaruhi Tokhmakhian, Operations Officer Sustainable Development Department Europe and Central Asia Region, 2012.

Balancing Past and Future Costs: As mentioned earlier, tariff levels should be set to cover not only 'cash needs' (operations and maintenance costs and, as appropriate debt service), but should rather aim to cover full 'utility/operator costs' as represented by operations and maintenance costs, depreciation of fixed assets and a return on fixed. However, given the nature of the water supply business in which investments as sometimes *lumpy*, there is a need to build up necessary cash reserves to deal with these lumpy investments throughout time. Not doing so would imply that every time a major investment would be required there would be need for a sharp tariff increase. Similarly, assuming that the demand decreases as the tariff increases, charging a tariff that simply recovers past costs will not give the proper signal to customer to adjust consumption behaviour to what is economically desirable.

Marginal Costs[10] *and Average Incremental Cost*: The economic theory tells us that resources are efficiently used when prices are set according to marginal costs. An efficient use of resources implies that supply and demand are balanced at a level of consumption that maximizes total producer and consumer surpluses. Accordingly, an increase of consumption by one m^3 would, in the short run, only result in an increase of pumping and treatment costs. But the marginal cost would rise sharply as the capacity of the infrastructure is being fully used up and decline rapidly after extension of the latter. Since it is not practical to reflect steep variations of the short run marginal cost (SRMC) in tariff levels each time the infrastructure is extended, it is more appropriate to set the latter according to the long run marginal cost (LRMC). But just using LRMC in water utilities is insufficient consideration should be given to the average incremental cost (AIC). This means the tariff calculated by dividing the net present value (NPV) of the sum of capital expenditure (Capex) and the incremental operation and maintenance expenditure (Opex), by the NPV of incremental sales. Doing this is a good proxy for LRMC of running a water utility. However if NPVs are being used, the discussion then centres on the appropriate discount factor as well as the planning horizon to be applied. Given the water sector is a long term business, it would seem appropriate to use a planning horizon of at least be 20 years to limit the influence of residual values on the result. Pricing water using AIC estimates would be more appropriate as, in theory, increased AIC should lead to increased tariffs which if applied, would lead to customer's adjusted consumption behaviour (sometimes referred to as elasticity of demand) which could defer or postpone capacity expansions in the system.

Customer Metering: There is an increasing debate about metering. Managing demand through pricing implicitly suggests that production and consumption are metered and that bills are established on the basis of consumption which has been accurately metered through verifiable readings. From a practical perspective consideration should be given to a number of factors involved in metering such as (a) the need to rebuilding house connections and tertiary distribution networks to enable metering installation which could be expensive and in some instances impractical, (b) the type of meter to be used to ensure accurate measurement despite intermittent service, (c) designing and installing meters to avoid tampering and or manipulation (eg anti magnetic meters), (d) ease of reading and (e) as appropriate, use of pre-paid meters.

4.5 PERFORMANCE INDICATORS

Monitoring and enforcing performance is at the heart of the successful implementation of any PPP agreement. In this section the subject is looked into in some detail, since this very important issue has to be dealt clearly and practically within the contractual arrangements.

Usually Developers have many ways of reporting on their performance in different areas. These include internal performance reporting for managers and other staff in the company, financial reporting for owners (shareholders) and creditors, as well as performance reports for regulatory purposes established in the obligations of the PPP agreement.

Since the provision of water and wastewater service functions are delegated to the Developer through a contractual structure – the PPP agreement – it is imperative that systems are in place to monitor and enforce performance. For example, according to Dinar (2000) it will be important that the Grantor is able to monitor whether the Developer is meeting his obligations with respect to:

- Obligations related to service coverage targets (if any) in terms of quantity and quality.
- Provision of water to required standards such as pressure, continuity of supply and so on (and for wastewater it may include specific treatments standards[11]).
- Upkeep, upgrade and rehabilitation of the assets (both above and underground).
- Adequate reporting by the Developer of its activities.

However, these more traditional metrics of service performance monitoring do not necessarily measure the customer's experience with the Developer[12]. It is to be suggested that for performance monitoring, enforcement and reporting to be effective, it requires access to performance indicators that are most relevant to the customer's experience with the Developer, namely, whether the service provided is reasonable considering the tariffs that customers are asked to pay. If metrics related to contract obligations as well as specific indicators related to the customer's experience in receiving a service are brought together, then PPP agreements and their resulting obligations are more likely to be understood by the public. Ultimately a better informed customer means that expectations in terms of service quality and reliability can be managed leading to an overall more sustainable PPP scheme[13].

An important question to answer is precisely what ***is*** effective consumer reporting? Cook and Stevens (2004) perceive it as: (i) when customers are knowledgeable about the quality of service they are receiving and the value for money, and; (ii) the collection of information that leads to improvements in service levels; (iii) an increase in transparency that increases public awareness of the balance between investments, tariffs and the service facing the utility.

Effective consumer-oriented reporting relies on a good set of easily-understood performance indicators. These will assist the Developer, monitoring units and governments in assessing the performance being provided and if made available in an easily-understood format, can assist the public in evaluating performance levels as well. Involving the public at this level can lead to improved decision-making, service delivery and accountability.

Basic service performance indicator categories for water distribution include (compiled from various sources including (Dumol, 2000; Ehrhardt *et al.* 2007, Environmental Resource Management, 2003b; Foster, 2001; Komives, 1998; Ofwat, 2003):

- Coverage of households or other potential customer bases.
- Quantity of water provided or consumed.
- Water quality.
- Water pressure and reliability of pressure.
- Customer service response times.
- Customer satisfaction with different aspects of service.
- Affordability.

Basic wastewater performance indicators include:

- Coverage of households and other potential customer bases.
- Service quality and reliability (frequency of sewage backups and overflows or frequency of collection of sewage from holding tanks).
- Customer service response times.
- Affordability.
- Level of treatment and quality of outflows of treatment plants to the environment (which is not a customer-oriented indicator, but one of more general consumer interest for water resource management).

Each of the above mentioned indicators can serve to inform decisions on service provision, policy decisions and customer decisions. The approach recommended is one that seeks to synchronise data from the Developer and customer surveys in order to generate a more meaningful perspective on the operation and management of the utility. At the same time, the advantage of this type of monitoring scheme within a contractual context is that it can be adjusted over time, along with operational and data quality improvements, and as such may be used to:

- Progressively tighten targets.
- Increase the number and sophistication of indicators at each review period.
- Progressively replace some of the surrogate indicators, as better data becomes available.

The benefits of this type of consumer based performance indicator may be summarised from the perspective of both the Consumer/Customer[14] and the Developer:

For Consumers: Information is power and information on actual service level provision can provide a powerful tool for customers to demand service improvements should they be required. Consumers will have three options under PPP arrangements to take action to obtain service improvements: (1) they can approach the utility, (2) file a complaint with the regulator (if one exists) or (3) approach their political representative with their issue. If a partnership arrangement is constructed between consumers and the utility, the information will stimulate discussions over service and tariff changes.

Critical decisions relating to the management of water and wastewater service provision are often based on inaccurate and insufficient data as discussed in Chapter 12. NGOs and consumer groups are often suspicious of the PPP process, and wary of the 'benefits' that will flow from such an arrangement. Insufficient communication between the Developers and its customers can breed misunderstandings which in the past, have led to the collapse of PPP arrangements[15,16]. It is important therefore that all stakeholders of a PPP process[17], and indeed all consumers, have access to information about the performance of the Developer responsible for their service provision.

For *Utility Managers* the greater accountability provided by the improved reporting mechanisms gives greater incentives for utilities to improve service performance, keep consumers happy and mitigates any pressure that the government may place on them in the event of inadequate service provision.

Some of the main barriers to improved public reporting may include:

- Secrecy of agreements in some cases, particularly in older PPP agreements, which are not in the public domain and prevent publizing information on performance reporting and performance target achievements[18].
- In relation to public reporting in the PPP agreement itself, lack of specification for information and processes requirements.
- Lack of definition of the role of monitoring units in the process of public reporting.
- Lack of Grantor interest in accountability and transparency in some countries (although in some instances this can be compensated by the interests of the NGO community[19] and the multi-lateral donors).

- Resistance/lack of incentive for some Developers to be transparent, since not all Developers feel that provision of information is in their best interest[20].
- Lack of a partnership approach in most cases in the relationship between Developers, customers and owners or regulators.

Accordingly, it will be important to develop strategies within the PPP agreement that allow improved monitoring of performance. It has been noticed that in some instances Developers have actively sought improved disclosure such as was the case in Buenos Aires Argentina[21] leading to the suggestion that NGOs may be co-opted to take on effective monitoring roles. Cases in Argentina, Colombia, Philippines, Zambia and Senegal[22], demonstrate that NGOs and Community Organizations can contribute to constructive dialogue for improved performance when given the task of monitoring tariffs and services and representing consumers and customers. Integration of NGOs to oversight committees has also taken place and can also be an effective method of introducing greater consumer-driven accountability.

A key recommendation is to ensure as great a level of transparency and information transfer as possible. Monitoring units that are housed in regulatory agencies tend to have a greater inclination to keep performance data confidential than those that are housed outside government bodies. Monitoring units are therefore recommended to use multiple channels to report performance standards to the public and develop mechanisms to obtain customer surveys on service levels and areas of improvement. Monitoring units are never completely independent and should be viewed as a useful watchdog, rather than a replacement for the Operator/Developer completing its contractually obligated tasks.

Various studies (Gentry *et al.* 1999 and Nichols, 2004) suggest that consumer-oriented reporting utilize multiple channels for public reporting, using the following components:

- Using media outlets such as the local press or radio networks.
- Produce annual reports available on the internet.
- Conduct customer surveys that request information about service satisfaction levels.
- Conduct regular dialogue roundtables with NGOs or Committees.
- Undertake road-shows to present performance information and host consumer feedback sessions.

4.6 SUMMARY AND CONCLUSIONS

The prime focus of the PPP schemes considered in the Water and Waste Water sector is the level of service provided to their customers.

An iterative process is used to establish effective scheme service standards, sizing, costs, funding and financing. If levels of service are taken as the prime concern of the PPP schemes, then this iterative process defines and sizes an affordable scheme for which effective operational, financial and contractual arrangements have to be developed.

Determination of the affordable tariff levels is key to ensure customers' ability to pay for the services, and hence the financial sustainability of the services. Generally some form of subsidy is necessary for schemes in developing countries, as affordable tariff levels are not high enough to give full cost recovery. The details of this, any subsidy arrangements and any progressive program for change in tariff levels as service improves, need to be clearly defined in the contractual arrangements.

There is a need to monitor and control performance parameters under the PPP arrangement. A variety of performance measures can be defined for many activities of a utility. However, it is proposed that measures should be output based and directly related to a customer focused approach.

There is a need to ensure that there is full stakeholder consultation at all stages of development and operation. Case history shows the positive effect of stakeholder involvement in design, development and implementation of PPP schemes (and also the potential negative effects when they are not involved). This leads to the premise that for long term effectiveness and sustainability it is necessary to formalize customer and stakeholder consultation throughout the duration of the scheme. Therefore, in order to reflect the importance of this issue, it is to be recommended that the contract terms include specific provision for stakeholder consultations to be established by the Grantor.

ENDNOTES

[1]Tariff setting is a highly complex issue. Tariff setting is a careful balance between costs, investment and affordability and social implications. These are interlinked with demand and consumer behaviour. Whilst tariff setting is vitally important in PPP Agreement, the information given here is simply to illustrate the importance of this issue for long term sustainability of the water and wastewater sector.

[2]An example of this is the divestiture contract in Tallinn, Estonia that specified the outputs in terms of coverage but left the Developer to plan in whatever way it saw fit to achieve that contractual obligation.

[3]Under the current financial conditions, we have seen a retraction of major lenders in infrastructure. This liquidity crisis has led to a delay of major investment in infrastructure (including water and wastewater). International Monetary Fund, internal briefing note to World Bank staff, 6th April, 2009.

[4]It should be noted that in some jurisdictions there are limitations imposed on the level of foreign currency that may be used for a project or capped imposed on the amount of profit that may be repatriated.

[5]See Financing WSS Investments – Estimating Revenue Requirements and Financial Sustainability; A Baietti and P Raymond, WSS Working Note no 7, World Bank (2005).

[6]Extract from Reforming Urban Water Utilities in Western and Central Africa: Experiences with Public-Private Partnerships. M. Fall *et al.*; Water Sector Board Discussion Paper, no 13; the World Bank, 2009. In Côte d'Ivoire the water service provider was financed exclusively on debt through a system whereby the water tariff was set to cover operations and maintenance costs (which involved paying a private operator SODECI, and repaying sectoral loans), with tariff resets every four years. The National Water Fund, responsible for servicing the debt, was replenished by the difference between the tariff paid by the customers and the tariff SODECI was allowed to retain on its collection from customers to cover the costs of its contract. When the exchange rate of the US dollar, in which a large share of the debt was denominated, steeply appreciated against the local currency, and a slowdown of the economy resulted in lower sales to large industrial customers whose water bills were the main contributors to the National Water Fund, revenues generated from operations were insufficient to replenish the Fund and to service the debt.

[7]For the purposes of this research the authors have looked at the point of view of a utility focusing solely on water and wastewater services and therefore do not consider any subsidiary income streams from outside this core business.

[8]The 1903 water riots in Port of Spain, Trinidad and Tobago resulted from an ordinance that increased the cost of water. A number of public meetings had been held to protest the increase, culminating in a demonstration on March 23 in front of the main government building (Brunswick Square) smashing windows and causing members of the legislature to hide under tables for protection. The protesters set the lower floor on fire, at which time police opened fire on the crowd, killing thirteen people, and injuring forty-two others. Another highly publicised event occurred in the last decade in Cochabamba, Bolivia.

[9]Conversation by the authors with Alain Locussol and Aldo Baeitti of the World Bank (2012).

[10]Water Pricing: The Importance of Long Run Marginal Costs; London Economics for OFWAT (UK), 1997; and Estimation of Long Run Marginal Cost (LRMC); Marsden Jacob Associates for QCA (Australia), 2004.

[11]These standards usually include specific metrics such as level of suspended solids (SS), Biological Oxygen Demand (BOD) etc.

[12]Customer-oriented performance reporting differs from these points, with the focus not on detailed internal or financial issues but on matters of direct interest to the end user. Such reporting will also provide important insights to others, such as consumer-advocacy groups, chambers of commerce, other nonprofits, regulators, and governments.

[13]Interestingly, experience in PPP schemes throughput the world shows that although the provision of water and wastewater services are a basic service, the interface with customers is often overlooked. Although an obvious statement, it seems that in many cases the introduction of the private sector into the provision of water and wastewater services has been used by governments to defray political attention and sometimes unpopular decisions. Examples abound (for example Trinidad and Tobago, Dar es

Salaam, Tanzania) of failed PPP schemes where the public was ill informed about the scope of the respective PPP agreement and customer's expectation were not managed appropriately. Developers need to ensure that they interface with the public

[14]The use of the word customer and consumer are used interchangeably. However it is noted that generally the distinction is that customers (unlike consumers) have rights, including the right to safe water, the right to be informed and the right to be heard, among others.

[15]One such example is what happened in Cochabamba, Bolivia, after a concession type contract was signed where there were a number of civil disturbances that ultimately lead to the early termination of the contract. Although not a single factor is attributable to this early termination, (for example there was an obligation by the Grantor for the developer to build an expensive dam), the PPP agreement allowed for a 35 percent increase in tariffs at the start of the contract and a 20 percent increase once the new dam became operational. Upon review of the causes of failure, lack of communication between stakeholders and no adequate mechanism for addressing concerns of these groups played a key role in the early termination.

[16]See Bechtel 2002 Company statement on 'Cochabamba and the Aguas del Tunari Consortium.'

[17]Stakeholders of a PPP process include the customers, potential customers, workers, private Developers and financiers and taxpayers.

[18]An interesting example of this is the PPP agreement in Trinidad and Tobago which was not publically available despite it being a management contract. PPP agreements in Guayaquil, Ecuador, Mexico City, Mexico, Bucharest, Czech Republic are also not publically available. The PPP agreement of Mendoza, Argentina (at the time in which it was won by a consortium of Saur and the failed US energy giant Enron) was not publicly available in Argentina. However because Enron was a NY listed company, there was a public disclosure obligation imposed by the Securities and Exchange Commission to make public such contractual documentation.

[19]NGOs, community-based organizations, and other civil society organizations should not be treated as a single group. They represent different stakeholders and different interests, and need to be engaged with a clear understanding of whom each organization represents and what its interests are. Some NGOs and community-based organizations will be effective conduits for dialogue with customers and consumers.

[20]Some of the big multinational water companies have felt that it is in their best interest not to make PPP agreements public as this allows 'greater flexibility' in terms of re-negotiations. Interestingly, at the time of writing this, accessing PPP agreements was difficult and in some cases special permission had to be sought from the existing Developer to use this document. In other cases, such as the case of Guayaquil, Ecuador, where the author was engaged as a mediator, specific permission had to be sought.

[21]This occurred during the preparation of the Second Five-Year Plan for the concession in 2000, and took more than a year to complete. The Concessionaire put together a series of presentations on service quality aimed at a variety of stakeholder groups. The information in these presentations was checked with the Regulator's office for validity

before they were made. Some of these presentations were made as part of a public hearing process, which was an outside-contract activity agreed upon with the Regulator for some municipalities and other stakeholder groups. This consultation process was intended by both the Regulator and Concessionaire to get 'buy-in' from the different stakeholders as the plans were adjusted to their needs. At the end of the process, the different municipalities gave their approval to the revised plan. This approval was not a legal requirement, but was critical to the political process.

[22]It has been noticed that NGOs and Consumer Organizations in particular, are playing an important and increasing role in water sector monitoring in many developing countries. However, there is a large variation in the approaches that are taken by NGOs in different countries and include some who are confrontational and do not contribute to a constructive dialogue for improving performance or getting more value for consumers. Other approaches can be very constructive and effective. NGOs and COs are often called upon to represent consumers on regulatory boards and other bodies. Examples of this are found in Zambia, where ZACA (a CO) was invited to be a member of the National Water Supply and Sanitation Council, in Senegal, where ADEETelS (a CO) has both a seat on the board of SONES the state asset-holding company in the water sector and has been appointed a member of the management committee for the next phase of water sector reform, in Metropolitan Manila, Philippines where CONSUMERNET is on the Board of Directors of the regulatory authority (MWSS).

Part II

Main Factors Influencing Effective Development & Operation of PPP Schemes

Chapter 5

Performance monitoring, legal and regulatory issues

5.1 INTRODUCTION

The design of PPP schemes requires a legal and regulatory framework within which the agreement can operate. Having determined the performance measures that will demonstrate the level of scheme effectiveness, the contractual and regulatory framework then needs to incorporate mechanisms to monitor and control these performance measures. Critical decisions relating to the management of water and wastewater service provision are often based on inaccurate and insufficient data both at design stages and during the long term scheme operation. This issue and ways of dealing with it are discussed in Chapter 12.

Effective legal arrangements are at the core of developing sustainable PPP schemes, therefore suggestions are made as to how the arrangements can be embodied in legally effective and enforceable frameworks. Key contractual recommendations are dealt with in Part III of this book, but also the overall legal regime of a given country is considered and how it may impact the manner in which a PPP agreement is developed and interpreted.

At the same time, and given that water and wastewater services are provided in a monopolistic fashion, this chapter considers those issues that are particularly relevant when establishing effective PPP arrangements which also protect customer interests through adequate regulation. As typically users are not in a position to choose who provides them with water and wastewater services the case for public intervention is quite clear. On the one hand, monopolies represent the best way to achieve universal access at the lowest investment cost. On the other hand, unsupervised monopolies may not set fair price or supply services in an efficient manner. Reconciling these conflicting interests requires regulation to allow for efficient investment and to compensate for the absence of market forces. Thus, the basic principles of economic and contractual regulation will be dealt with in this chapter. Developing adequate regulatory provisions and details of the precise regulatory functions to be carried out and by whom will be crucial.

Adequate regulation will be determined by currently existing institutions, how effective they are and whether it is necessary to create new institutions to complement those already existing.

It must be noted that although traditional advice suggests that creating an independent regulator is a panacea to all problems (Bertolini, 2004), experience collected by Groom *et al.* (2007) suggests the situation is far more complex than a one-stop solution could possibly handle. Instead, this chapter argues that governments may find that assigning regulatory functions to different organizations may provide the most effective means of improving standards and service provision. Governments should ascertain whether there are existing organizations under its auspices that have regulatory capacity and whether existing legal and administrative institutions are capable of managing utility regulation. Alternatively, regulation could be achieved through specific contractual provisions within the PPP agreement. This is all discussed in this chapter.

5.2 PERFORMANCE MONITORING

Successful public reporting systems make use of a partnership approach between the main stakeholders in a PPP arrangement – namely the Operator/Developer, regulator/monitoring unit and the customers. The principal aim is to reduce any potential for confrontation between the key partners by taking a more inclusive approach. The design of future contracts should contain provisions that oblige the government and the Operator/Developer to divulge more information to the public as part of the partnership approach. However, this recommendation is complicated by the lack of specific guidelines on the responsibilities involved in public reporting to be granted to monitoring units or other performance monitoring bodies. Additionally, where regulation occurs through the contract (instead of via an agency) more thought must go into how to create a continuous dialogue between the Operator/Developer and the customer. This type of regulation will require higher levels of customer participation created in the design of the monitoring arrangements which may open up avenues, for NGOs to play partnership roles. Developing a partnership between the Operator/Developer and its customers, particularly in a developing country, requires considering the political landscape and the resources available. Additionally, the size of the utility and the population it serves will also determine the parameters of the partnership. The political landscape is critical as the more the politicians are involved in the issues, greater the need for a solid partnership with effective service performance reporting. Greater transparency can help to fend off politicizing the arrangement and information sharing can help dispel suspicions and distrust surrounding private Operator/Developers.

The details of the partnership should be stipulated in a contract and the relationship between the utility Operator/Developer, the customers, NGOs and any monitoring units, should be defined in the PPP Agreement (Environmental Resources Management, 2003b).

Contract and non-contract agreements are required to create a partnership agreement between the key stakeholders. Using standardized clauses for contract and non-contract responsibilities in consumer-oriented reporting, will give the partnership structure the degree of flexibility required to accommodate the intricacies of local political and social landscapes. Where contracts are not appropriate, it is recommended that a publicly available charter with clearly defined goals and responsibilities be created. Cook and Stevens (2005) detail the decision-responses to each of these areas of interest and the key stakeholders most affected by the information.

Standard Performance Reporting could follow the schedule governing the publication of the Developer's annual report and would build on Cook and Stevens (2005) and recommendations by Nickson and Vargas (2002). This would also be appropriate for information that requires a longer time-frame to gather, for example, average hours of service, extent of coverage and so on. Some elements of performance reporting can be gathered in shorter time-frames (monthly or quarterly) such as water quality, shortages or service interruptions. Table 5.1 provides some of the possible consumer-oriented performance indicators.

Table 5.1 Decision responses for different stakeholders involved in a PPP arrangement.

Consumer-oriented performance measures	Goals / performance indicator	Decisions	Stakeholders involved
Operational Service (coverage, hours of service, pressure, reliability)	–Expand service to all –Receive high quality reliable service	–Increase number of connections –Increase service quality and reliability	Operator Regulator Public Trust Customers/NGOs
Customer Service (response to complaints and requests)	–Respond effectively to customer requests	–Increase the effectiveness of responses to customer complaints and requests	Operator Regulator Public Trust Customers/NGOs
Customer Satisfaction	–Ensure customers are satisfied with service	–Improve service in areas important to customers	Operator Customers/NGOs
Affordability	–Receive fair and equitable pricing	–Improve affordability of tariffs	Operator Regulator Public Trust Customers/NGOs

Source: Adapted from Cook and Stevens (2005).

In some of the PPP arrangements reviewed here, Operator/Developers are required to post monthly reports to their regulator (e.g. Guayaquil, Ecuador). Clearly, the decision on the frequency of reporting will be based on the capacity of the relevant parties to gather the information and disseminate it in an accurate and understandable manner. Additionally, the cost of more frequent reporting will have to be borne in mind. Cook and Stevens (2005) suggest a frequency timetable for such performance reporting as detailed in Table 5.2.

Table 5.2 Suggested frequency of performance reporting.

Indicator category	Performance indicator	Recommended frequency
Financial	Financial indicators	Yearly
Operational efficiency	Efficiency indicators	Yearly
Operational service	Coverage	Yearly
Operational service	Water pressure	Quarterly
Operational service	Water quality	Monthly
Customer service	Customer complaint Response time	Quarterly
Customer satisfaction	Satisfaction with service	Yearly

Source: Cook and Stevens (2005).

Allocating responsibility

The partnerships should detail precisely which organizations are allocated the responsibility for each element of the reporting process (Ehrhardt & Irwin, 2004). The general flow of responsibility will start with the Developer from its own internal performance information requirements. The customer satisfaction information will then have to be collected by another stakeholder. The monitoring unit could be granted the job of filling in the blanks and processing the information, if a complete information package is not delivered.

Additionally, the partnership would require a method of verifying the information gathered, particularly information gathered by a private Developer. An independent auditor could be contracted to perform this role. Grantors may wish to consider using an independent agency to gather and report additional information, particularly in difficult political landscapes where distrust of private Developers may run high.

The monitoring unit, where it exists, would be responsible for reporting performance levels to the public; otherwise the local government must take on this responsibility.

Customer Committees would be responsible for providing information from the customer's perspective on the Developer's service performance and detail areas where customers believe service levels could be improved.

Key elements to keep in mind when determining the indicators include:

- Ensure the list of indicators can be gathered consistently and uniformly.
- Choose indicators that fit specific goals and needs.
- Adapt past experience to local landscape.
- Be realistic with what indicators can be assembled, not what one might wish could be assembled.
- Gathering credible data is the most important criteria.
- Ensure stakeholders each possess the technical capacity to gather and process the relevant data.
- Use dialogue with Customer Committees and NGOs to pilot indicators.

In a practical sense, and using as an example the suggested core indicators, Table 5.3 illustrates how such indicators 'evolve' for the duration of the PPP Agreement.

Looking at the evolution of targets, and using the monitoring indicator of 'Increasing the number of connections in wastewater' as an example, it can be observed that depending on the actual situation on the ground, there is no actual target in the initial period from 0 to 5 years in the PPP Agreement. Once basic information has been collected and reasonable and realistic data is available, specific targets can de set in terms of coverage making an assessment of the investment needs required to reach such targets. These targets need to be made consistent with the Grantor's overall sanitation policy. The following keys steps are recommended for implementing a consumer-oriented performance and monitoring regime:

- *Establish Objectives*: Once the performance indicators have been agreed upon and the system of reporting negotiated, an important next phase is to establish performance objectives for operating the reporting system as well as developing performance indicator targets. As the quality of data improves, the targets should be re-assessed. Baseline standards should be established for each indicator and performance targets should be set over a period of years.
- *Develop an Action Plan*: A comprehensive action plan should defined setting out each step to be taken to achieve the desired functioning reporting system that each stakeholder can use to review its obligations.
- *Develop a Self-Sustaining Reporting Process*: Achieving financial and political sustainability will be critical to the long-term success of the reporting system. However, long term financial and political sustainability will be dependent on the success of the reporting system. A reliable source of funds must be procured. A Developer will not incur large expenses in

Table 5.3 Potential monitoring indicators and corresponding targets*.

Goals	Improvement actions and inputs	Potential monitoring indicators	Recommended targets year 0–5	Recommended targets year 5–10	Recommended targets year 10–15
		% households with direct access to water services – [nos households connected to water services/total house count]	Performance targets: depending on local current figures	Performance targets: depending on local current figures and operation parameters	Performance targets: depending on local current figures and operation parameters
			Milestone targets (year 0–2): identify and map – number of connections, nos households within 250 m of a shared connection; total house count – can perhaps be done by aerial photograph		
	Increase the number of connections	% households with direct access to sanitation services – [nos households with toilet or latrine/total house count]	No performance target	Depending on local current figures and sanitation policy	Depending on local current figures and sanitation policy
			Milestone targets (year 0–2): design and acceptance of publicity programme of sanitation measures. All new water connections to undergo sanitation education training.		
		% households served by non-piped delivery systems – [nos households/total house count]	No performance target	No performance target	No performance target
			Immediate action (year 0–1): Procure mapping/aerial photographs for all urban areas over 10,000 estimated population Milestone targets (year 0–2): Identify and map – location and estimated population; house count without direct access to piped supply, total house count		

Expand service to all			No performance target	No performance target	No performance target
		Number of connections [Nos connections total or by category – e.g. domestic, non-domestic]	Milestone targets (year 0–1): Enumeration survey of all urban areas and review and complete consumer billing database		
		Number of areas not supplied [estimated population not supplied, or number of areas greater than 1000 population not supplied; or percentage of population supplied]	Performance targets: depending on current figures	Performance targets: depending on local current figures and operation parameters	Performance targets: depending on local current figures and operation parameters
	Expand service area		Milestone targets (year 0–1): Identify and map – location and estimated population; census data; population growth rate Other milestone targets (year 0–2): water resources assessment, demand forecast study		
		Extension of pipe network (km) [km of pipe to be laid by period, total km of pipe by period/original km of pipe as beginning of period]	Performance targets: depending on local current figures	Performance target: depending on local current figures and operation paramaters	Performance target: depending on local current figures and operation paramaters
		Available production capacity (mld) [Sum of design capacity of each operational works/predicted capacity in 2010]	Milestone targets (year 0–2): identify and map – location and estimated population; demand forecast, water source, pipeline route and length		

*Author's suggestion based on Cook and Stevens (2005), Ehrhardt and Irwin (2004) and Estache *et al.* (2004).

obtaining the required data and can agree to continue collecting the information without additional arrangements. Public utilities can reach an agreement for support from the local government and all other stakeholders must have secure funding provisions determined from the start of the PPP contract. Political sustainability comes with government support and results. The partnership strategy recommended in this chapter seeks to ensure that political support occurs more rapidly and with broad levels of consensus.

5.3 LEGAL FRAMEWORK

No one project or its jurisdiction will be the same. It is essential that when entering into a PPP Agreement the government or government's entity that enters into a contract has the right or the power (*vires*) to enter into the obligations involved in the project and is able to undertake its roles and responsibilities. All actions which the Grantor takes must be permissable by law (*intra vires*). Acts which are outside the law or outside or beyond the scope of the party performing their duties (*ultra vires*) must be avoided. All decisions and acts taken by the Grantor must be enforceable and should not be subsequently rescinded or invalidated under applicable laws. In some instances there will be particular administrative or legal requirements to be satisfied before the obligations set out in the PPP Agreement can become binding. In many cases, major infrastructure projects will require special permission to be granted, either by central government or the governing authority. These entities may differ from the actual entity that enters the contract. In other words, the Grantor may be different from the authority that provides the right to enter into the PPP Agreement.

The legal design of the PPP Agreement must provide a sound legal basis for the transaction (Delmon, 2009; EconOne Research, 2003; Gómez-Ibáñez, 2003 and Mandri-Perrott, 2009). Furthermore, the legal environment under which the PPP Agreement is to be procured needs to be specified clearly. It is essential that the legal code governing private participation is well understood, as this will dictate the legal instruments that may be used. In this way mechanisms can be developed to ensure compliance with obligations.

The choice of instruments in any particular case depends on the legal system, the practice of the country and the purpose the instrument is intended to serve. There is no absolute structure or rule for choosing legal instruments. A PPP Agreement may rely on several legal instruments to underpin its validity. For example, the PPP Agreement may be authorized by a statute and spelt out in an executive order, but the specific conditions of effectiveness are provided for in the PPP Agreement itself.

The PPP agreement is an arrangement that seeks to bind both the government (through the Grantor) and the Developer. At first sight it may appear that the government may be restricting its ability to take action and thus reduce its flexibility. Yet one party's flexibility often equates to the other party's risk. If the

Grantor were to have complete freedom to change the PPP Agreement within the duration of the contract, it is highly unlikely that a Developer would agree to such arrangement without assuring full protection. In order to obtain the benefits of private participation, the government has to commit to standards of behaviour and procedures to be regulated through the contract.

Civil law[1] is a legal system inherited from Roman law and pertains to the private rights of citizens. The term comes from the Latin Jus Civile, or 'citizens' law', the Late Imperial term for its legal system (Neubauer & Meinhold, 2007). The principle of civil law is to provide all citizens with an accessible and written collection of the laws which apply to them and which judges must follow. It is the most prevalent legal system in the world and is used in France, Spain, and most of their former colonies, as well a number of other continental European countries. A slightly different legal system called Common law is used in the United Kingdom, most of its former colonies, and the United States. In civil law countries, a separate administrative law usually governs relationships and contracts between the government and a private partner[2]. In common law systems, that is not the case, and there is no legal distinction between contracts that involve only private firms and contracts that involve a public authority and a private firm. The same law and the same courts govern private participation contracts that business people rely on for their transactions with each other. Common law contracts are very flexible, and almost any agreement can be put into a contract and can be enforced.

The difference between civil law and common law lies not just in the codification itself, but in the methodology used in the approach to codes and statutes. In civil law countries, legislation is seen as the primary source of law. By default, courts thus base their judgements on the provisions of codes and statutes, from which solutions in particular cases are to be derived. Courts thus have to reason extensively on the basis of general rules and principles of the code, often drawing analogies from statutory provisions to fill lacunae and to achieve coherence. In contrast, case law in the common law system, is a prime law source, while statutes are often seen as supplementary to judicial opinions and thus with a narrow interpretation. Under a civil law tradition, some important administrative rules seem to be common in many countries (such as the rights of the grantor to unilateral modification and unilateral cancellation[3]). Even if some of these issues are part of the background law that applies to a PPP Agreement, it may still be necessary to spell out the relevance of these rules because the law may not be explicit enough about their application. Thus, writing clear rules into the contract is the safest solution (Delmon, 2009). It is generally a good idea to specifically outline in the contract the grantor's rights (for example, to demand unilateral changes in services) and to include provisions within the PPP agreement that explicitly address the circumstances under which the developer will be compensated. However, the legal validity of explicitly spelling out how an administrative law principle will be applied within the PPP Agreement will need to be checked. For example, some civil law codes contain mandatory notice periods before termination for breach of contract; such rules cannot be avoided or overridden within the PPP Agreement.

In some countries, there has been a trend toward creating general laws for private sector participation. These special statutes could provide a clear framework for a PPP Agreement. A well-designed law addresses such issues as identifying which agencies have the power to initiate the introduction of private participation and set out the mechanisms that will allow competitive and transparent procurement processes. At the same time, poorly drafted PPP laws may hinder the PPP process (Harris, 2003). Even with specific laws (either general or sector specific), there are often limitations. Successful PPP Agreements will include appropriate methodologies and techniques that make optimal use of the approach described in this chapter.

5.4 BASIC PRINCIPLES OF REGULATION

As towns and cities developed, the provision of water and wastewater services passed from individual, to collective responsibility. Citizens began to organise themselves and form structures that were in charge of providing services to the community as a whole. Since these early beginnings, water and sanitation networks have grown in both size and scope. Associated costs and resource requirements have increased according to the complexities involved. Modern water and wastewater systems require large and continuous capital investment. The need for realising a return on this investment would intuitively seem to conflict with policy goals including universal availability and affordability. Left to its own devices, a naturally competitive marketplace for water and wastewater services could result in redundant infrastructure in some areas and no infrastructure in others. Profit incentives for private investors and customers' ability to pay would probably not result in a socially optimal water and sanitation network. Because these services are vital to quality of life and economic development, the case for public intervention is quite clear.

In most countries, specific regions are served with water and wastewater services provided by a single entity. Structuring utilities as 'natural monopolies' helps to achieve universal accessibility and affordability and concentrate investment into one single infrastructure network. Inevitably, citizens have no choice but to receive service from this sole provider. The resulting combination of vital services and monopoly pricing power gives obvious cause for concern. Public ownership and control would represent the simplest method for protecting the interests of citizens. However, as discussed in previous sections, public ownership is not without limitations and failings.

Many cities require financial resources and implementation capacity that public sources simply cannot supply. This, of course, is one reason why private sector participation can add value and result in net social benefits. Nevertheless, unchecked private monopolies would be undesirable. Economic theory suggests that monopolists could set prices above a socially desirable level where the quality of service supplied may also be below par. Rather, citizens and policy makers would prefer pricing to be equivalent to the 'marginal cost' of supplying services.

Overall economic efficiency is tied to the marginal cost associated with producing one more unit of output. Wise allocation of limited resources brings with it the implication that customers should pay the true cost for their consumption (Mandri-Perrott, 2009). Pricing below marginal cost leads to overconsumption because the benefits of each additional unit of service outweigh its expense. Pricing above marginal cost leads to under consumption because the expense of each additional unit outweighs its benefits. Competitive market forces naturally compel firms to price near marginal cost where the corresponding level of public consumption is also optimal. Unfettered monopolies prevent this phenomenon from occurring.

Providing capital intensive public services presents an interesting challenge. On the one hand, monopolies represent the best way of achieving universal access at the lowest investment cost. On the other hand, unchecked monopolies may set inefficient prices or supply services. Reconciling these conflicting interests requires regulation which will allow for efficient investment but at the same time compensate for the absence of market forces.

However, the preceding discussion should not imply that regulation exists *only* to ensure economic efficiency. According to Ballance and Taylor (2001), other goals for sound regulation include:

- *Financial viability*: Private firms expect reasonable returns on capital invested. Intelligent companies take risks only in accordance with the potential for rewards that may follow. When anticipated rewards are insufficiently small for some amount of risk, firms will not commit scarce resources to a project. Continuous investment is critical for maintaining quality water and wastewater services. Return on capital is also an important incentive when justifying system expansion to meet the goal of universal access. Effective regulation therefore, should attempt to preserve investment incentives at an appropriate level. This threshold should approximate the 'opportunity cost' of scarce funds. For certain levels of risk, investors must receive a given level of return in accordance with fair market rates. Allowing firms to recover normal returns from their investments is a critical regulatory goal because it provides the incentives for corresponding investment required to maintain good service quality and expanding the water and wastewater network to all citizens.
- *Operational efficiency*: As previously discussed, water and wastewater services are socially beneficial and should ideally be priced at their respective marginal costs to optimize both consumption and supply. Reductions in the marginal cost of services allows for additional, yet optimal, consumption. There is, therefore, clear benefits to reducing costs and increasing the efficiency with which a regulated monopoly supplies

these services. Creating incentives for private sector efficiency requires some potential for reward, in order to incentivize progress to make improvements and further cost reductions. Sound regulation should allocate some portion of the benefits from reduced service costs back to private partners as an incentive to improve operations and make them more efficient.

- *Dynamic efficiency*: Rivals in competitive markets will undertake investment and formulate plans to accommodate future demand in order to retain or increase market share. In the absence of competition, regulation should stimulate private monopolies to still plan and invest for future growth as again, unsupplied demand would be undesirable.
- *Distributive justice*: Access to clean water and adequate sanitation is a fundamental human right. Accordingly, policy makers should assure affordable access to all those who seek access to such services. In many cases, achieving this goal may require some kind of subsidy. These subsidies may often come from customers with greater capacity to pay for services (either directly through a 'cross subsidy' arrangement or indirectly through tax monies). In this way the structure of prices is consistent with paying capacity. Balancing the inherent conflict between distributive justice and economic efficiency is another goal of good regulation.

From the outset, it must be stated that there is no one-size fits all regulatory design that can be reliably applied and guaranteed to solve performance and efficiency problems. Regulatory design must be conceived through the lens of the wider specific contexts of the country in order to reform the way in which water and wastewater services are delivered and accounted for.

Economic regulation

As discussed above, the provision of water and sanitation is an essential service, and has strong natural monopoly characteristics. This makes it important to control service providers and make them accountable in some way, regardless of whether the provider is publicly or privately owned and operated. Private providers are generally regulated by autonomous agencies or contracts, or both. Regulation is considered necessary because, without some form of constraint, private providers may act in a way that is not in the public interest or make unreasonable profits by overcharging and under delivering (Groom *et al.* 2007).

Economic regulation can be defined as the legal controls placed on service providers in order to overcome the problems inherent in an essential, monopoly service. Economic regulation sets the formal rules and associated organizations that set, change, monitor, and enforce permitted tariffs and permitted service standards for water providers (Smith, 2003), and further defines economic regulation as mimicking the pressures that competition provides in other markets. In other words, it can help stop tariffs from increasing above the level required to recover reasonable costs. It will also help to set and enforce service standards and ensure that those standards are achievable against set tariffs.

It is important to point out the need to balance the tariff and the associated level of service, or the standard that one seeks to achieve. There is no point in setting service standards which require massive investments that cannot reasonably be recovered by the service provider. If it is a private provider, tariffs will need to cover its reasonable costs that is, the cost of operation and maintenance required to provide the service, as well as the cost associated with borrowing capital for making the necessary investments, and providing a reasonable return (or profit) to the shareholders of the service provider. Alternatively, if the service provider is a public entity, recovery of reasonable costs should in principle, seek to set tariffs which would cover these costs (Public Private Infrastructure Advisory Facility and Water and Sanitation Program, 2002a). However the key difference is that the cost of capital for governments is generally significantly lower than that obtained by the private sector, and those public borrowing benefits could be passed on to the customer.

Public provision should not per se imply that the sector is being subsidized in an un-transparent manner. Even if there is a public provider, the surplus, after operations and maintenance, as well as government's cost of capital, should be retained by the service provider and reinvested. After all, one of the key characteristics of the water and wastewater sector is the fact that it is highly capital intensive. For this reason economic regulation should seek to set rules to ensure that a service provider is able to invest continuously to achieve sustainability of service (Public Private Infrastructure Advisory Facility and Water and Sanitation Program, 2002b). It is important that tariffs are set at adequate levels: not too high so as to allow excess profits or surpluses, and not too low so as not to allow adequate cost recovery. Note that it is important that government should not have to bear the fiscal burden of providing either implicit or explicit subsidies to the sector.

Furthermore, specific national legislation may prescribe ownership rights related to water, assets and so on, including who is responsible for service provision, and how the owner exercises control over the utility's management. Country governments need to ensure that these factors and the regulatory design align with each other. Furthermore, the regulatory regime must be coordinated with other bodies involved in water sector activities. It is important that each country brings together the different bodies that deal in a direct or indirect way with the water sector. It must ensure adequate coordination of public expenditure priorities. Consistent and appropriate policy decisions and implementation plans are key to achieving a balanced approach to the development of the water sector (Rama, 1999).

The traditional approach to regulation for water and wastewater services has proscribed the use of an independent regulator (or similar model) with the intention that this body should solve the issues that usually beset the sector. This solution has had a mixed level of success with numerous examples of public disputes due to uncertainty over the relationship between contracts and the regulator[4], often leading to appeals and government intervention on tariffs. Regulation needs to be developed within a variety of legal instruments and organizations and must be implemented with an understanding of the specific context of the country. Crucially, the following questions must be asked:

- What are the real problems in the sector and which can be dealt with by regulation?
- What are the specific objectives that regulation will attempt to achieve?
- Which combination of instruments and organizations is best suited to achieving these objectives?

Answering these questions is part of the first step towards creating a regulatory system that is credible, predictable and legitimate. Regulatory design must take into account the state of existing laws and institutions. Good regulatory design must reflect the specifics and should demonstrate the following key attributes (Ehrhardt *et al.* 2007):

- *Coherence*: a well designed regulatory system will ensure the alignment of interests of the principal parties for an agreement (customers, Developer, government and investors). The relationship between tariffs and service standards is clear and correlated – the higher the service standards, the higher the costs – the regulatory system should ensure Developer recover his costs and people receive the services they are willing to pay for.
- *Predictability and credibility*: when operating in an environment with little risk of changes in standards and rules for cost recovery and tariffs, attracting a competent Developer is easier and will also serve to encourage the Developer to invest in the water services infrastructure. Predictable and credible regulation is best achieved within a sound policy and governance framework. Regulatory decisions should be time-consistent, and made on clear precedents and rules. In turn, this will minimize risk to Developer, encourage good service provision and sufficient asset maintenance and capital investment.
- *Legitimacy and accountability*: the regulatory process should be as open and accessible to the public as possible in order to eliminate suspicions of corruption, ensure greater transparency and provide accountability for the customers who bear the impact of tariff and service provisions. Regulatory institutions require incentives to make decisions for the long term interests of consumers and therefore robust accountability levels would assist in this regard.

Working with the existing institutional framework

Generally, regulation is introduced into an environment of existing institutions with well established legal and political mechanisms and interactions with the public.

Regulatory design does not have to be imposed upon existing structure, instead it can emerge from the existing institution framework and legal instruments and utilize the strengths of the existing system (e.g. precedents for contract enforcement) whilst introducing specific tools to address the weaknesses in the system (such as a contract enforcement unit if this is currently lagging). The familiarity of existing institutions is an asset for obtaining public acceptance for changes and designing regulatory regimes around this notion can be advantagious in winning the public's support.

Research suggest (see Ehrhardt *et al.* 2007; Estache *et al.* 2000; Saltiel, 2003; Shirley, 2002 among others) that typical strengths and weaknesses of institutions within each country may vary, with specific characteristics for each. No institution is perfect in respect to all the necessary tasks required for a successful PPP scheme, so Governments must choose between institutions with strengths in some areas and weaknesses in other – or create new institutions. Table 5.4 (World Bank, PPIAF, 2006) shows the typical strengths and weaknesses of institutions.

Table 5.4 Typical strengths and weaknesses of institutions.

Institution	Typical strengths	Typical weaknesses
Operator	Knows the business well—has much relevant information. Has many relevant skills.	May lacks independence and won't have the right incentives for some tasks as a result of being one of the main parties to the agreement.
Ministry	Knows aspects of the business well. Already functioning. Will have legitimacy in some countries.	May lack independence. May not have all the expertise required for some tasks, such as applying tariff-adjustment rules.
Independent regulator	Can be independent of the operator and somewhat insulated from political pressure on the contracting authority. May have more legitimacy with customers than other more independent bodies such as international arbitration panels.	May not be effectively independent from the government in practice. May not be trusted by the operator or the contracting authority.
Joint decision – making (bipartite committeet)	Represents the interests of the operator and the contracting authority. Incorporates expert knowledge of the business and	May fail to reach an agreement if both parties hold out, for a better deal. May not operate transparently. May not directly

(*Continued*)

Table 5.4 Typical strengths and weaknesses of institutions (*Continued*).

Institution	Typical strengths	Typical weaknesses
	has many relevant skills. Has incentives to take into account the interests of both the operator and the contracting authority.	involve customers in decision making.
An independent expert	Is independent and likely to have technical expertise.	May lack legitimacy in the eyes of customers.
Arbitral panel	Is independent and can have expertise in resolving disputes.	May lack legitimacy in the eyes of customers.
Courts	Already exist. Generally independent of the operator and the contracting authority. Often have legitimacy.	Typically reactive, rather than active, limiting their role. Lack expertise for certain tasks. May be slow or corrupt. May lack legitimacy.

Source: PPIAF World Bank, Approaches to Private Participation in Water Services, 2006.

Regulatory designers should also be mindful that a country's institutional capacity and credibility can change over time and the regulatory institution created should be able to reflect this change by taking on a stronger or more extensive role. Such changes can reveal themselves over time as government policies change or obvious faults in regulatory design become apparent. For a regulatory system to gain the trust of the public, it must demonstrate long term commitments and contain lasting structures. Accordingly, regulatory schemes must:

- *Choose the correct legal instrument*: As discussed previously, past experience has long favoured the use of independent regulatory bodies with broad enforcing powers, rather than regulation through the contract. Lessons from experience now suggest that the contract is more often than not, a good base from which to establish clear and easily enforceable regulatory rules[5]. Civil-law countries are more used to using contracts for regulatory control over service standards and Developer remuneration. At the same time, it must be made clear that contract (or sole contract) regulation may not be suitable in all circumstances. In situations where governments have to deal with multiple regulated water service providers, it is advisable to create a single legal instrument, such as a statute or uniform license, that details the rules of the game for all the players. This will also serve to reduce regulatory risk as investors will take comfort from the presence of national laws backing up contracts entered into with local governments.
- *Be aware of parallel processes*: Common regulatory models that were created to interact with fully privately owned utilities. In water and wastewater PPP

projects where the PPP Agreement is between public and private entities, governments should be mindful of this when designing regulations to ensure that there is a seamless relationship between economic regulation and the PPP arrangement it is designed to support.

- *Create an appropriate role for politics*: Attempting to separate politics from regulatory independence cannot always be guaranteed and under certain specific circumstances may not always be unwelcome. Politics can play a positive role within a framework that permits politicians to make sound decisions and honour contractual commitments. To achieve this, regulatory designers must specify and define a role for politics within the regulatory structure which encourages sensible political intervention when required. A clear and defined role for politics in regulation provides a source of collective decision making, accountability and a tool for government action when faced with public scrutiny. A key provision is to ensure coherency and to avoid the short-termism that is endemic to most political arenas[6]. Contract provisions that depoliticize the usually sensitive issue of tariffs, by removing the link between consumer tariffs and operator revenue, can provide the government with an ongoing role without being too detrimental to the Operator's performance or risk management structure.
- *Limit the discretion given to regulatory decision makers*: Quality decision-making requires the government to make predictable, coherent and transparent decisions. It is important to select the right individuals and establish clear written rules which set out the level of 'discretion' that decision makers are granted. Within this context, regulation by contract is associated with low discretion and independent regulation with high discretion. In reality, the actual relationship is far more complicated and beholden to other factors[7]. Low discretion rules are also likely to be best utilized in countries with limited institutional capacity and regulatory traditions due to their ability to make the outcomes of regulatory decisions clearer and more predictable and are achieved using fewer resources. This serves to increase the credibility of the regulatory system in the eyes of the public and the Developer. Increasing the transparency of decision making can reduce the impact of external pressures that influence decisions.
- *Provide simplicity over sophistication*: Designing a regulatory framework that takes into account country contexts and regulatory experience is the starting block. The next critical stage lies in the specific regulatory details (tariff formulas, rules for calculating costs). The more room for misinterpretation of the rules, the greater the scope for disputes and messy interventions. Keeping rules simple increases their enforceability, promotes their legitimacy and require less expertise to manage. The more complicated the rules and regulations, the more difficult they are to enforce, making decision making more difficult and allowing greater influence of external pressures (Groom *et al.* 2006).

5.5 DEVELOPING A REGULATORY FRAMEWORK FOR EFFECTIVE PPP AGREEMENTS

The water and wastewater sector has strong (natural) monopoly characteristics which heighten the importance of developing robust accountability measures in any regulatory structure. Usually, private Developers are regulated by autonomous agencies or by contracts or a combination of both. Without sufficient regulation, private Developers may lack sufficient incentives to avoid overcharging and under delivering. A key element in the approach to regulatory design should be on how Developer can be given incentives to deliver services at a reasonable price, solve the regulatory issues and interface well with existing institutions and systems. The risk of failure can be reduced through careful planning, assessing where and how regulation is needed:

- Define the underlying problems and objectives in the sector.
- Assess whether existing institutions can move the sector forward and address the underlying problems.
- Detail the specific regulatory institutions required to address the underlying problems.
- Decide on the cocktail of legal instruments and which institutions are best suited to fulfilling regulatory functions.

The use of water resources and the manner in which wastewater is treated and disposed of into the environment needs to be well managed. When introducing the private sector into the provision of water and wastewater services, it is important to define clearly the roles and responsibilities of each party involved. The suggestion is that the private sector should be responsible for providing the service, that is, the operation and management (including sometimes financing the sector's needs), whilst the public sector's role shifts to one of ensuring protection of this vital resource and regulating the manner in which the services are provided.

It is also important to recognise the potential harm from bad regulation. In particular, regulation that oversteps it boundaries can severely hinder investment, limit development in the sector and could ultimately amount to expropriation. This may occur when regulatory changes deprive private firms of the potential benefits that would otherwise accrue from an asset or other investment. It is essential that the framework under which the provision of such services are to be rendered is clearly defined beforehand within the context of existing property rights legislation, so that risks involved may be allocated to the appropriate party. This regulatory framework for including the private sector in the provision of water and wastewater services effectively sets 'the rules of the game' and its form may vary depending on the type of contract structure.

Typically, when the private sector is involved in the provision of water and wastewater services, there is some form of contractual relationship that enshrines the roles and responsibilities of each party (Guasch, 1999). Notwithstanding the contractual form chosen, the framework needs to create a business approach clearly focussed on meeting the requirements of its customers, that is, quality of product and customer satisfaction. Similarly, regulation must not only protect the customers from monopoly abuses but it must ensure some level of risk protection for the private sector, particularly in relation to investment. In PPP contracts such as BOTs, Concessions or Divestitures, the private sector has a long-term investment commitment that requires assurances that the project's structure will allow adequate recompense for repayment of the loan principal and interest. Private sector participation needs an enabling set of rules not only of a regulatory nature but also from a political, legal and economic perspective.

Process for developing the most appropriate PPP agreement

The first step in designing the most appropriate PPP Agreement and associated regulatory provisions must start from a thorough identification of the common causes of poor sector performance. These can be summarised as: (adapted from Budds & McGranahan, 2003)

- Inadequate investment in and maintenance of system asset. Caused by –
  - Inadequate cash flow, tariffs set below cost-recovery level, or higher external costs.
  - Poor long term planning or disagreements between government and Operator over investment plans.
  - Lack of confidence in credibility of regulatory arrangements.
- Poor operating and maintenance standards.
- Vague or unclear service standards which are not enforced.
- Absence of service expansion policies.
- Barriers to entry for small private providers.

Assessing the inadequacies of the existing water and wastewater system is the first step towards its reform. Having identified the main problems within the sector, the government needs to develop a strategy for reform that can get the sector moving in the direction of actually providing the services that people want, at a price they can afford. The combination of legal instruments and institutions will vary from case to case, country to country. From previous analysis of the work on policy setting, it can be concluded that good starting points for designing the framework for the regulatory system to suit the characteristics of the country and its water and wastewater sector, include:

- Understanding the main problems, the main objectives and the performance improvements that are desired.
- Understanding how the rules and institutions required will monitor performance, manage tariffs and service standards and achieve the performance improvements desired.

- Delineate the specific regulatory actions required to achieve the above objectives. This will be framed by the inherent performance issues hampering the efficient running of the existing system. With this understanding, decisions can be made on whether the focus should be on improving cost recovery levels or extending services.
- Make a decision on which legal instruments will best achieve the regulatory goals and which institutions will best serve their regulatory purposes.

However, research also suggests (see Camdessus & Winpenny, 2003) that care must be taken when designing water and wastewater sector reform. Governments should be wary of setting out to reform their sector with overly ambitious objectives that may not fully consider the reality of the country context. For example, insufficient service levels create a vacuum into which small local private suppliers may emerge that provide water services that may be unreliable and that do not meet basic quality and quantity criteria. Increased investment in infrastructure to expand coverage and obtain new customers may result in poorly managed utilities being forced to charge tariffs below cost recovery and are thus unable to either fund maintenance operations or invest in assets to increase coverage. This evidence suggests that reform in the sector should occur progressively and in a realistic manner. Efficient management systems that reduce costs should be established as well as creating the necessary conditions to allow a tariff at cost recovery levels to ensure the operator can provide basic services to all.

Regulation, national policy within the EU legislative context

As discussed above, economic regulation focuses primarily on setting, monitoring, and enforcing rules on tariffs and service quality. Yet depending on the legal structure of a given country, its legal tradition and other factors, economic regulation may overlap with other areas of regulation and policy. In this section we are only concerned with the main principles of regulation and their implications with respect to EU policy and particularly how economic regulation can be ‘woven into’ a PPP regime. In this context therefore, it is up to each Member Country to apply its own rules as to the specific manner in which it seeks to undertake its regulatory functions. Each Member Country must decide whether a particular regulatory function should be considered part of economic regulation or dealt with in another way. For example, setting drinking water standards may be considered part of economic regulation, or it may be dealt with as an aspect of health or food safety regulation. Service coverage may be governed by regulatory requirements to the provider to connect more customers, or by subsidies for service expansion, or both.

Another form of exercising regulatory influence may be through the use of powers granted through law to instruct providers to do certain things, and to enforce these instructions through penalties or providing other forms of incentives. All in all, what is important in the EU context, and in particular for EU Accession countries, is that whatever form of regulation is chosen is designed in such a way that it is consistent with EU law. After all, the EU is explicit in allowing each Member Country the discretion to implement its own sector laws so long as they are coherent with EU law and specifically in the water and wastewater sector with the EU Framework Directive, EU Competition Law, principles for PPP and the EU Utilities Directive, among others. Of particular relevance are the implications of State Aid, as EU Competition Policy recognizes in some instances the need for State Aid. A Government may choose to influence providers through subsidies such as output-based payments for extending service to poor households, or by providing specific tax incentives. All this needs to be set out in a comprehensive set of national rules together with adequate institutions to ensure these rules are enforced. This is all discussed in the next chapters.

The EU context allows for each country to develop its own approach to the water and wastewater sector. Accordingly, each Member Country has to develop a set of national rules that provide for the correct overall implementation of the EU Water Framework. However, setting these 'rules of the game' should combine not only regulation but also a national framework that works hand in hand with economic regulation. For example, the overall policy making function should be developed to ensure that the Member Country sector objectives and principles, as well as the actions to achieve these, are clearly set and consistent with EU policy. Uniformity with EU Utilities directive, as well as the principles of good economic policy, should seek to define the extent to which consumers or taxpayers should pay for water services. Furthermore, specific national legislation may prescribe ownership rights related to water, assets and so on, including who is responsible for service provision, and in what way the owner exercises control over the utility's management. Member Country governments need to ensure that these factors and the regulatory design are aligned with each other. Furthermore, the regulatory regime must be coordinated with other bodies involved in water sector activities and ensure adequate coordination of public expenditure priorities. Consistent and appropriate policy decisions and implementation plans are key to achieving a balanced approach to the development of the water sector.

Governments embarking on water and sanitation sector reforms need to decide how providers will be given better incentives to deliver the services that people want, at a reasonable price. This is where regulatory design becomes important. As mentioned earlier, the regulatory system should include a combination of rules and organizations (including appropriate Member Country institutions) that are able to help deal with the country's key regulatory challenges, the inter-linking nature of Member Country law with EU law, and how these factors affect the manner in which national institutions can work. Brown *et al.* (2006) suggest that the starting point for any national regulatory framework should start with a thorough statement of the underlying problems and objectives in the sector. Key issues should be determined, for example whether existing institutions could adequately enforce regulatory provisions or whether these should initially be enshrined within the PPP Agreement. This assessment would help a Member Country to decide, in the light of existing institutions, which legal instruments are best suited to embody the regulatory rules, and which organizations are best suited to perform the regulatory functions.

5.6 SUMMARY AND CONCLUSIONS

As discussed above, economic regulation focuses primarily on setting, monitoring, and enforcing rules on tariffs and service quality. Yet depending on the legal structure of a given country, its legal tradition and other factors, economic regulation may overlap with other areas of regulation and policy. We are only concerned here with the main principles of economic regulation and their implications in respect to national water sector policy and particularly how economic regulation can be 'woven into' a PPP regime. In this context, it is up to each country to apply its own rules as to the specific manner in which it seeks to undertake its regulatory functions. Moreover, it is up to each country to decide whether a particular regulatory function should be considered part of economic regulation or dealt with in another way. For example, setting drinking water standards may be considered part of economic regulation, or it may be dealt with as an aspect of health or food safety regulation. Another form of exercising regulatory influence may be through the use of powers granted at law to instruct providers/developers to do certain things, and to enforce these instructions through penalties or incentives. A Government may choose to influence developers through subsidies such as output-based payments for extending service to poor households, or by providing specific tax incentives. All this needs to be set out in a comprehensive national set of rules together with the presence of adequate institutions to deal with the enforcement of these rules. In this regard, Governments will have to decide whether existing institutions can adequately enforce regulatory provisions or whether these should initially be enshrined within the PPP agreement.

In all this it is important to recognize that there are problems that economic regulation cannot solve, or at least cannot solve alone. A good example of such limitations is the need for a country to develop an integrated approach to ensure services remain affordable to low-income households, or that there are sufficient incentives given to the service provider to increase efficiency, capital expenditure planning, and achieve low payments by government departments and agencies. Thus, even when governments choose to expand the role of regulation beyond simply setting, monitoring, and enforcing rules on tariffs and service quality, supporting regulation with other policy initiatives is generally necessary.

The design of PPP schemes requires a legal and regulatory framework within which the agreement can operate. The contractual and regulatory framework needs to incorporate mechanisms to monitor and control minimum performance measures related to quantity and quantity of service. These mechanisms have to operate effectively throughout the duration of the PPP arrangement, and institutions have to be established to be responsible for maintaining and operating the monitoring and control activities, and for the regulatory activities. It is important to ensure that these institutions are funded and resourced in the long term, as well as being endowed with the necessary enforcement and remedial powers. Critical decisions relating to the management of water and wastewater service provision are often based on inaccurate and insufficient data both at design stages and during the long term scheme operation. This issue and ways of dealing with it are discussed in Chapter 12, including ongoing use of monitoring and control information.

ENDNOTES

[1]An example of civil law would be the Code Napoleon (1804), named after French emperor Napoleon Bonaparte. The Code comprises three components: 'Persons,' 'Things and Different Forms of Ownership,' and 'Different Ways of Acquiring the Ownership of Things'. Rather than a catalogue of judicial decisions, the Code consists of abstractly written principles as rules of law which are to be applied by judges.

[2]The primary source of law is the law code, which is a statute grouping rules and standards concerning a particular subject matter and arranged in classified order. Law codes are usually created by a legislature's enactment of a new statute that embodies all the old statutes relating to the subject and including changes that may be needed and which result from court decisions. In some cases, the change results in a new statutory concept.

[3]It must be noted that not all countries under the civil law tradition grant the right of unilateral termination.

[4]Examples include the regulatory dispute in Guayaquil, Ecuador, Cochabamba, Bolivia, Buenos Aires Argentina amongst others.

[5]A good example of this was the case of Tallinn, Estonia where the Grantor, unable to create a full independent regulatory body prior to the PPP agreement, decided to set out the monitoring and regulatory provisions within the PPP Agreement. Subsequent to contract award, the creation of a regulatory body set out the joint contractual obligations of the Grantor and Developer, Three years after contract award, an independent regulatory body was created and the regulatory functions in the PPP Agreement migrated to the regulatory entity.

[6]Politics can often interfere in the delicate contractual ecosystem that balances cost recovery and consumer protection. They may attempt to force through tariff reforms that only recoup short-term operating costs, to the detriment of long term investment and asset maintenance. Alternatively, unrealistic demands on service provision expansion, without a commensurate increase in tariffs, can often be experienced. Key service and tariff provisions should be ring-fenced within contracts but politicians should be granted avenues to influence tariffs.

[7]For example, provisions for how to handle resets under contracts or with agencies are often vaguely defined and provide a source of risk and uncertainty. Under a contract system, provisions to deal with resets may contain guidance on how resets are to be agreed with Operator and details on how disputes are to be handled. With an independent regulator, resets may be imposed upon the Operator. In either case, it is preferable to have clearly delineated rules with binding restrictions on how price resets are to be handled, limiting regulatory discretion and reducing sources of uncertainty.

Chapter 6

Understanding and managing risk

6.1 INTRODUCTION

There does not appear to be a consistent definition of risk as it is dependent on innumerable factors (Landsburg, 2003; Hubbard, 2009; Flyvbjerg *et al.* 2002; Flyvbjerg *et al.* 2003). One definition is that risk is an issue which could be avoided or mitigated as opposed to a potential problem that needs to be fixed immediately. In the context of PPPs for the water sector, risk is described both qualitatively and quantitatively as a situation which could potentially lead to negative consequences. Therefore an essential planning tool in the successful delivery of major infrastructure projects is the proactive identification and allocation of responsibilities and risks. Whilst the contractual arrangements allocate responsibilities to the parties, it is also necessary to establish who takes the risks associated with the scheme. Commercially viable and cost-effective risk allocation between the parties is an integral part of a project's success and is one of the most complex elements of a project arrangement. Even though the allocation of risks has significant efficiency and distributional implications, the real challenge lies in how to balance the private sector risks so that they no longer constitute a significant impediment to private financing of infrastructure projects.

Chapter 3 introduced some of the typical models of PPP arrangements, showing the differentiation between the various models by looking at the way the Grantor and Developer shared risks. Some major Developer Risks in each model are shown diagrammatically, as can be seen in Figure 6.1.

This chapter will explore methods of allocating the risks and responsibilities between the Grantor and the Developer along with a discussion on specific risks in managing water and wastewater. Clearly, the roles and responsibilities between the Grantor and the Developer are allocated within the PPP contractual form that is chosen (as discussed in Chapter 3). So for example, demand risk affects the developer's risk profile to varying degrees in all contractual models, except it does not have a direct impact on the Developer's remuneration for management and service contracts. This section will look at the most complex forms of risk typically present in concession and divestiture type PPP arrangements. Recommendations are made on how to identify the key risk allocation issues, as an important and practical tool for effective decision making in complex PPP arrangement design.

> *"Risk is the chance of an event occurring which would cause actual project circumstances to differ from those assumed when forecasting project benefit and costs."* (State of Victoria, Australia, Department of Treasury and Finance, 2001)

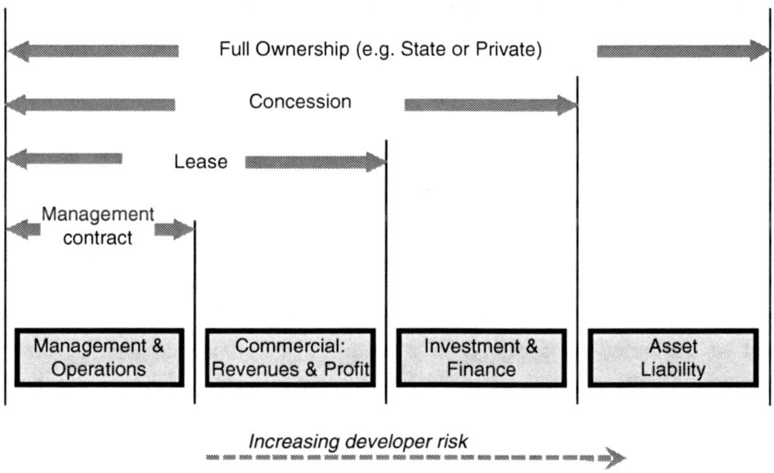

Figure 6.1 Contract risk areas for each type of PPP model. *Source*: Authors, from a presentation produced by them for the Water and Sanitation Toolkit, 2009.

6.2 ANALYZING RESPONSIBILITIES AND RISKS

All parties involved in the PPP arrangement have an interest in identifying any risks that a project may face and will have an equal stake in ensuring that these risks do not jeopardise the project. After the consideration of a number of factors, the precise allocation of risks among the various parties involved will be defined and would include (UNICITRAL, 2001) public interest in the development of the project, as well as other investors and lenders (and the extent of their ability and readiness to absorb those risks at an acceptable cost).

Financing large infrastructure projects requires a good projection of capital costs, revenues and expected costs, expenses, taxes and project liabilities[1]. Effective risk allocation is premised upon the notion of allocating responsibility for dealing with the consequences of each identified risk to either the Developer or the Grantor, or through a system of shared responsibilities. Risks and responsibilities are usually allocated on the basis that: (Ringskog, 2000; Smith, 1997a; Torres de Mästle & Izaguirre, 2008; Woodhouse, 2003).

- Threats to project costs and risks to the project are minimized by allocating particular risks to the party in the best position to reduce the *probability* of the risk and in the best position to *manage the consequences* of the risk after it has materialized.
- One party may be in a better position to diversify or absorb the risks than the other parties involved in the arrangement.
- Allocating a risk to a party should provide incentives for that party to spend time and resources in delivering the expected outcomes.
- Not all risks within a PPP Agreement can be foreseen and therefore the agreement needs to allow for mechanisms to deal with unpredictable, unforeseen or unmanageable events. This is best dealt with in the contract for example, 'force majeure' and other clauses for unforeseen events.

It must be noted however that risks are sometimes allocated on the basis of commercial and negotiating strength in which the stronger party seeks to allocate the weaker party an unwanted risk. In such circumstances, the Grantor must ensure that when developing the PPP Agreement it foresees such a situation and plans accordingly. The Grantor must not find himself with a counterparty unable to mitigate a particular risk which may jeopardize the PPP project (Gupta *et al.* 2002).

An example of the format for a risk analysis matrix developed for a recent major PPP scheme is given in Annex 4. This matrix was developed at an early stage in the design and development of the PPP scheme to identify key risk allocation issues, and was used as an important and practical tool for effective decision making in a complex situation. In practice, risk allocation is often a product of policy considerations and the negotiating strength of the parties. In allocating project risks, it is important to analyze the strengths of each entity undertaking a specific risk and their ability to shoulder the consequences of the risk should it occur.

6.3 COMMON RISKS AND RESPONSIBILITIES

Risks can be divided into three broad categories (Mandri-Perrott, 2009):

- *Political and Macroeconomic Risks* – Country risk refers to the possibility that changes could occur in the business environment that may reduce the profitability of doing business in a country. These changes may affect operating profits adversely as well as the value of the assets. This includes political risks and macroeconomic risks.
- *Sector Risks* – The risk that the sector may be affected by economic or other factors which specifically pertain to that sector.
- *Project Risks* – Refers to those circumstances which may have an effect on the responsibilities of each party in the PPP agreement and the benefits they may achieve from the project. These include risks related to financing, design and construction and operation and maintenance.

- *Counterparty Risks* – Relate to specific risks arising from the counterparties in the PPP agreement not being able to fulfil their responsibilities under the PPP agreement (this would mean in a typical PPP agreement, the Grantor as the main counterparty but there may be other counterparties, such as a particular national agency that grants special permits for water abstraction to the developer).

The magnitude of the risks varies depending upon the phase of the project. It may be that there are greater risks in the early stages such as in the bidding process but there will be other risks that will continue throughout the life of the project. Obviously this will affect who will assume those risks and to what degree they will be shared or not. There are four distinct periods in a project in which risks occur and are allocated accordingly:

- Before bid submission.
- Between bid submission and financial close.
- During construction.
- During operation.

Each risk category contains a number of specific risks within the responsibilities connected to the project environment. Variations in specific risks can have positive or negative effects on the cash flow of the project and on the total value of the business. The Grantor should not transfer all risks to the Developer because the Developer will seek to recover the cost of such risk (i.e. price the risk).

The major responsibilities and risks involved in delivering water and wastewater services will now be discussed in greater detail, but broadly speaking the following key risk allocation rules should be kept in mind to balance full risk transfer by creating the right incentives between the parties in the PPP Agreement:

- If a Developer perceives a risk to be outside his control, he will tend to price this risk on a 'worst case basis'.
- Economically it does not make sense for the Grantor to pay the same price for a certain risk as for one that only *might* happen.
- For such risks, if the Grantor capable of managing those risks itself, the emphasis should be shifted towards incentives (e.g. 'cost + fee + pain/gain' approaches).
- The optimal incentive would be that the 'pain/gain' is commensurate with the cost of doing one's best.

6.4 POLITICAL AND MACROECONOMIC RISKS

Political risks

The Developer and the financiers face the risk that certain actions by the Grantor, other Government agencies or the legislature may have a negative impact on the project. 'Traditional' political risks include nationalization, new tax regimes and other events that impact debt service and profits. 'Regulatory' risks include the imposition of new standards or the introduction of competition, while 'quasi-commercial' risks include breaches by the Grantor or interruptions due to changes in the Grantor's plans (UNCITRAL, 2001). Other political risks include acts of war, rebellion, default or failure of public sector entities. Normally, the Grantor is the participant in the project who has the greatest ability to manage risk of change in the political climate, and so often takes on this responsibility.

In some cases, risks of disruption in construction or operation by individuals or groups opposing the project will be borne by the Grantor unless the disruption was caused by a specific act or omission of the Developer himself. The same would normally apply in the case of legal challenges against the Developer preventing him meeting project objectives[2].

Change of law risk

Changes in law including any adoption, modification, or repeal may happen at any time after a PPP agreement has been signed and become effective (the Effective date). Developers will be particularly wary of future changes in any law which might affect standards of water and wastewater quality for example, and for which there may not be initial adequate planning[3]. It is essential that there is an effective mechanism to deal with the consequences of any change of law after the bid date (not only after the effective date[4]), particularly where this may require the Developer to incur costs or which may result in a decrease in profits. Accordingly, change of law provisions included in the PPP agreement should address the issue of who should be responsible for the costs arising from changes in law and how such costs should be funded. This is discussed further in Chapter 7.

Significant changes in law might include:

- Currency or capital repatriation limitations.
- Nationalization of developed assets.
- Import/export prohibitions.
- Deprivation of Developer rights.

Contingent liabilities

The United Kingdom's HM Treasury (2003), states that contingent liabilities represent commitments to future expenditure if certain events occur. Many of the risks associated with private sector participation in infrastructure create sizeable contingent liabilities for public institutions. Because such liabilities are uncertain and do not correspond to definite cash flow events, simply relying on cash based budgetary analysis will fail to consider their potential impact on affordability. This is an important issue for the Grantor, and a more detailed review of contingent liabilities and their impact is given Chapter 7 on Financing.

Risk of change in interest rate

Private investors and local or provincial governments have almost no control over prevailing interest rates, all of which are affected by central government actions. Generally loans are quoted in relation to a floating interest rate (based on some reference value such as LIBOR[5] or EUROBOR) and such interest rates change over time and are not something that bidders or Developers can control. Usually, Governments are not prepared to compensate Developers for changes in interest rates during construction or operation.

Since generally revenues do not change according to variations in interest rates, equity holders, lenders and Governments usually prefer that the winning bidder–Developer sources a significant portion of debt at fixed rates through an *interest rate swap*[6]. This is a financial derivative product by which the developer seeks to manage his exposure to the interest rates charged for the project loans. The swap is to exchange a floating rate loan to a fixed rate loan.

However, the bidder/Developer cannot enter into a swap agreement since he does not know the precise date of the financial closure of the PPP Agreement. Bidders must therefore use calculated assumptions of applicable swap rates in their bids, even though they have no control over the swap rate until they enter into a swap contract. More than likely, the swap rate at financial close will differ from that at the time of bidding and this variation can act positively or negatively for the bidder/Developer. Who should then take the risk related to this underlying variation in interest rate swaps? This question is true of any country and is a typical risk associated with project financing. Countries experienced in PPP transactions have generally opted for the Grantor to take over the risk of changes in swap rates between the time of bid submission and financial close.

Risk of change in inflation rate

Under PPP arrangements, the construction and operational risks are mainly borne by the Developer. However, inflation can have a serious impact on the costs (both construction and operational) of a project. To lenders, covering this risk is extremely important in order to limit their financial exposure, maintain the project cover ratios and also the net benefit anticipated from the revenue stream. For a typical PPP project, the time period between bidding and operating is long and the cumulative impact of inflation over time will be significant. Box 6.1 shows an example of how exchange risk has been dealt with in some sample contracts.

BOX 6.1 EXCHANGE RATE RISK IN MANILA, BUCHAREST AND BUENOS AIRES

Exchange rate risk in the Manila concession contracts is managed via extraordinary price adjustments. The contract allows for a Currency Exchange Rate Adjustment (CERA) to be added to tariffs when exchange rate movements are in excess of 2%. Maynilad, the operator in the west zone of Manila, was required, through the concession fee, to service 90% of the public authority's previously incurred U.S. dollar denominated debt. The Asian financial crisis led to a large devaluation in the peso against the dollar, leading to a significant increase in the debt service burden. As the tariff adjustment process allowed for only a slow exchange rate pass-through to customers, this led to serious financial problems for Maynilad. Both operators have since successfully negotiated an amendment to the agreement establishing a new mechanism called the Foreign Currency Differential Adjustment, which allows faster recovery of exchange rate losses from customers.

A different sort of problem arose in Bucharest. Foreign exchange devaluation led to significant local currency inflation. Indexation was linked to local currency inflation, not to the exchange rate, but the increase in local inflation led to significant tariff increases, which were highly unpopular with customers.

The Buenos Aires concession contract of 1999 allowed the concessionaire Aguas Argentinas under the US dollar to Peso parity law, to increase end-user tariffs to compensate for the resulting increase in foreign currency cost. This resolution also included provisions for annual tariff adjustments based on variations in US produce and consumer price indices. After the extreme devaluation of the Argentine Peso in December 2001, Government enacted a public emergency law in January 2002 that ordered a suspension on all tariff indexation in accordance with US indices. Shortly after that law was passed, Government issued Resolution 38/02 suspending any tariff increases (this resolution was only applicable to national public utilities). Both the law and the resolution were in direct contravention of the provisions in the concession contract. All public utility contracts in Argentina were then subject to a renegotiation process, which has not yet been concluded.

Sources:

- Concession Agreement between MWSS and Maynilad.
- Bucharest Concession Agreement.
- Various legal documents related to the Buenos Aires water concession, including the original concession contract, its amendments, as well as resolutions issued by ETOSS, the regulator.
- Ducci (2008).
- Interview with Ondeo and Mr Raul Pedroni (former adviser to Aguas Argentinas).

Since bidders have no way of managing inflation risk, if asked to bear such risk they will either refuse to bid (the risk is too high) or they will make very conservative inflation assumptions (risk pricing) which will push up project costs significantly. In the latter case it is very likely that it would have been far cheaper for the Grantor to have compensated the Developer for actual cost inflation. For this reason inflation risk is typically passed through to the end user or the Grantor through the indexation of capital grants and other contract payments (e.g. availability payments, fares).

Generally a transparent indexation formula is applied: it may simply refer to general inflation (e.g. Consumer Price Index/Retail Price Index), or preferably, in many cases, a basket of inflation indices which better reflect specific project costs (e.g. construction materials, power, labour). It is important that the indices used are from public sources to ensure transparency and minimize bias.

Risk of change in foreign exchange rates

According to Gray *et al.* (2003), almost all developing country currencies have a history of depreciation against the US dollar. The implications of this is that the true local currency cost of servicing debt denominated in US dollars (or other international currencies) will be higher than the interest rate on the loan[7]. Thus, estimating the cost of finance by looking only at the international currency interest rate will systematically underestimate the cost of finance, and hence underestimate the total cost of service[8].

It should be noted that PPP schemes in the water and wastewater sector may involve funding and procurement structures that depend on a variety of currencies:

- PPP projects are often financed with significant amounts of foreign capital, for example, in the form of syndicated bank loans, bond issues, bridging and stand-by facilities, with multilateral and export credit agency loans and guarantees.
- The project participants may have revenues in one currency, but costs in several currencies.
- Some of the capital and operating costs may be denominated in a currency other than that of the country in which the project is being constructed. Exchange rate fluctuations directly affect the price of imported parts or other inputs required to maintain plants and equipment or construct or rehabilitate new assets.
- Project financing involving more than one currency exposes the project to changes in exchange rates.
- Changes in exchange rate of local currencies will have an effect on the level of planned revenues and profit taken offshore by the Developer.

As a result, foreign exchange is a significant issue. Grantors and Developers may mitigate this risk by aiming to reduce reliance on imported inputs or foreign currency borrowing. However, in practice, foreign exchange risk is a significant part of most PPP schemes in the water and wastewater sector.

In some instances, it is generally impractical for the Grantor to assume all the foreign exchange risk. Additionally, the fiscal effects of assuming this risk need to be carefully considered. It is possible for the Developer to purchase protection against variations in project costs caused by foreign exchange fluctuations. This may be achieved through a Currency Swap which involves exchanging principal and interest payments on a loan in one currency for principal and interest payments on an equal loan in another currency.

6.5 SPECIFIC RISKS IN THE WATER SECTOR

Demand risk

In the economic appraisal of any PPP project in the water and wastewater sector demand forecasts form a key input. Customer demand is required in order to calculate revenue income and assess operational costs, maintenance, rehabilitation of assets and investments. Future demand forecasts are a fundamental input to any economic appraisal. Demand is vulnerable to a number of factors such as (Mackie *et al.* 2005) which require careful management. Box 6.2 shows an example of the effect of these factors:

- Economic shocks, including fuel price shocks and economic boom or recession.
- Changing demographics and usage habits.
- Political intervention.
- Random error in the forecasts.

BOX 6.2 THE CASE OF MANAGING DEMAND RISK IN BOLIVIA

A good example of the potential impact of demand risk in the water and sanitation sector is the 1997 La Paz – El Alto concession in Bolivia. The developer Aguas del Illimani, took responsibility for operating and managing the water and sewage systems and agreed to specified service expansion targets. The arrangement provided for tariff revisions every five years, based on a set of formulas covering the cost of repayment of capital for existing investments, the cost of operation and maintenance, the cost of expansion to meet targets set out in the contract, and commercial costs. In between tariff reviews Aguas del Illimani bore normal business risks, including demand risk. Demand growth, including demand for new connections, has been substantially lower than initially forecast. Despite requests from Aguas de Illimani for an extraordinary tariff adjustment to reduce the impact of this shortfall on its financial position, the regulator refused to review tariffs before the first five-year revision. Ultimately this PPP arrangement failed. In Argentina, the PPP agreement relied on a law which provided for direct conversion on a one to one basis between the Argentinean peso and the US dollar. This protected the Developer against exchange rate fluctuations as a 'change of law' event in the PPP Agreements. Unfortunately the end of the peso–US dollar parity in 2001 and the economic crisis that ensued, led to an early termination of the Agreement.

Depending on the availability of accurate and reliable historical usage data, customer numbers, and economic and demographic trends the extent to which demand risk is shared between the Grantor and the Developer will be determined. However, there seems to be a tendency for the Developer to be reluctant to fully assume demand risk and for him to pass, at least partially, part of this risk on to customers in tariffs or service levels through developer mechanisms.

In discussions with sector specialists[9] it is clear that the developer has limited ability to influence the demand for water services. This is because of the typical

Developer cost structure whereby a large proportion of its costs are fixed. So when demand falls, the average cost to the Developer of delivering each unit of water rises (or the cost of treating wastewater effluent rises). It is unlikely that a Developer will be allowed to increase tariffs if demand is substantially below forecast levels[10]. There are however, some limited technical, operational and financial actions open to the developer to help mitigate the effects of demand risk which could include cutting back on capital investment if demand is reduced, or bringing forward investments or leakage reduction plans if demand increases.

Operation & maintenance risk

Operational risks refer to the responsibilities associated with operating the existing and new assets and maintaining them at required standards. Depending on the type of PPP Agreement, the Grantor may take less or more of the demand and revenue risk. Usually, the Developer is responsible for operational and maintenance costs, including any costs related to latent defects in the construction for which he is responsible.

Furthermore, and also related to foreign exchange risks described above, the exchange rate may affect the price of imported inputs required for the Developer to provide operational and maintenance activities. For example, a change in the exchange rate may change the cost of imported fuel oil, which may affect the domestic price of electricity used to pump water. Similarly, changes in foreign exchange with local currency will affect the supplies such as chlorine and other chemicals.

Risk of increased project costs

Given the high capital cost involved in water and wastewater PPP schemes, errors in forecasting can have a major effect on the real economic cost of infrastructure, financial planning and project management. Unanticipated escalations in construction costs and higher than expected inflation rates are factors that may lead to potential overruns[11]. Clearly, the party who bears this risk depends on the type of PPP Agreement. For example, under a Net Cost Contract with Investment also known as a Concession (or some form of Build Operate contract), the Developer would be exposed to such escalations which in turn may affect the Grantor. According to the United Kingdom's HM Treasury (2003), the two main causes of bias in capital cost estimates are:

- Inadequate definition of scope and objectives of projects in the business case – Grantor risk.
- Inadequate management of the project during the implementation phase so that costs are not controlled and contractual risk mitigation instruments are not adhered to – Developer risk.

Bias in capital costs can be dealt with by improved estimation of the capital costs of each option, adjusting these estimates on the basis of empirical evidence, and reducing these adjustments based on confidence in the capital cost estimates and the risk management and mitigation systems.

Demand forecasting has a direct effect on establishing capital and operating costs. When evaluating bids, care must be taken to evaluate the forecasting techniques and costs used by the Developer to establish project costs.

Development risk

The Development phase involves the preparation of the project up to procurement of the financial close[12] of the PPP Agreement. This phase includes the invitation to tender, the bidding, the negotiation of the PPP Agreement and various documents pertaining to the project, its debt and equity funding. Given the nature of water PPP projects, both the Developer and the Grantor will spend significant time and resources in negotiating the PPP Agreement. The costs during this phase are generally understood to be normal development costs and are borne by each party at his own risk. Occasionally, and depending on the way the bidding process is designed and the terms in the tender documentation, the Grantor may cover some of the Developer's bidding costs.

It is important that the Grantor is aware of the possible changes in costs that may arise from initial estimates to actual costs once the PPP Agreement is implemented.

Design and construction risk

Responsibilities associated with the design, procurement, engineering, construction, completion, testing and commissioning of assets (if applicable) fall within the design and construction risks. Depending on the type of PPP Agreement chosen, the Grantor will sometimes stipulate such outputs as water quality standards, number of connections and so on. Developers will base their bids on these specifications. The risk allocation during the design and construction phase is complex but can be summarized as follows:

- *Design* – whichever party takes responsibility for the design, generally takes the risk of errors in design that may lead to the failure of the project to satisfy contractual requirements or laws. The risk of faults or changes in design, latent defects and asset life expectancy need to be specified and responsibilities allocated. Should any variation be required by the Grantor, he will generally bear that risk.
- *Permits, licenses and Access* – the Grantor is responsible for providing the Developer support in obtaining the permits and licenses (e.g. abstraction licenses, wastewater discharge licenses) necessary for the construction and operation of the water project. However, the Developer is ultimately responsible for obtaining such permits and licenses. The Grantor is also responsible for guaranteeing the timely delivery of land required for the

development of the project, free of legal or physical encumbrances and will bear the costs of dealing with these encumbrances.

- *Construction* – depending on the type of PPP Agreement, the Developer is responsible for constructing the project. The Developer carries the risk of increases in construction costs (price of labour, materials). The Developer is also responsible for the performance of all sub-contractors. The risk of delays is generally borne by the Developer.

Financing risk

Depending on the type of PPP Agreement and the corresponding level of financing, the Developer will be responsible for raising the private financing necessary to complete the project. The risk of increases in interest or inflation rate on the privately financed portion of funding will be borne by the Developer after financial close. This is dealt with in further detail in Chapter 7.

Government guarantees

Guarantees are provided by the Grantor to support infrastructure investments and transfer risks from the Developer to Grantor in order to make a more attractive arrangement for the private sector/Developer. During the construction phase of a project, Developers may be utilizing inputs whose costs depend on the exchange rate when the value of these inputs may rise and fall with the local currency. In such cases the Government may provide an exchange rate guarantee to mitigate the effects of depreciation in the currency, for example, tariff dollarization. However, Irwin (2007) warns that Developers face reduced incentives to improve performance levels when government guarantees cover a risk which the Developer is capable of managing and controlling better than the government. Furthermore, when government guarantees are provided, the contingent nature of the guarantees makes valuing them difficult and raises issues over how they are to be accounted for within the government's financial and budget reports (Ter-Minassian, 2005).

Environmental risk

Generally the Developer will be considered responsible for meeting environmental norms and standards, for example, water discharges, over abstraction and pollution. However, the Grantor will retain certain responsibilities related to specific issues such as pre-existing conditions or special compliance waivers[13].

6.6 RISKS ASSOCIATED WITH MANAGING THE PPP AGREEMENT

The mechanism for implementing and monitoring risk allocation rules has to balance the required future certainty of the PPP Agreement with a need for possible adjustment if there are unanticipated changes in laws, indexation and value testing – all critical factors in operating cost risk. Developers will be more willing to take on risks related to unforeseen changes in the operating environment provided they are given value-testing assurances that will not be disadvantaged (HM Treasury, 2007).

The type of PPP Agreement will set out its respective financial implications. Put simply, the payment mechanism contained within the PPP Agreement details the financial arrangements made between the Grantor and the Developer and also sets out the framework of incentives used to encourage the Developer to provide an efficient service at a cost that provides value for money. The way in which a Grantor attracts private sector financing and promotes the success of the project will dictate the financial stimulus it gives to the PPP arrangement (Torres de Mästle & Izaguirre, 2008). Some of the mechanisms used by the government aimed at reducing the risk of private financing for water and wastewater schemes include:

- *One-off subsidies* which usually involve capital subsidies in which a large portion of the subsidy is paid after the targeted beneficiaries are connected to the water and wastewater network and connections are verified.
- *Transitional subsidies*[14] could be a useful instrument given the current financial crisis for users who could afford services before the crisis but are now out of a job or household resources are so low they are no longer able to afford the tariff. These subsidies could also help ensure that the project remains financially viable by filling the gap between what is deemed that the user is able and/or willing to pay and the Developer's cost-recovery level (e.g. long-term marginal cost) of the tariff.
- *Ongoing subsidies* would be required in cases where there is a continuous gap between affordability and cost recovery – including for consumption costs.

Capital Grants – to cover part of the infrastructure construction costs. By reducing the privately financed construction costs the project could be more financially attractive to the private sector and should be considered when customer revenues would not be enough to recover the full construction cost of a project.

Furthermore, a Grantor could use bonuses and penalties as part of a mechanism to enforce the risk allocation rules. Additionally, they would enhance the Developer's incentives to fulfil their general responsibilities under the terms of the PPP Agreement and to meet agreed performance targets. Penalties and bonuses should reflect the economic costs and benefits of the behaviours they are trying to prevent or promote. Without such incentives, the Grantor reduces his ability to influence the Developer and also his ability to demand improvements should they be required.

Price and tariff variations

The regulations for adjusting the payment structure are a critical component in the risk allocation architecture. Throughout the life of contract a number of variables, such as inflation, input costs and legal regulations are likely to change in unpredictable ways. To reflect these uncertainties the PPP arrangement should allow for payment adjustments over time to ensure the Developer continues to earn a reasonable rate of return and that incentives are maintained (Kerf *et al.* 1998). The Developer should always be encouraged to control costs but mechanisms to control unanticipated increases can go a long way to reducing an excessive contingency risk pricing in the Developer's bid.

Tariff indexation formulas reflect changes in price indexes but not necessarily changes in a Developer's costs. Tariffs should be adjusted at regular intervals (e.g. every six months) instead of responding to specific events. The indexation process should aim to compensate the Developer for the effect of exogenous cost increases on the Developer's inputs. Importantly, indices reduce the risks faced by the Developer without blunting the performance incentives (Hudson Institute, 1999). Specific indexation formulas can adjust tariffs according to changes in either the rate of inflation, consumer price indices or a consumer price index (related to changes in the system's likely costs such as a basket of prices, exchange rates or specified inputs).

A chief concern for the Developer is the risk of costs inflating during the life of the PPP Agreement, rendering payments made insufficient to cover the operating and financing costs. The tariff, as well as other specific payments made by the Grantor to the Developer, should be indexed and the proportion of such tariff or payment components should be determined at an early stage. The choice of indices or proportion should be determined by the Grantor. Indexations which reflect the underlying cost faced by the Developers could reduce the cost risks and provide considerable savings through the duration of the contract (HM Treasury, 2007).

Cost pass-through is usually used to cover the cost of risks over which the Developer has no control. Depending on the type of PPP Agreement and when applicable, if input costs rise, the tariff adjustment rules agreed upon between the Developer and the Grantor will allow for these changes in costs to be passed through to customers in the tariff. Changes in PPP project inputs may include tax structure changes or regulations governing the operation of the water system.

Residual value risk

Where an asset still has value (not fully depreciated) after the PPP Agreement has expired, the Grantor should agree upon the way this residual value is to be dealt with. Normally a PPP financed, asset will depreciate according to a depreciation schedule. In most long term PPP Agreements there will be a residual value at termination, especially if significant investments were made towards the end of the contract period. If the Grantor wishes the assets to be transferred at zero value upon expiry, then an accelerated depreciation programme would have to be carried out resulting in increases in the availability payments to cover this.

Alternatively, if the assets are transferred at net book value at expiry (i.e. the residual value) then the Grantor should pay the residual value to the Developer at expiry and would not suffer large availability payments in the final years of the PPP Agreement. This residual value could be easily audited from the Developer's financial accounts of the project (i.e. at a fully transparent auditable transfer value). Other issues that the Grantor should consider include how the residual value transfer impacts any termination payments upon the termination of the PPP Agreement. Other risks related to the implementation of the PPP Agreement such as early termination and associated triggers are dealt with in the chapter on contractual recommendations – Chapter 9.

6.7 SUMMARY AND CONCLUSIONS

Once the appropriate service delivery scheme has been chosen, identification and allocation of responsibilities and risks between the private and public sector parties is an essential step in the successful planning and delivery of major infrastructure projects. The chapter looked at the importance of analyzing and allocating these risks in light of the need to balance the private sector risks to a level in which they offer optimal benefits to the Grantor, but no longer constitute a significant impediment to private financing of infrastructure projects.

The main PPP arrangement types can be analyzed to show how the level of risk to the developer increases according to type of arrangement chosen. Within these types of PPP schemes the Concession (or BOOT model) and Divestiture type arrangements generally involve the most comprehensive array of risks and responsibilities to be taken by the private Developer. The comprehensive nature and the extent of allocation of risks and responsibilities is one of the main reasons why these contract forms have been used as the starting point to develop recommendations for contractual issues (Chapter 9).

A variety of ways are used for allocation of responsibilities and risks and, given the importance of this issue, a systematic approach for registering this allocation, such as the Risk Matrix, is recommended.

In allocating risk, four basic tenets are followed:

- Allocating particular risks to the party in the best position to reduce the *probability* of the risk being realized and in the best position to *manage the consequences* of the risk after it has materialized.
- One party may be better placed to diversify or absorb the risks than the other parties involved in the arrangement.
- Allocating a risk to a party should provide incentives for the party to spend time and resources in delivering the expected outcomes.
- Not all risks can be foreseen within a PPP Agreement and therefore the agreement will need stipulate mechanisms to deal with unpredictable, unforeseen or unmanageable events.

Variations in specific risks can have positive or negative effects on the cash flow of the project and on the total value of the business. The Grantor should be selective as to which risk he transfers to the Developer, as often the Developer will increase his rate to cover the perceived cost of the risk, adding to the cost of the project.

These risks and others, and their risk allocation and means of mitigation need to be dealt with effectively in the PPP contractual arrangements, and recommendations for detailed ways to deal with this contractually are give in Chapter 9.

ENDNOTES

[1]It must be noted that all these projections are not an exact science. Developers and lenders will make assessments of the various capital costs, revenues and expected costs, expenses, taxes and liabilities for the project based on their experience and their appreciation of country, sector and other risks associated with undertaking the project.

[2]There may not be any conventional insurance available to cover the Developer from this risk or may be so costly that it may not be practical.

[3]In Eastern European countries this has been of particular concern, as some of the PPP arrangements were made prior to their admission into the European community. In the cases of Sofia, Bulgaria and Tallinn, Estonia both countries entered their respective PPP arrangements prior to being full members of the European Union. In particular, the Estonian government pointed out that any obligations related to the European's Water Framework Directive's water quality and associated standards were already enshrined within the PPP Agreement at the time of contract award.

[4]The effective date is the date in which the contract obligations become effective.

[5]The London interbank offered rate (or LIBOR) is a daily reference rate based on the interest rates at which banks borrow unsecured funds from banks in the London wholesale money market (or interbank market). The euro interbank offered rank (or Euribor) is the rate at which euro interbank term deposits within the euro zone are offered by one prime bank to another.

[6]A swap is an agreement between two parties to exchange future cash flows according to a prearranged formula. They can be regarded as portfolios of forward contracts. The streams of cash flows are called 'legs' of the swap. Usually at the time when the contract is initiated, at least one of these series of cash flows is determined by a random or uncertain variable such as an interest rate, foreign exchange rate, equity price or commodity price.

[7]For example, if the interest is 10% in US dollars, but the local currency depreciates 5% per annum against the US dollar, then the local currency cost of financing is around 15%.

[8]This research suggests that this approach has been the norm, as represented for example by the cases of Manila, Jakarta and Buenos Aires and elsewhere, governments and operators agreed to tariffs which were only viable if the exchange rate remained more stable than history would suggest was likely.

[9]Discussions were had with Alain Locussol former lead water specialist at the World Bank, Antonio Vives, former Vice President of the Inter American Development Bank, Chris Jennings senior water specialist of the Inter American Development Bank, Dr Geoff

Thorpe Director of Regulatory Affairs for Interagua (concessionaire company for Guayaquil, Ecuador), and Alain Mathis, Director for Suez Lyonnaise de Eaux.

[10]If tariff setting rules leave demand risk largely to the operator rather than customers, the operator's overall risk exposure increases significantly and the sustainability of the arrangement may be threatened.

[11]For example, in a United States government study (Pickrell, 1992) found that most cost overruns were the result of critical errors in forecasting either the volume of materials and services required to build and equip projects, or the future costs of purchasing inputs.

[12]Financial close means that the loan agreements are in place and the first draw-down can be made on the loans.

[13]In some instances the Developer may be granted special waivers allowed the Grantor, for example, to meet European Environmental Directives. This does not mean that the Developer does not have the ultimate responsibility to, against a pre-agreed timetable, meet such standards. In such situations the PPP Agreement must specify a schedule for such waivers to phase-out.

[14]Transitional subsidies are often used to support tariff reform. Typically the subsidy is transitioned out after a specified period of time (e.g. months or years) as the user contribution increases (and possibly as tariff levels required for cost recovery decrease with efficiency gains). Under the current financial crisis, it will be important to devise a strategy deal with the period when tariffs are raised to 'normal' levels of full cost recovery.

Chapter 7

Funding and finance

7.1 INTRODUCTION

This chapter summarizes the more important funding and finance issues likely to arise in PPP water projects. PPP Water projects are expensive and developing strategies for sourcing funding (both public and private) need to be an integral part of the PPP strategy that is devised (Nickson & Franceys, 2003). Generally there are three primary sources for funding PPP schemes for projects:

- *Debt*: through commercial bank loans (local and international banks), publically traded bonds, private debt placement, loans from project sponsors, supplier credit, export credit financing, loans from international financial institutions (IFIs) or development organizations[1].
- *Equity*: usually sourced from consortium partners, passive investors, investment funds (e.g. emerging market, infrastructure or sector specific equity funds), domestic public entities, local property developers, international financial institutions (IFIs) or development organizations.
- *Grants*: often provided by public institutions (local, national or supranational), local property developers, international donors or other groups benefiting from project implementation.

Decisions made on policy, management, technology and commercial issues during the development of PPP water projects may affect the availability of some of these funding sources. For example, grant funding may only be available for water PPP projects that aim at accomplishing certain policy goals. Similarly, large amounts of private debt financing may not be available in cases where private concessionaires have to assume the entirety of certain major risks (e.g. full commercial and collection risks). A critical component of effective planning is to understand the implications that the structure of the project has on funding and finance.

7.2 CONSIDERING BANKABILITY

Water PPP planners face a daunting challenge when planning project funding and financing requirements. Finalizing a project's financial structure usually occurs during the later stages of planning and procurement – after many key decisions have already been made. Revisiting public approval processes along the way can result in substantial delays and loss of confidence should earlier decisions need amending. Therefore, it is imperative that those early decisions take into account such requirements and their impact on a project's future financial and commercial structure. 'Road shows' and other events designed to test market interest can help planners get a sense of the market's perception of proposed risk allocations and other project features. The term 'bankability' summarizes investor/lender sentiments and their willingness to commit debt or equity capital toward a project. It is important to realize that considering bankability goes beyond financial analysis alone and should include much wider project considerations.

Bankability generally depends on four broad criteria:

- Credit worthiness.
- Legal viability.
- Economic viability.
- Technical feasibility.

Table 7.1 summarizes many of the key questions investors/lenders will ask in order to assess a project's bankability and determine their level of interest in a project. Concepts discussed in Chapter 6: *Understanding and Allocating Risk* relate to many of these issues.

Table 7.1 Factors influencing project bankability.

Project aspects	Questions for determining bankability
Credit Worthiness	Are project cash flows sufficient to support envisaged levels of debt? How risky are project cash flows? How certain are project revenues? Who bares revenue collection risk and how realistic are demand forecasts? Is there potential for regulatory 'claw back' if actual numbers exceed estimates? Will the regulator be isolated from political pressures? Can the Grantor meet his financial obligations on the project? Does the project benefit from any Grantor or sovereign guarantees; will the project benefit from guarantees/insurance on its debt (e.g. partial risk or credit guarantees, political risk insurance)?

(*Continued*)

Table 7.1 Factors influencing project bankability (*Continued*).

Project aspects	Questions for determining bankability
	Is there sufficient equity cushion to protect lenders if the concession's value decreases? Do project developers have sufficient 'skin in the game' (i.e. value at risk)? In the event of termination, what mechanisms guarantee debt repayment and what proportion of the debt will be covered? Do project developers have adequate capacity and incentives to deliver sustainable long term operational performance? Do they derive significant value from ancillary activities outside of the concession company (e.g. local property development, turnkey construction contracts, etc.)? Do the project's financial ratios meet lender expectations (e.g. principal and interest cover ratios, debt service cover ratio, loan life ratio, debt:/equity ratio)?
Legal	Does the Grantor have the authority to grant the concession? Will the project require any additional legislation (e.g. sector law, PPP law)? How strong are the project's contractual arrangements with input suppliers (i.e. rolling stock suppliers)? What legal protections/channels for recourse do investors have in the project's jurisdiction (e.g. access to international arbitration)? Are legal decisions enforced in the project's jurisdiction (rule of law)? How strong are property rights in the project's jurisdiction?
Economic	Is there a market for the project's services? Does regulation protect against the threat of new market entrants? How stable is that regulatory environment? Are project inputs (e.g. electricity, rolling stock, etc.) available at reasonable prices? How stable are input supplies? How stable is the project's macroeconomic environment? How would changes in inflation, foreign exchange, interest rates, etc. impact project cash flows? How will such risks be mitigated in the PPP contract? Have any standby credit facilities been arranged to deal with potential lags between financial shocks and tariff adjustments?
Technical	Are construction costs reasonable and realistic? Is the construction and commissioning timetable realistic? Does the project rely extensively on proprietary technology? How flexible is the systems design? Is the proposed technological solution appropriate for local conditions and the availability/scarcity of skilled labour?

7.3 STRUCTURING FINANCE FOR WATER PPP PROJECTS

Financing PPP water projects may take the form of either corporate or, more commonly for large projects, project finance structures. The primary difference between corporate and project finance is the means by which lenders derive security for repayment. Project finance structures will involve a Special Purpose Vehicle (SPV, an independent legal entity expressly established for the purpose to undertake the project) into which lenders and project sponsors contribute debt and equity to fund project costs. As Special Purpose Vehicles are legally separated from project sponsors, lenders are said to have 'limited or no recourse' to sponsors regarding their debt investments. Limited recourse is the more usual variation and is used in financings, where recourse is limited to either a fixed monetary amount (e.g. USD 50 million) and/or is subject to certain performance criteria (e.g. cost and time overruns during construction, revenue/cash shortfalls during operations when developers commit further capital toward debt service). For project financings the loan structures rely solely on project cash flows for repayment, with the project's assets, rights and interests held as secondary security or collateral. Project finance structures are very common for water PPPs that require substantial upfront private investments; such investments exceed the capacity of developers' balance sheets.

Project finance has a number of key characteristics:

- *It requires certainty in cash flow*: this underpins the structure and risk allocation.
- *Credit intensive*: with a multitude of stakeholders and types of risk, project financing requires a multi-disciplinary and rigorous approach to risk allocation and management.
- *International*: most major projects in developing countries tend to have extensive international involvement (lenders, investors, developers, contractors, operators, suppliers etc.).
- *Long-term*: repayments are usually over extended periods (10 years or more) to accommodate high capital costs and subsequent operating margins.
- *Highly structural*: project financing risk allocation means rigorous and comprehensive structuring among the parties.
- *Large*: high development/transaction costs can usually only be supported or justified on large transactions.
- *Mature*: particularly in developed countries there are many experienced participants in the international PPP markets. Projects can, therefore, be structured smoothly by experienced financiers and advisors. For developing countries the challenge is often to attract such experienced participants and financiers.

Other characteristics of project finance structures include (Vives, 1997):

- Financing is provided through a special purpose legal entity whose sole business is the project.

Table 7.2 Project finance vs. corporate finance structures[a].

Financing structure	Advantages	Disadvantages
Project Finance	• Off balance sheet. • Separates project credit profile from developers' credit profiles – the project company is insulated from developer default. Similarly, developers are insulated from project liabilities[b]. • Can achieve high leverage ratios. This can provide greater rates of equity returns from smaller cash flows. • Can create greater tax shields because of greater leverage – which can reduce the overall cost of capital. • Reduces managerial discretion over free cash flows because of lender imposed constraints. • Mitigates the 'underinvestment bias'[c]. • Provides additional mechanisms for risk spreading through syndication and securitization. • Additional lender due diligence can improve project quality. • Lender reporting and control requirements can help achieve much greater transparency for the Grantor with regards to project company operations and financial performance. This can have important implications when assessing performance related payments and when setting or adjusting tarif levels.	• Greater transaction costs compared to corporate finance • In developing markets only available for larger projects • Requires additional due diligence and associated time when structuring. • Bid prices can change when final price is dependent on prevailing credit market conditions (i.e. interest rates) as final financing terms may not be set until after preferred bidder selection. • Given the complexities of project finance structures and documentation the time between bid award and financial close/project commencement may be lengthy (generally in the order of 12 months, but may be much longer if the project structure is not bankable). This can conflict with the expectations.of Grantor/political project's timetable.
Corporate Finance	• Lower transaction costs because credit risk assessment is based on sponsor credit rather than complex project credit risk • Simpler, easier to obtain and faster to structure than project finance. • Requires smaller amounts of due diligence and associated time because of additional security provided by sponsors' balance sheets. • Depending on sponsor's credit rating, may offer lower margins on debt. • Offers greater flexibility to accommodate changes (such as renegotiations).	• Places sponsors' assets/balance sheet at risk. • Lower debt to equity ratios. • Typically available only for smaller investments (sponsor not willing to accept larger risk). • Project is not isolated from sponsor credit profiles and vice versa. • May limit public control over refinancing activities and prevent public sharing in refinancing gains. • Public authorities may realize fewer benefits related to project transparency compared to project finance structures.

[a]Adapted from various sources but principally Yescombe, (2002).

[b]The degree of protection afforded by project finance structures varies depending on legal jurisdictions.

[c]*Underinvestment Bias*: High proportions of debt increases a company's management reluctance to finance low risk positive NPV projects as smaller returns almost entirely benefit lenders only. Similarly, other high risk positive NPV projects may be foregone because of the risks they pose to managers. Project finance can achieve higher leverage to enhance returns of smaller NPV projects and can also protect management's interests by partially insulating them from potential downsides of riskier projects (Yescombe 2002).

- Usually raised for green-field development or extensions to existing systems.
- Project capital structures normally involve greater leverage (typically in the range 60–90% debt).
- Debt is generally 'non-recourse' or 'limited-recourse'; meaning that lenders rely primarily on future project cash flows for repayment.
- Lenders derive security from the concession contract or PPP Agreement.
- Financing reflects the time-restricted nature of the concession contract. The tenor of project debt will be less than the life of the concession, but usually linked to it.

In contrast to project finance, lending for corporate finance projects derives security for debt repayment from the sponsors' or shareholders' balance sheets or other non-project security. Corporate finance lenders typically would have rights to all the borrower's assets in the event of default. Characteristics of corporate finance include:

- Contractual arrangements between a Grantor and a legal entity which derives its financial strength directly from the balance sheets of the sponsors or shareholders.
- Are generally smaller investment amounts compared to project financed alternatives.
- Lender due diligence on the project absent or substantially reduced, since the lender is taking the sponsors' credit risk, as opposed to project risk – lender security is apart from the financed investment. It is important to note, however, that this characteristic only applies if the sponsor is considered to be a good credit risk – if not, due diligence type project finance will still be needed.

The relative merits of either structure will depend upon the size and risk profile of the project in question. Larger, green-field water investments will almost always recur to some form of project finance in order to limit the liability of developers and raise sufficient capital. Nevertheless, smaller water investments (e.g. minor extensions or refurbishments) may draw on corporate finance depending on their size and scope. Different financing techniques have their relative merits and limitations summarized in Table 7.2.

7.4 TYPICAL FINANCING STRUCTURE OF A LONG TERM PPP AGREEMENT

As discussed above financing for water PPP projects usually take the form of project finance structures. Project finance structures involve a Special Purpose Vehicle (SPV, an independent legal entity specifically established for the project) into which lenders and project sponsors contribute debt and equity to fund project costs (Yescombe, 2002). Figure 7.1 shows a representation of a typical financing structure for a long term water and wastewater PPP scheme. Characteristically, the Special Purpose Vehicles are legally separated from project sponsors. The

Developer of a concession type or any other form of long term PPP scheme will be a legally separate company from the international parent company. If things go wrong, the extent of the parent company's losses is the equity it invested in the Developer.

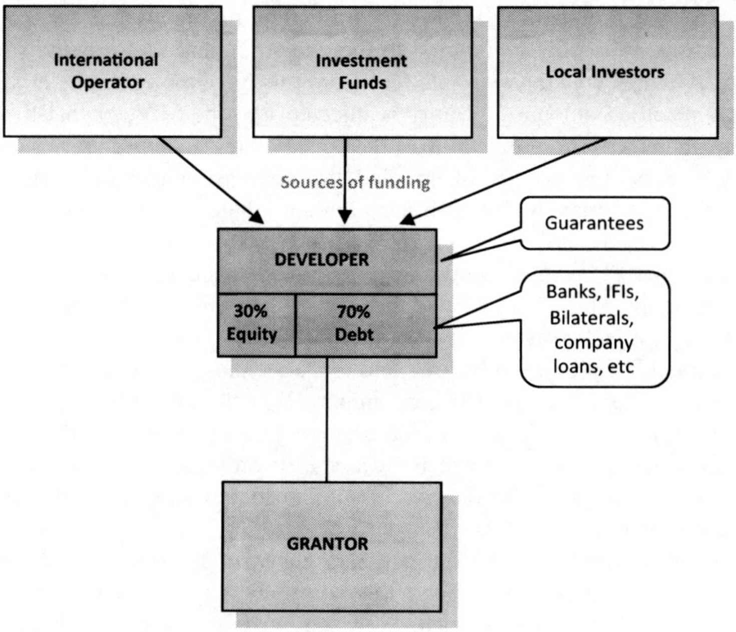

Figure 7.1 Typical financing structure.

Initial lender due diligence and subsequent monitoring can enhance project transparency and quality. Since banks or other lenders may require significant changes in a project's structure or agreed risk allocation before committing financing, consultation with them is recommended as early as possible in the procurement process. Although these changes can help rationalize risk allocations, they may also take considerable time to negotiate, as well as damage project and grantor credibility, which is why it is so important to focus on developing a project structure that is bankable, from project inception. In this respect, it would be invaluable for public authorities to hire experienced transaction/financial advisors familiar with taking PPP projects to market and financing them (usually investment/project finance banks or in some cases IFIs).

Project financiers will look for a balance between sponsor equity and debt financing (gearing) in a particular project. The project's gearing will reflect:

- How much debt can the project cash flows support?
- How risky or uncertain are the project's cash flows (market risk)?

Although many project financings aim to maximize gearing (debt is generally cheaper than equity), equity requirements are nevertheless usually substantial to ensure that the sponsor's commitment is too costly to withdraw support when the going gets tough. Debt equity ratios of 75:25 or 70:30 have been common, although some more conservative export credit agencies have sought lower gearing (60:40). Lenders and Grantor will want to know that the parent or sponsor company is truly committed to the operation, and will provide sufficient funds to fix things which go wrong. To ensure this, governments may require that at least a specified amount of equity is injected into the SPV, or that the parent company guarantees the obligations of the concession Developer up to a specified level. It is good practice to ensure that the concession company has enough equity, or parent company guarantees, to remain viable[2] even if things do not go as expected (see Ehrhardt & Irwin, 2004). More risky project structures (e.g. full concession type PPP Agreements) are also likely to induce lenders to insist on lower gearing, as well as to increase the cost of debt – a 'double-whammy' effect on increasing project costs.

The certainty of project cash flows and risk allocation will determine the nature and amount of debt that projects can support. Spreads on project debt will also reflect perceptions of riskiness in addition to general market conditions at the time of issuance. Higher leverage for water PPP projects with riskier cash flow characteristics is a recipe for disaster, assuming that sources of debt and equity are even available on this basis.

International commercial lenders are also increasingly conscious of the social and environmental impact of projects to whom they lend. Over 60 of the world's leading financial institutions have now adopted the Equator Principles, a voluntary set of guidelines in the financial industry to determine, assess and manage environmental and social risks in project financing. Consequently these Equator Principle Financial Institutions (EPFIs) have adopted these Principles in order to ensure that the projects they finance are developed in a socially responsible manner and reflect sound environmental management practices. Therefore, in order to attract financing from these institutions, Water PPP projects will, need to have been subjected to environmental and social impact studies and public consultation prior to launch.

7.5 PUBLIC FUNDING, SUPPORT AND INTERNATIONAL FUNDING INSTITUTIONS

As discussed, depending on the type of PPP structure chosen, there will be a lesser or greater level of support required. As the provision of water services are public in nature, are of high political sensitivity and are vital for human subsistence, most water PPP services require some direct or indirect form of public support. This section provides a non exhaustive list of some possible forms of public support.

Capital grants and Public 'soft' loans. The upfront capital costs required to construct new water systems are typically in the hundreds of millions of dollars. Moreover, the tariff revenues generated from customers are often too small to service the costs of project debt required to construct a system's infrastructure. Accordingly, the Grantor may opt to incorporate an element of capital grant designed to offset the private sector's initial construction costs and associated ongoing debt repayment obligations. Reducing debt service during operations enhances the stability of Developer concession companies and helps to prevent default, which is often costly to public partners as well. Determining the appropriate level of capital grant and defining appropriate construction bench marks is a challenge for project planners. Too much capital grant can reduce risk transfer to the detriment of value for money incentives, when private partners do not have sufficient 'skin in the game'. Conversely, insufficient capital grants can result in potentially unstable concession companies and correspondingly large risk premiums that is, poor project bankability.

Lending public money to private partners at concessionary interest rates can help overcome disadvantages resulting from the private sector's higher cost of capital. Preserving appropriate risk transfer to the concession company requires careful consideration when projects incorporate public financing. To protect public interests and ensure rational risk allocations, there is an argument for senior public loans in the project's capital structure which should benefit from appropriately large junior debt/equity cushions. Private lenders may, however, have an issue with such an approach to fund structuring – in this respect it is important to clarify and agree whether the intention is that the public loan will be treated as a 'quasi-grant' (subordinated to senior debt) or not. Ideally, substantial amounts of private financing should still accompany public loans, to take advantage of private lenders' due diligence and subsequent oversight.

Exemptions from taxes and import tariffs

Exempting developers from selected import and customs tariffs and local and national taxes (e.g. tax holidays) can reduce costs to public institutions and customers. In most cases such waivers will require legislative acts or close coordination between public agencies. Early planning for exemptions and their associated legal requirements is essential for ensuring timely project completion. In addition, public institutions should carefully consider the 'value for money' associated with long term tax exemptions (especially on corporate income). Existing tax codes and accounting practices may already provide substantial tax shields in the contract's early years through depreciation and interest expense deductions. Providing tax throughout the full life of the contract may not be necessary and may in some instances reduce incentives for further capital investments as well as reducing potential public revenues.

Loan guarantees

Public guarantees (sovereign, municipal etc.) aim to improve project bankability and to reduce the cost of debt for private concessionaires. In some cases, lenders may not finance a project without the provision of such guarantees, especially where the creditworthiness of the Grantor is deemed deficient. However, 'wrapping' project debt in the full faith and credit of public institutions implies a measure of caution regarding risk transfer and private incentives. Planners should ensure that public guarantees on senior debt benefit from adequate equity capital cushions at risk. Contingent liabilities associated with guarantees should also be priced into any 'value for money' analysis and should be disclosed transparently.

Equity contributions

Contributing public equity capital to water concessions aims to align public and private interest in a system's financial success. Sharing common equity can also provide a mechanism for public influence on operating and investment activities. Public equity is often contributed 'in kind' through public asset transfers for Brownfield concessions, or through capitalization of capital grants. However, the drawbacks of public equity ownership often outweigh its benefits. Reduced risk transfer and increased likelihood of political influence can harm water operations and compromise private incentives to perform. Therefore, the shareholders' agreement must address these issues specifically and explicitly in a situation where such a structure is being contemplated. Differences between public and private shareholders can result in conflicting strategies or incoherent company objectives.

Export credit financing

Export credit agencies (ECAs) exist to promote and facilitate the foreign investment and export of goods and services of the companies of their particular nation. In water projects ECAs have commonly supported the financing of imported rolling stock and associated signalling/communication systems. A number of ECA's can provide project financing through banks or directly to the buyer, via guarantees to the buyer's bankers. ECA financing can take the form of loan guarantees, political risk insurance, concessionary lending or working capital guarantees. In some instances, ECAs require matching contributions from private lenders. Most ECAs from OECD countries abide by the 'Arrangement on Officially Supported Export Credits' which sets upper limits on the amount of assistance that foreign governments can offer to support their exports. ECA financing may be particularly useful when local credit markets are underdeveloped or when sovereign risks reduce the attractiveness of private finance.

Where local public resources are lacking, support may be available through multilateral or bilateral development institutions. Historically the World Bank,

IFC, ADB, EIB, MIGA and EBRD have been active MLA project financiers. Most of these organizations operate exclusively within developing countries. Multilateral and bilateral financial institutions may offer support through:

- Loans (potentially to both projects and governments).
- Grants.
- Equity investments.
- Guarantees on debt and equity.
- Advisory services.

Multilateral and bilateral support can help to make water PPP investments more financially viable and provide developers and lenders with additional comfort on political/regulatory project risks. However, working with these institutions may also lengthen the project development process or constrain policy discretion. They may be very demanding in terms of environmental and social safeguard requirements, or impose stringent procurement procedure requirements. Multilateral and bilateral financial institutions with strong development mandates may also be able to help planners determine whether a Water PPP would be a sound policy consideration, as discussed in Chapter 1.

7.6 CONSIDERATION AND EVALUATION OF CONTINGENT LIABILITIES

Risks associated with private sector participation in PPP Water projects invariably create contingent liabilities that can influence the amount of public support services may require. Contingent liabilities usually relate to the provision of public guarantees, but also reflect other potential public sector obligations that do not appear or are not accounted for in public budgets and auditing. Intuitively, public institutions should plan to cover the expected value of such liabilities as part of their efforts to budget for system affordability. This could be accomplished either by verifying in later years sufficient budgetary 'cushion' or by setting aside appropriate reserves for future uncertain budgets. In practice, accurate budgeting is challenging to achieve, as contingent liabilities often involve uncertain cash flow events and associated opportunity costs to public funds. Allocating excessive reserves to cover such expenses may divert public funds from other important uses. Conversely, allocating insufficient reserves may strain future budgets in periods where expenses occur. This can prevent public authorities from meeting contractual obligations or from making necessary investments to ensure quality is maintained. Therefore a critically important funding consideration for public institutions is to value contingent liabilities correctly. Methods for valuing contingent liabilities include:

- *Actuarial or Statistical Techniques*: these techniques use historical data and current trends to estimate future losses related to uncertain events. The application of actuarial techniques is limited by the availability of historical data and their failure to explain the patterns behind losses which they predict.

- *Econometric Models*: unlike actuarial techniques, this method identifies factors which drive future losses. By forecasting future changes in those underlying factors, econometric models endeavour to predict future loss distributions. For example, econometric models can be used to calculate government exposure on a direct loan to a private Developer by forecasting those elements which potentially drive default and the corresponding risk of loss, or early repayment and associated reinvestment risk.
- *Contingent Claims Analysis*: this method can be particularly useful when historical data is lacking or when planners seek to evaluate a single specific element within a larger collection of risks. Contingent claims analysis employs many of the same techniques used to value financial options by identifying elements within larger risks that have options?? like cash flow features. For example, the value of a direct loan with risk would be deemed equivalent to the value of a risk free loan, less the value of a 'put' option on the borrower's default (Lewis & Mody, 1997).

7.7 HEDGING

Hedging instruments provide mechanisms that help public authorities and private developers manage financial risks associated with specific liabilities (both contingent and otherwise). However, such instruments provide counterparties with certain rights and/or obligations at some future date. Hedging is a common feature of Water PPP projects and is often used to offset exposures to:

- Foreign currency movements (e.g. swap and forward contracts/options).
- Large purchases of raw materials and other project inputs (e.g. forward contracts for steel and concrete).
- Interest rate movements (e.g. swaps).
- Counterparty risks (e.g. credit default).

Deciding when and how to hedge certain project risks requires careful consideration of associated costs and benefits. Whenever projects procure hedging instruments from international markets, planners should ensure that:

- Prices are fair and competitively obtained (see case example below).
- Purchased hedges offset a risky exposure.
- Hedging instrument covers are purchased in appropriate amounts.

It should be noted however, that hedging comes at a significant cost, with commercial lenders often earning as much from hedging instruments as they do from the project loans. Planners should, therefore, endeavour to construct natural hedges that eliminate risky exposures altogether when circumstances allow. For example, when lenders can provide loans in local currency, projects can match revenues with debt service payments avoiding the need for expensive currency swaps or the need for the public sector to bear foreign exchange currency risk. Chapter 4 discusses many of the risks that planners may want to consider hedging.

7.8 REFINANCING

The capital and debt structure of a project can affect both investor returns and a concession's financial risk profile significantly. Striking the balance between stability and appropriate rates of return usually occurs towards the financial close when lenders conduct extensive due diligence and also specify restrictive covenants on project debt to minimize default risk. However, market conditions and a project's credit profile are rarely constant throughout the allotted time. Favourable changes in either of these factors may provide additional capture value by opportunistically altering elements in a project's capital or debt structure. For example, a concession company may be able to extend the maturity of project debt (within the constraints of the term of the PPP Contract) and/or reduce interest rate margins and/or replace the hedging instruments. This could reduce the debt repayment profile, leaving additional cash free for paying dividends or making investments in service improvements.

The length of most water concessions allows ample time for potentially valuable refinancing during a project's life as risk profiles or credit market conditions change. Post-construction refinancing is relatively common and usually planned, and recently is often built into the original financing as a margin ratchet. However, other opportunities for refinancing might be less predictable. This may be especially true in rapidly developing countries where sovereign risk spreads contract over time hopefully to reflect improving governance, increasingly stable macroeconomic conditions, or developing local capital markets. Financing terms may similarly become more favourable when systems demonstrate several years of sound operations and timely debt service or when contracting authorities show competence at managing privately financed infrastructure projects. In practice, the opportunity for later stage refinancing often arises from a combination of public and private efforts. Therefore when not originally planned, it is reasonable to include in a well crafted PPP Agreement, some structure for sharing these unexpected 'refinancing gains' between public and private partners. Profit maximizing developers and investors would naturally pursue refinancing whenever gains exceed associated costs sufficiently (e.g. transaction fees, additional financial risk, hedge breakage costs etc.). However, the benefits of refinancing decisions for public institutions may be less clear given different policy objectives and reduced appetites for risk. In addition, refinancing may limit a project's flexibility or may create additional termination liabilities for contracting authorities. Value for Money analysis should precede any agreement to changes in a project's capital structure. Well designed contractual arrangements should therefore also include mechanisms for dealing with refinancing opportunities that may be in both the public and private interest. Box 7.1 provides an example of how refinancing gain is dealt with in the United Kingdom.

BOX 7.1 REFINANCING GAIN – HM TREASURY

The United Kingdom's HM Treasury Department provides detailed guidance on refinancing and arrangements for sharing in version 4 of its *Standardization of PFI Contracts* guide. Specifically, in the October 2008 Amended Refinancing Provisions the Treasury states that the grantor ('Authority') is entitled to between 50% and 70% of such refinancing gains depending on the amount and kind of gains generated, and subject to the Authority's approval and value for money analysis. Transactions where sharing may be warranted include:

1. Reduction in interest margins.
2. Reduction or release of reserve accounts.
3. Release of contingent junior capital.
4. Extension of the maturity of debt.
5. Increase in the amount of debt.
6. Refinancing undertaken without the direct involvement of the Contractor.

HM Treasury also mentions that certain transactions should not require an authority's consent and likewise would not require any form of sharing. These include:

1. Disposal of junior capital, which, in terms of rights, is equity in all but name.
2. Refinancing agreed in the project's base case financial plan.
3. Transactions originally taken on a corporate finance basis.
4. Gains on interest rate hedging.
5. Changes in taxation or accounting policies.
6. Qualified banking transactions such as syndication or securitization of loans.

Considerable analysis should precede an agreement to refinance any portion of a project's capital structure. HM Treasury recommends that contracting authorities should diligently analyze how refinancing can potentially increase termination liabilities or otherwise impact the operational/policy flexibility of contracts. More information from HM Treasury can be found at: www.hm-treasury.gov.uk

7.9 SUMMARY AND CONCLUSIONS

This chapter has examined the different contractual arrangements for private sector participation that can effectively prescribe the funding and financing mechanisms for Water PPP projects. The majority of project financiers will look at the proposed transaction structure and try to find a funding structure that best fits. This in turn can have significant impact on other important considerations such as affordability, value for money, risk transfer and overall project feasibility. As financial close usually occurs later in a project's development cycle, planners will inevitably be challenged to anticipate the implications of their decisions prior to receiving final investor feedback. Indeed, it should be borne in mind that sometimes the project may need to be restructured to suit funding requirements. Understanding the basic financial considerations of Water PPP projects is therefore essential when seeking to align financing implications with public interests during early planning stages.

It is also important for public authorities to appreciate that private capital comes with an expectation of reasonable return. Rational, profit maximizing developers and investors are prepared to take risks only if they expect to earn commensurate rewards. Planners must consider how private investors will recover normal returns throughout project development and implementation. Public-private partnerships are never free, and failing to understand the need for reasonable rates of return can render projects financially non-viable or result in underinvestment during subsequent stages of operation and maintenance.

It is recommended that policymakers understand the implications that project structure has on funding and finance and underline that this is a critical component of effective planning. We discuss the concept of 'bankability' in reference to investor/lender sentiments and their willingness to commit debt or equity capital toward a project. We emphasized the importance of realizing that considering bankability goes beyond financial analysis alone and should include much larger project considerations. Further, we explored how to structure finance for Water PPP projects and discussed the merits and limitations of the different financing techniques available. In addition, we examined the role of public support and bilateral and multilateral financial organisations and the possible forms this may take, depending on the type of PPP structure chosen. Finally, we end with a discussion on the consideration and valuation of contingent liabilities along with other risk management techniques and refinancing.

ENDNOTES

[1]Terms such as 'leverage' or 'gearing' describe the relative proportion of debt in a project's capital structure. Highly 'geared' or 'leveraged' projects involve greater proportions of debt to equity – this has the effect of increasing lender project risk and reducing sponsor/equity investor project risk.

[2]It is important to note that in interviews undertaken with Severn Trent Water International (UK), Veolia (France), Ondeo (France), Biwater (UK), Cascal (UK-Netherlands), Odebrecht (Brazil), and Gelsenwasser (Germany) within a period from 2006 to 2009, these international player are not willing to effectively 'bet the (parent) company' on a single contract. It is therefore important to take into account the willingness of Developers to bear risk, and accept that Developers will need to limit their risk to a specified level. It is not good practice to seek unlimited parent company guarantees, since few Developers would be willing to bid in such circumstances.

Chapter 8

General procurement issues of PPP's

8.1 INTRODUCTION

This chapter is an overview of the key issues to be addressed in the selection and award of a PPP Agreement to a suitable Developer to implement the chosen long term Water PPP Agreement. Given the size and complexity of PPP Water projects, the chosen procurement method has not only to meet local procurement standards but will generally need to be adapted to satisfy international norms and standards, to ensure effective involvement of financing institutions and developers. The overall aim is to establish an effective method to select a long term Developer who is financially, technically and operationally capable of developing an effective water system and long term operation under the PPP Agreement. Although there is a focus on a competitive bidding approach, there is reference to some key issues of other procurement approaches. The selection criteria and assessment methods are issues that should be established at an early stage in the process. Furthermore, given the nature of development of PPP Water schemes, it is beneficial to maintain a continued stakeholder consultation throughout the procurement process, with the aim of developing an optimal scheme. This chapter shows how from the initial survey of interest the procurement process can be managed, through to final bidding, negotiation and award of a PPP Agreement and achieving financial close. Finally it must be noted that this chapter provides the backdrop and general principles for the recommendations that are made later on in Chapters 12 and 13 related to dealing with imperfect data and improving the overall sustainability of long term PPP Agreements.

8.2 PRINCIPLES FOR GOOD PROCUREMENT AND MANAGING THE BIDDING PROCESS

"Proper planning prevents poor performance"

The selection process is intended to achieve efficiency and maximize value for money within a dynamic and flexible environment. Conducting a selection process requires balancing control, flexibility and efficiency. The major parties involved in the tender procedures, the Grantor, the bidders and the public will require assurances that the process has been carried out with transparency and that competition amongst bidders obtained the best price for the project to deliver value for money and quality service. The winning bidder will likely present the most economically advantageous offer, provide adequate price/service certainty and appropriate risk allocation.

According to Brook & Cowen (2002), key considerations governing the award of public contracts are as follows:

Competition: amongst bidders injects efficiency into the process by allowing market forces to mould the procurement approach selected by the Grantor. The aim of a bidding competition is to reduce the price of the project and to induce firms to offer best technical and financial solutions. Bidders are aware of other competing firms and each firm or consortia will compete by reducing its price and improving its technical and financial proposals in order to be selected for final negotiations. Price is not the only consideration. The competitive process can spur greater innovation on the part of the bidders and this can only benefit the Grantor by providing him with technical and financial solutions that otherwise may not have been presented in non-competitive tender.

Economy and Efficiency: Competitive bidding allows the selection of the Developer best able to deliver a high quality service (including all investments required) at the most cost effective or most commercially attractive proposal. Promoting competition between bidders is the most effective method to achieving this. The competitive approach can also be used to develop most effective technologies and operational methods[1].

Integrity and Fairness: The selection procedure should be clear and transparent and should ensure equal treatment of all bidders. Bidders can spend substantial amounts of time and money on preparing bidding packages and most will do so with confidence in the integrity of the selection procedures. Bidding firms are likely to decline to participate, or will price in the higher risk of participation into their bidding package, if they not sure of the fairness of proceedings. The confidence of participants will be bolstered by an evidently fair approach with confidentiality mechanisms to protect the information detailed in the bids. Assurances should be made that no information will be disclosed to competing bidders and that all negotiations or discussions will be confidential (UNCITRAL, 2001). Any form of abuse, either by firms participating in the process, or by the parties administering it, should not be tolerated and the integrity of the approach should be demonstrated with a specific system of sanctions.

The *transparency* of the bidding process, the way that it is administered, the scope, requirements, pricing and technical specifications should be clearly detailed (Dinar & Subramanian, 1997). Not only does this reduce the need for clarifications during the process, but allows the bidding firms to generate a clearer view of the risks, resource commitment and costs involved in participating in the process. By giving every potential bidder the same information of the Grantor's requirements, the costs of bidding and the probability of winning become much more calculable, increasing the number of potential bidders and increasing the competitive environment. The Grantor should ensure that all procedures governing the selection of the Developer are transparent and fully disclosed. Unexpected costs can be reduced and competition improved if there is transparency in tendering procedures and will also serve to mitigate potential corruption. All selection procedures should be clearly recorded to help reduce potential disputes and assist in transparency and accountability. This in turn may attract interest from non-traditional investors, thereby increasing potential finance sources for bidders.

Managing the bidding process

A competent and efficient procurement management team, linked to careful planning and co-ordination of the procurement process, will ensure the success of the selection proceedings. It would seem that there is a need for the Grantor to establish a strong bid management structure able to handle the complexities of procuring the project and managing the bidding process. The Grantor should ensure that suitable administrative and personnel support is provided to undertake the chosen selection procedure. I would strongly recommend that representatives from key public agencies should form a steering committee to oversee the execution of the process[2]. Furthermore, I would recommend:

Appointment of Award Committee: The Grantor should establish an award committee that will be responsible for evaluating the proposals and making recommendations on selection. The committee should ensure a fair, transparent and efficient procurement process, executed in accordance with local and international laws and regulations. The composition of the committee should be decided by the Grantor, with members of suitable standing and skills. These will be supported by suitable technical, operational, legal and financial staff. Outside expertise will be critical in providing knowledge and expertise to establish the Water PPP scheme. The advice of independent experts or advisers to help determine the evaluation criteria, performance indicators and specifications and to prepare the bid documentation may be sought[3]. The responsibilities of the committee would include overseeing the bid management team to: draft bid documents, identify short listed bidders, and identify the winning bid. However, it must be noted that a separate team may be required to finalize negotiations with the winning bidder.

Use of Advisers: As mentioned before it will be important that the Grantor has appropriate financial, technical, operational and legal support. Demand analysis is a key area where external technical advisor participation is often crucial. Technical support is required as well in:

- Preparation or review of construction costs and assumptions for the feasibility review.

- Drafting output specifications and risk analysis.
- Structuring the technical aspects of the bid documentation.
- Evaluation and negotiation of the technical aspects of the bids.
- Design review.
- Construction supervision.

External legal advisors can assist on specialized legal aspects of the project and particularly in drafting the PPP Agreement and all bid documentation, and, together with the Grantor, ensuring that the bidding procedure fits with relevant procurement legislation (Basañes *et al.* 1999).

Market Sounding is the process of assessing the reaction of all potential bidders to a proposed project and procurement approach before formally initiating a procurement process. Market sounding helps to establish the existence of a market for the proposed project, level of investor interest and project feasibility. It helps to obtain feedback on basic fundamentals of a project and its key constraints (Yescombe, 2002). If the project's financial model is shared with the potential pool of bidders, insight on commercial viability can be sought[4]. However, the focus of market sounding should be to obtain information on: the following key areas: *Feasibility* – whether the proposed project is feasible, experience from other similar projects; *Capability & Capacity* – whether potential bidders are able to execute the project; *Maturity* – whether there is an established market for the requirement and whether a competitive procurement could be held.

It is recommended that the market sounding process be open and that all participants are treated with equality and fairness. Giving any potential bidder an inside advantage is to be avoided at all costs to preserve a level playing field. Market soundings can range from a request for comments, to a full market-testing questionnaire, through 'road shows' or meetings with potential bidders to address key issues that may help optimize project design and implementation.

8.3 SELECTION CRITERIA & BID EVALUATION

Contracting authorities seek bidders with proven experience and capability in the service being tendered. As discussed above, this suggests that the award of public contracts is best achieved through methods that promote competition among a range of bidders within structured, formal procedures. Competitive selection procedures generally provide optimum conditions for competition, transparency and efficiency.

Competitive selection is generally carried out in two stages:

(a) Pre-qualification of suitable bidders.
(b) Formal bidding, and selection of the winning bid.

Contracting authorities should decide the method, principle, scoring process and weightings before starting the selection process. Contracting authorities need to indicate the broad evaluation framework to the bidders in order to maintain transparency and increase bidder's confidence. It may also be necessary to train staff in reviewing bids in order to maintain consistency in the evaluation process.

In this chapter we consider evaluation at both pre-qualification and bidding stages.

Selecting criteria for competitive bidding

The complexity of many infrastructure projects requires contracting authorities to design evaluation systems to compare proposals from different bidders. Bidding packages which have passed the threshold of quality and technical aspects can be judged solely on a single factor, such as a global price offered for the construction works and long term operation, or some basket of financial measures (which could include lowest level of public support, greatest revenue sharing between the public and private parties, etc). However, price alone should not be the sole determinant. Privately financed infrastructure projects are expected to be financially self-sustainable, maximizing recovery of the development and operational costs from the project's own revenue, and optimizing amounts of subsidies needed. The Grantor will therefore need to analyze the suitability of the technical elements and the commercial and financial feasibility of the project, the soundness of the financial models and of the bidding proposals (Vives, 1997). The manner in which the winning bid will be selected must be determined prior to the tendering process. The following evaluation procedures are most commonly used:

- *Pre-qualification* – Firms interested in the project are required to establish technical competence and financial viability. The purpose is to limit the field of applicants to those who meet the minimum requirements to participate in a competitive award process. The Grantor reviews the information and selects a short list of firms invited to participate in the next stage.
- *Evaluation of technical proposals* – Technical proposals on how the Developers would execute the project are submitted, together with further technical and financial viability information. Technical evaluation criteria will be established by the Grantor in order to assess the proposals. To be chosen to participate in the next stage, bidders have to achieve a minimum score according to the criteria.
- *Evaluation of financial proposals* – Financial evaluation criteria are then used to assess the bidders' financial proposals. The bidder with the best financial proposal is then awarded the PPP agreement.

The *technical evaluation criteria* determine the fitness of a bidder to undertake the project if selected. The criteria are used during the pre-qualification and the technical evaluation stages. As a pre-qualifying test, a bidder's operating capacity may be evaluated by investigating the bidder's previous experience in similar projects, its human resources and technical capability. Financial capacity can be established by analyzing the bidder's financial statements and whether other financial institutions are supporting his bid. The financial capacity of the bidder evaluated at the initial stage is only to see whether he has the resources to carry out work of the size and complexity of the proposed Water PPP scheme (Klein, 1998b).

Common quality criteria include:

- Technical skills.
- Personnel skills.
- Management team.
- Supply chain management.
- Methodology.
- Environmental criteria.
- Relevant experience.
- Past performance.

At the final bid stage more specific technical evaluation criteria are required, specifically related to the detailed water scheme and the type of PPP Agreement. Depending on the type of PPP Agreement being bid, the emphasis of the technical evaluation differ – for a PPP Agreement the Grantor will be more concerned with the construction, operation, maintenance and extension plans included in the bids; while for a Management Contract, the evaluation will focus more on the quality and expertise of the management personnel, management style and systems. Operational and capital works investment issues are the main focus in the technical proposal. Financial information on the bidder's working capital and investment plans may also be included to ensure sufficient resources will be available. Technical evaluation of proposed business plans involves a great deal of discretion and judgment, and care needs to be taken to ensure continuing transparency of procedures. The Grantor can liaise with bidders on the technical and service requirements before finalizing a bid on a technical package[5].

The Developer's *financial proposal* will need to assure the Grantor that sufficient financial resources will be made available to ensure that the technical proposal can be carried out. This will be evaluated on the strength and deliverability of the financing plans, equity and debt providers' levels of commitment, and the risk allocation terms. Criteria required for financial evaluation could include:

- Customer/Developer tariff required/Subsidies payable.
- Upfront fee, periodic lease payments, or concession payments to the Grantor.
- Price for shares or assets to be sold.
- Capital investment committed by the Developer.
- Service Quality targets.
- Service or management fees payable to the Developer.
- Subsidy payable to the Developer.
- Subsidy payable by the Grantor.

8.4 PRE-QUALIFICATION

With the size and complexity of privately financed infrastructure projects, the Grantor may decide to limit the number of bidders to only those who satisfy pre-qualification criteria. Pre-qualification is the selection of suitable bidding firms from among all interested parties, involving evaluation of bidders' qualifications and resources, as well as technical and financial capacity for handling the project (Klein, 1998b). Those bidders who qualify may be selected to take part in the tendering phase.

The pre-qualification process is also beneficial in helping to simplify the final bid selection process, since assessment of ability and capacity to perform will not be required again. The pre-qualification process has the benefit of helping the Grantor focus on development of the project details in a clear and effective way in which industry input can be used in the process. Key documents issued to potential bidders at the pre-qualification stage could include:

- Background information memorandum on the proposed project.
- Bidding process instructions and bidding selection criteria.
- Pre-qualification questionnaire for all potential bidders.

Pre-qualification Questionnaire: A key purpose of the pre-qualification questionnaire (PQQ) is not only to ensure the potential bidders' capacity, but also to be used as an evaluation tool to reduce the bidders invited to tender down to a manageable number. The Grantor will evaluate the completed questionnaires and exclude the bidders who do not meet the minimum requirements (Office of Government Commerce, 2005). The pre-qualification questionnaire explores:

- *Business Activities* – the bidders' previous experience and expertise in the field.
- *Financial Information* – the bidder's financial health and whether there are any risks in completing the contract satisfactorily. Information on the bidders' capacity to raise finance and the type of debt to be raised.
- *Legal Information* – the legal status of the bidding firm.
- *Resources* – Quality of the personnel involved in the project.

Road Shows: Before launching a pre-qualification process, Governments (aided by their advisors) generally undertake a 'road show' to promote the potential transaction and review the level of investor interest. By meeting potential bidders, contracting authorities can set the criteria to ensure that there will be a sufficient number of bidders based on their knowledge of investor interest and the technical and financial characteristics of potential bidders.

Public Notification and Pre-qualification: In the interests of transparency and competition, the invitation to the pre-selection proceedings should be handled in a manner which attracts the broadest possible interest to ensure a competitive selection process[6]. The invitations should be published domestically and internationally in order to foster interest at home and abroad. The procurement process generally begins with a public notification in electronic and traditional media outlets stating the details of the tender opportunity. The details of the notification will adhere to the requirements of the country's procurement laws and will contain information on the project with a request for qualified companies, groups, or consortia to participate in the tender process. Companies expressing interest in the invitation are then sent pre-qualification documents.

The pre-selection documents should contain clear and concise information detailing the precise infrastructure to be built and the specific requirements expected of the selected Developer, the financial structure and a summary of the terms of agreement envisaged by the Grantor.

Pre-selection Criteria: Bidders should be required to demonstrate that they possess:

- The technical and professional qualifications.
- Financial and human resources.
- Managerial capability.
- Reliability and experience.
- The specific criteria required for large scale infrastructure projects such as a water scheme include:
    - Experience managing financial aspects and experience operating infrastructure services under public regulatory oversight (proven through quality indicators of their past performance in previous projects).
    - The experience of the key team members to be involved in the project.
    - Organizational ability (experience in construction, operation and maintenance).
    - Financial sustainability (demonstrated by the amount of equity the bidder is willing to provide and the support of financial institutions defending the financial standing of the bidder).

The challenge is to identify the right parameters by which to judge quality. Performance criteria may also be utilized which allow bidders to demonstrate a minimum level of efficiency in their relevant operations elsewhere.

Dealing with bidding Consortia: The pre-qualification process should be designed in such a way to ensure the participation of the broadest range of local and international companies with relevant experience in the field. The number of firms experienced in building and operating water and wastewater services is relatively small and in most cases contracting authorities will be faced with consortiums of real estate developers, rolling stock providers and construction companies. Developer consortiums seek to offer the full range of expertise and experience. Grantors should be careful to evaluate the full financial capacity of the consortium and whether there is a track record of the component firms working together in the past. There is a need to assess the forms of any joint ventures that have been established to make the bid and the likely effectiveness of their varying types of equity or sub contract structures.

One firm should be designated as the spokesman for the entire consortia for the purpose of efficiency. It is recommended that the Grantor oblige the members of the consortia to sign an agreement committing them to remaining in the consortia in the event of their selection. The Grantor may also stipulate that the members of the winning consortia create a special purpose vehicle to carry out the project.

Evaluation of Pre-Qualification Responses: The Grantor should create a framework to evaluate the responses to the questionnaire using a pre-determined scoring matrix that determines the scale by which each component of the questionnaire will be graded and the level of importance to the Grantor (the weight) to be applied.

Generating a Shortlist for Bid Selection: Potential bidders prefer shorter lists for an increased probability of winning the competition. Past experience evaluation showed that some bidders may withdraw from the process if too many other firms are pre-qualified (more than five prequalified firms is the general threshold before firms reconsider their decision to proceed with a bid)[7]. At the same time, contracting authorities will seek to ensure greater competition between bidders by pre-qualifying more bidders. Past experience suggests that at least three bidders are necessary to ensure competition with four chosen as an insurance option should one firm drop out. Quantitative pre-selection criteria are more easily applied and transparent than qualitative criteria

- *Pre-Qualification Threshold* – the Grantor determines a quality threshold and all the bidders that pass that threshold are placed on a shortlist. The disadvantage is that too many or too few bidder will be shortlisted.
- *Set a Number of Short Listed Bidders* – the Grantor decides upon a maximum number of bidders that will be shortlisted. The top-ranking bids evaluated are then selected until the 'quota' is reached. The disadvantage is the probability of including low-quality bids and excluding high quality bids.
- *Hybrid* – interface the above two approaches by creating a maximum quota of five firms to pass the threshold on a shortlist but at the same time the Grantor can ring-fence the right to choose the top two or three ranked firms if less than five bids meet the threshold.

8.5 CHOOSING THE BIDDING PROCESS

Contracting authorities have at their disposal a broad range of options for the bidding and award procedures. The three main approaches are:

- Competitive bidding.
- Competitive negotiations.
- Direct negotiations.

In practice, most award processes contain an interface between competition and negotiation. In this section we look at the competitive bidding and negotiation approach as being the most common, and generally considered the most appropriate for Water PPP schemes.

After the pre-qualification stage, it is recommended that the Grantor review its original feasibility study and project/performance specifications in the light of information and comments gleaned from the pre-qualification process. The design of the bidding and award process can have a significant impact on the economic efficiency, transparency and quality of the outcome. At this stage of proceedings, the Grantor should then decide whether a single or a two-stage procedure or two-envelope bidding system will be used to request proposals. Figure 8.1 represents graphically the single stage bidding process.

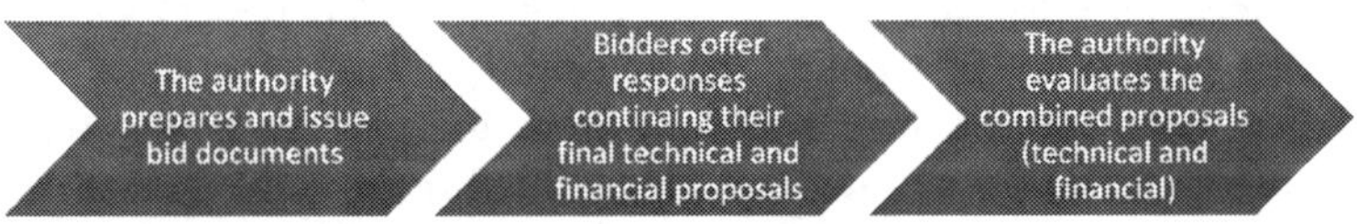

Figure 8.1 Single stage bidding process.

Single stage bidding

If the Grantor is confident that the final project specifications and performance indicators meet the required standards and designs, the selection process may be structured as a single-stage process. After pre-qualifying bidders, the Grantor would then proceed to issue a final request for proposals.

Two stage bidding

For complex projects, such as a Water PPP scheme, the two stage bidding process is generally the most effective (Figures 8.2 and 8.3). The process includes a stage for dialogue between the Grantor and the eligible bidders to ensure that bids are more likely to meet acceptable standards. The purpose of this initial stage is to determine what the market considers feasible and bankable, and allows the Grantor to take advantage of this information to refine the bid requirements. At this stage there are no negotiations on the terms of contracts.

The process includes the following steps:

- Grantor issues request for proposals which contains technical specifications and bidding documents. The request for proposals should also include details on the risk allocation envisaged by the Grantor.
- Bidders submit a first round bid containing only preliminary technical proposals based on the information contained in the request for proposals for comment and discussion.

- The bidders' responses to the proposed risk allocation can assist the Grantor in assessing the feasibility of the project. The Grantor consults with each eligible bidder to discuss the bids and what changes are required to improve them. The Grantor must avoid disclosing details of any discussions to competing bidders and must treat all discussions as confidential.
- After these discussions, the Grantor should review, and if necessary, revise the initial project specifications. The bidders should be advised of any changes to the project specifications in the invitation to submit final proposals. At this stage, bidders no longer interested in contesting should be allowed to withdraw from the process.

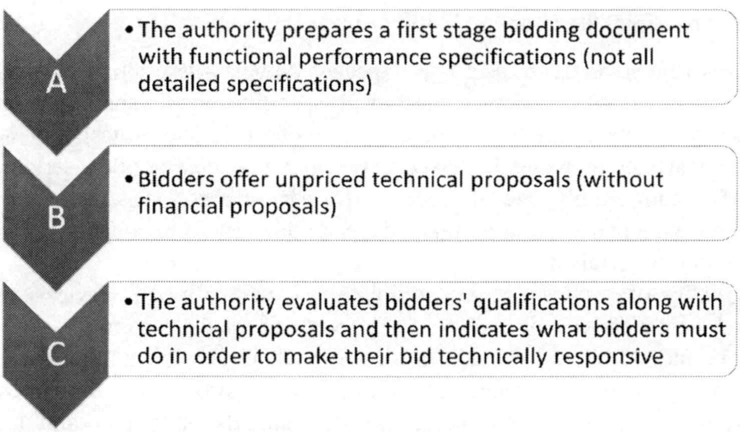

Figure 8.2 Stage one of a two stage bidding process.

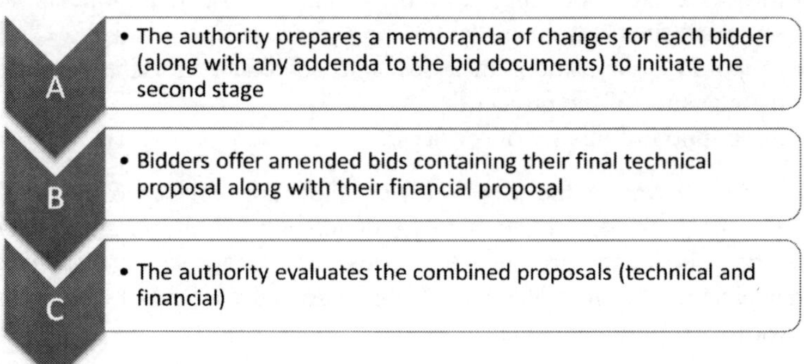

Figure 8.3 Stage two of a two stage bidding process.

Final request for proposals[8]

The Grantor's chief responsibilities at this stage will include: providing the bidders detailed information including the tender rules and procedures; consulting with the bidders and delineating the precise format and content of the proposals. Each of these responsibilities will ensure the maximum amount of transparency within the competition bidding process.

The Grantor should ensure that it fully discloses all information regarding the potential project and precisely what the expectations are for the winning bidder. This will minimize accusations of unfairness and corruption. In addition, full disclosure will allow bidders to prepare the best possible bids according to the outlined requirements and given the same information, bidders are likely to produce easily comparable bids. Developers who receive more relevant information will be more open to taking on the associated risks.

Where possible, the Grantor should provide information on:

- Description of the project and expected outputs – description of works and services to be performed, technical specifications, plans, drawings and designs, time schedules for the execution of works and provision of services, technical requirements for the operation and maintenance of the service.
- Minimum applicable design and performance standards, including environmental standards – the desired quantity and quality of service should be detailed.
- Quality of service – general obligations, availability of services, and the relevant standards of quality to be used to assess the system.
- The bidders should provide information where possible so that the Grantor can evaluate the technical soundness of proposals; the operational feasibility and the responsiveness to standards of quality and technical requirements, including:
  - Preliminary engineering design, along with the proposed schedule of works.
  - Project costs, operating and maintenance cost requirements and the proposed financing plan (debt to equity ratios).
  - Proposed organization, methods and procedures for the operation and maintenance of the project.
  - Description of quality of services.

The contractual terms of the proposed risk allocation envisaged by the Grantor should also be included in the bidding documents. If the Grantor details his preferred risk allocation (provided it is bankable) it can ensure that Developers will bid against a common standard. Other essential elements to be included in the request for proposals include:

- Information on the duration of the agreement.
- Formulas and indices to be used to calculate tariffs.

- Government support and investment incentives.
- Bid bond requirements.
- Regulatory agency requirements.
- Monetary rules and regulations.
- Revenue sharing agreements (if any).
- Transfer of assets at termination.

Information should be shared through the following channels:

- Bidding documents and information memorandums.
- A data room.
- Meetings with Grantor.

A data room provides a one-stop location for bidders to receive further information about the proposed project. All bidders should have equal time and access to the project data.

Interaction with Bidders: Dialogue between the Grantor and bidders can be beneficial to both parties. Consultation with potential bidders prior to formally beginning the procurement process can increase investor interest in the contracts and can be tailored to increase the attractiveness of the project. The main approaches for bidder interaction are bidder conferences and arm's length consultation. Bidder conferences are arranged by the Grantor to meet all the prospective bidders, explain the bid process and consult them on their ideas for the project. The Grantor may provide the conference attendees with draft documentation before the conference and bidders are given the opportunity to raise questions or request clarifications of specification. It is to be recommended that these questions and clarifications are distributed to all potential bidders prior to the conference. The aim is to ensure fairness and transparency of information. However, it must be noted that there are disadvantages to this approach:

- Bidders may not wish to share innovative plans with their competitors.
- Bidders may not answer questions honestly during the conference.
- Bidders may seek to manipulate the transaction terms through their comments on the draft documentation.
- Bidders may seek to collude and persuade the Grantor to take on more project risks than necessary.

An alternative to this is the use of 'Arm's length' discussions, where bidders submit written and independent comments on the draft bidding documents and the proposed PPP Agreement. The Grantor reviews the comments and is made aware of the key issues and concerns held by potential bidders. Whilst this written process does not suffer from the competitive issues of bidders conferences, it does not offer the advantages of potential dialogue at a bidders' conference.

The technical proposal

The Grantor must clearly specify in the request for proposals (RFP) the precise information and format required in the technical proposals of the bidders. The format and content should be directly related to the evaluation system being employed by the Grantor as this will facilitate easier comparison and scoring of bids. Mott Macdonald (2008) recommend that each bid package should contain a technical package with details as to the Bidder's design, construction, commission, operational, maintenance and handover (if contract stipulated) proposals for the project to the Grantor. Bidders must respond to the performance specification and construction specification as detailed in the invitation to bid. The marking scheme and weighting allowances that the Grantor will use to evaluate the bids should not be revealed to the bidders before bid closure.

The bidding procedures evaluated here show that technical proposal formats differ but their general purpose is to demonstrate the bidder's ability to execute the proposed project according to the standards specified in the request for proposals. Bidders should provide sufficient details to enable the Grantor to obtain a good understanding of the essence of the development work that the bidder would undertake should they be awarded the PPP Agreement[9].

In evaluating the proposals, the Grantor should ensure that the bidders' plans fulfil the terms of the bidding documents and are able to ensure the service levels stipulated in the proposed PPP Agreement. In general, the evaluation process will look at:

- The scope, clarity, quality, robustness and deliverability of the proposals.
- The technical effectiveness of the proposals in relation to the design reference and performance specification.
- The superiority of the operations and maintenance proposals.
- The inherent quality, reliability, availability and maintainability of the proposed system.
- The robustness of the program methodology, schedule and delivery plans.
- The technologies to be employed for the supply, construction, billing and commercial systems, operations and maintenance.
- Compliance with environmental and sustainability requirements.

The financial proposal

The financial proposal will demonstrate the Developer's detailed financial approach to providing the technical solution detailed in his technical proposal. This generally includes details on: Financial Plan (funding as well as operational), Financial Model (provided by the bidder) demonstrating key financial parameters and assumptions, additional proposals made by the bidder and the required levels of State Financing (for example level of subsidy to be paid by the Grantor to supplement connection or tariff charges for poorer customers, Grantor contributions to the capital expenditure program etc.).

The structure of the financial proposal is critical as it can have important effects on the funding and the operation of the project, and ultimately on consumers. The financial plan within the bid is used to demonstrate the adequacy and feasibility of project financing proposals. The Grantor should follow the following basic principles:

- The structure should be as simple and transparent as possible so that the bid award is automatic (avoiding complex formulas or anything requiring subjective judgment).
- The structure should promote economic efficiency on the part of the Developer and users of the proposed system.

The financial proposals will be evaluated by the bid evaluation committees to check whether they are compliant with the requirements of the invitation to bid and whether they correspond to the legal and technical parts of the bid criteria. Generally, evaluation panels will award a coefficient to each financial part, and consider the quality of the financial plan and the model regarding its maturity and feasibility.

8.6 SUBMISSION, OPENING AND COMPARISON OF BIDS

Given the complexity of the tendering procedures for large infrastructure PPP projects and the multitude of criteria to be evaluated, it is recommended that contracting authorities opt for a two-step evaluation process. The technical proposal is evaluated separately from, and generally before, the financial proposal evaluation in order to avoid placing greater weight on price considerations.

The two-envelope system has been used in many water and wastewater tenders. The process occurs as follows:

- Grantor pre-qualifies bidders based on their technical and financial capacity.
- Bidders then submit technical and financial proposals in separate envelopes at the same time.
- The Grantor will evaluate the technical and financial proposals separately and will not open the financial proposal until the technical evaluation is completed.
- Using a suitable scoring method, the technical and financial scores are combined and a winning bid is selected.

This approach eliminates the influence of price during an evaluation of the technical proposals and therefore the technical evaluation will be price more subjective and less vulnerable to manipulation. However, care should be taken to ensure that technical criteria do not exclude solutions that are technically adequate and might offer more competitive financial advantages. International institutions such as the World Bank do not generally support such procedures due to the difficulty in evaluating proposals objectively under such a system.

An alternative to the two envelope system involves the Grantor receiving both technical and financial proposals in one single proposal but the evaluation is structured in two stages. The Grantor establishes minimum quality and technical score thresholds and would rate each proposal according to the degree to which the requirements specified have been met. The weighting and scoring mechanisms will be detailed in the request for proposals so as to ensure transparency. The Grantor will then proceed to evaluate the financial proposals of all the technical bids that have scored at or above the threshold.

The overall approach to selecting the winning bidder should be decided in the initial stage of the tender process. The specific evaluation criteria and their use for determining the winning bidder should be detailed in the Bid Procedure documents.

Compliance: The Grantor should review each bid submission to ensure that they comply with the procedures and requirements set out in the Bid procedure documents. Elimination may be mandatory at this stage under certain jurisdictions. Any form of material non-compliance may lead to either the elimination of the bidding firm from the evaluation process or a request for a revised proposal.

The Grantor should review any changes made to the proposed PPP Agreement. Any changes to the proposed PPP Agreement should be marked clearly so that suitable risk adjustments can be made so that all bids evaluated are using the same risk parameters. Bids should fulfil the requirements of any engineering design, and should include the core commercial, technical, operational, maintenance and financial details.

Financial and Technical Evaluation: As discussed in previous sections, the technical and financial evaluations are performed sequentially but not simultaneously to ensure unbiased analysis and to ensure that cost considerations are not factored into the analysis of technical proposals. Public opening to financial proposals are commonplace in which the primary benefit is to avoid allegations of impropriety in the evaluation process. Arrangements are frequently made for opening of major bids in public, with attendance by the media. Financial criteria will depend on several factors:

- The type of arrangement proposed.
- The level of cost recovery achieved by existing tariffs.
- Predictability and value of future tariff-based revenues.
- Grantor's objectives.
- The rules for setting future tariffs.
- The need to prevent bidders from deliberately under-pricing their bids (low-balling).
- Whether the proposed terms by the developer contained in the bid can be taken to financial close.

Bid evaluation methods available

After assessing both the technical and financial viability of bids, the Grantor must then select the best overall bidder based upon the conclusions formed from those assessments. Some technical and financial criteria can firmly be established as crucial, but other elements of bids, such as personnel, are not governed totally by objective criteria, yet can form a key element to judge if a company is able to execute a project successfully A common approach therefore is to choose weights for the technical and financial scores and then combine the scores to produce an overall score. Pakkala (2002) suggests that the weighting of these factors should be appropriately defined and measured as objectively as possible in order to be most effective. Each factor will be weighted differently and this will depend on the preferred combination of quality and price of the contracting authorities. Considerations include:

- The bid weighting should reward bids that match the quality and price criteria. Therefore, appropriate percentages for quality aspects should be considered in order to improve innovation and have the organization best qualified to provide maintenance.
- Technical proposals are usually evaluated first, the bidders are informed of the scores and then the financial proposals are assessed. The final scores are calculated using the pre-determined formula and then the winning bidder is chosen.
- The technical score is subjective and is decisive in determining the winning bid. However, an avenue for corruption may open since the weighted average score can be manipulated in cases where an unwarranted higher technical score might be granted, with little chance of being detected, thus allowing a favoured bid to win.

Another possible bid evaluation method is to, within the invitation to bid document, the Grantor indicate the percentage of total assessment marks that will be awarded to the different aspects of the bidding packages – the financial, legal and technical sections. The technical assessment and marking will allocate a percentage of the marks to each of the sections of the performance specifications and the code of construction practice and marks are awarded to the Bidder's response to each section of the documents. The marks are recorded on bid marking sheets which are then forwarded to the Grantor by his technical assessors. The completed sheets should be kept securely for a period of years after the completion of the bidding process. The detailed marking scheme and weighting of the marks should not be revealed to bidders or any other parties before Bid Closure. After Bid Closure, bidders may privately request access to examine the bid marking sheets related to their own bid after the bid assessment process has been completed.

Other weighting methods include:

- *Technical threshold, highest financial score* – this decision method is used for concession contracts. All bidders are evaluated on their technical proposals and a 'technical threshold' score is set. All bids that do not meet this score are rejected. The remaining bids are then judged on their financial proposals and the best financial offer wins the PPP Agreement.

  - This method is simple and transparent, based on the objective criterion of price, thereby making it harder to manipulate.
  - However, this method eliminates higher priced bids that dould provide higher quality solutions. In many cases the cheapest bid does not necessarily provide the best value over the long term.

- *Fixed budget, highest technical score* – Grantor sets a budget and informs the prospective bidders. The bidders are then asked to submit their best technical proposal based upon the set budget. For management contracts, the bidders would then compete on the amount paid for the management fee. This method encourages creativity towards quality rather than focusing on minimizing cost.

Table 8.1 provides a comparison between a price and quality based evaluation system for bids received.

Table 8.1 Weighting bids – price vs. quality features.

	Advantages	**Disadvantages**
Price Based	• Easily administered • Widely used and accepted • Clear and objective selection process	• Lack of innovation • Tendency for change • Orders and cost overruns • Tendency for mediocre quality
Quality Based	• Quality potential • Potential for innovation • Best qualified organization • Better people skills • Best methodology potential • Potential for project success	• Some criteria are subjective • Requires more effort and skill for tender evaluation team

Source: Pakkala 2002

Final selection

After completing the evaluation of the technical and financial proposals, the Grantor may proceed by selecting the proposal that contains the highest combined scores in price and non-price criteria. Alternatively, the tariff charged over a given period (usually five years as in Buenos Aires and Tallinn) may be used as a deciding factor. The winning bidder will likely present the most economically advantageous offer, provide adequate price/service certainty and appropriate risk allocation. The award committee is advised to provide written explanations for their selection beyond simply accepting the lowest priced bid. Negotiations then commence with the best rated bidder. If there is more than one high score bid or there is a minimal difference between bids, more than one bidding firm or consortia can be invited to further negotiations over the transaction documentation and to answer any queries from the bidders' lenders. If negotiations are unsuccessful it may be necessary to restructure and rebid the whole selection process.

Negotiations may occur in two ways *Non Competitive Negotiations* where the best rated bidder is invited to a final round of negotiations to finalize certain elements in the project. If final agreement cannot be reached, then generally the next rated bidder enters into negotiation. However, if the consultative selection process suggested is used and combined with the acceptance of contract terms at the bid stage, the risks of non-agreement at this stage should be unlikely. *Competitive Negotiations* involve the Grantor negotiating simultaneously with two or more bidders in order to enhance the competitive nature of the negotiated transaction. Competitive negotiations are suited to complex, non-standard projects. However, the process is less transparent and more subjective.

The negotiated bidding process involves:

- The Grantor details its requirements and requests expressions of interest through a request for proposals (RFP).
- The Grantor reviews the proposals and selects those that are technically responsive to the RFP.
- The Grantor then negotiates the PPP agreement terms with the selected bidders.

The advantages for selecting competitive negotiations are:

- Induces bidders to submit innovative and creative project proposals.
- Reduces 'optimism bias' in bids and discourages bidders from deliberately underbidding in order to win PPP Agreements.
- Allows for greater dialogue between contracting authorities and more opportunities for the Grantor to evaluate bidders on criteria other than just bid prices.

The disadvantages are:

- Non-standard bids make comparisons difficult.
- Increased level of closed dialogues between contracting authorities and bidders decreases transparency (increasing the potential for corruption).
- The prolonged bidding process and the costs involved may deter some firms from engaging in the process.
- Bidders may attempt to make changes which fundamentally alter the price or risk allocation originally contained in the proposal, thereby distorting the criteria upon which the bid was originally rated.

At this stage, the Grantor should inform the remaining bidders that they may be contacted for negotiations should negotiations with the highest rated bidder fail to end in agreement. If this scenario occurs, the Grantor will inform the bidder of his decision to terminate negotiations and then advance with negotiations with the next highest bidder on the basis of its ranking. It is recommended that contracting authorities do not re-open negotiations with previously failed bidders.

Negotiating with preferred bidders

After completing the financial and technical evaluations, contracting authorities have various options available to them. The Grantor can:

- Choose the firm obtaining the highest score according to the bid procedure criteria at the price stated in the firm's bid.
- Enter into negotiations with the selected firm.
- Hold another round of bidding due to:
  - Additional issues remaining to be agreed with the preferred bidder,
  - The opportunity for a better offer.

Negotiation options

- *No Negotiation* – in the case that all technical and financial consultations have resolved any issues over their bids, thus permitting the bidding teams to submitting their bidding packages and associated documentation including the signed and unchanged PPP Agreement. The Grantor then selects the preferred bidder and accepts the signed PPP Agreement.
- *Negotiation with Preferred Bidder* – the Grantor negotiates outstanding issues with the preferred bidder. Contracting authorities should make sure that they have the adequate resources to negotiate on equal terms with the more experienced bidders to ensure that modifications requested by the bidders are justifiable and are not merely attempts by the bidder to take advantage of the Grantor's lack of experience.
- *Additional Round of Bidding* – all submitted bids are shared amongst the bidding teams and an additional round of bidding is then held.
- *Competitive Negotiation*[10] – the Grantor negotiates simultaneously with two or more bidders. The bidders compete against each other with the aim of generating a better deal for the Grantor. A major criticism of this approach is that it encourages 'private' negotiations between the Grantor and the bidders and the transparency of such negotiations is questionable.

Competitive negotiation may be formalized to include the 'Best & Final Offer' or BAFO approach (Box 8.1 gives some of the downsides of the BAFO approach). This is becoming common in some European projects. After bids have been evaluated, the bidders are called back to give a 'best and final offer', with the aim of increasing benefits to the Grantor.

BOX 8.1 THE DOWN SIDE OF BAFO PROCESSES

Best and Final Offer (BAFO) processes can both benefit and harm the public interest. Reduced bid prices or increased design value can clearly benefit the value for money that public authorities derive from PPP structures. However, BAFO processes can add time and complexity to the procurement process while also introducing excess subjectivity into the selection of preferred bidders. Compromises made during BAFO processes can potentially harm the economic viability of concession companies which may lead to insolvency or renegotiation at a later date. Planners must be aware that sometimes the most accommodating bidder is not necessarily the best choice.

The main features of the EU competitive dialogue system are:

- Dialogue is permitted with selected suppliers in order to identify and develop solutions to the Grantor's project requirements.
- Dialogue occurs in successive stages, at the end of each stage the number of bidders/solutions is reduced.
- The Grantor makes its decisions based on which bid provides the most value for money.
- No changes are allowed in the post-final tender or post award stage of bids.

The procedure is as follows:

- Contracting authorities publish a PPP Agreement notice which explains their needs and requirements.
- A prequalification questionnaire is completed by the interested parties and is evaluated by the Grantor's bid evaluation team. Prospective bidders are then selected.
- Prospective bidders are invited to begin a dialogue to identify and develop the best solution to the infrastructure required. The Grantor must describe his needs clearly, in order to assist the process and ensure bidders do not use the dialogue to market test ideas.
- Provided there is sufficient competition, the Grantor can reduce the number of prospective bidders with whom he is in dialogue. If there is not a sufficient number of prospective bidders, the authority can continue its dialogue with the remaining bidders.
- Contracting authorities can discuss all aspects of the PPP Agreement with the chosen candidates, but care must be taken to ensure that all candidates are treated equally fairly.
- The prospective bidders will then specify their proposals in writing and the award criteria contained in the PPP Agreement notice will be used to evaluate the proposal. The number of bidders may then be reduced further at this point.

- The Grantor will then decide which of the proposals meets his requirements and the proponents of these proposals are then requested to submit their final offers based on dialogues with the Grantor. The final tenders should be as complete as possible as there are various limits on post-tender discussion and alterations to bids.
- The bids are then evaluated against the award criteria and the most economically viable proposal is selected as the winner.
- Post-tender discussions with candidates who submit final tenders are permitted to allow clarification, fine-tuning and additional information. No changes to the fundamental nature of the bid are permitted.
- Dialogue with the preferred bidder will promote further design details, the finalization, of the PPP Agreement and financial due diligence.

The intention of the dialogue phase is to encourage and, promote more value for money and innovative solutions from the bidders. While competitive dialogue may extend this phase, it should shorten the closure phase of the preferred bidder and lead to a shorter procurement timetable[11].

Unsolicited proposals and direct negotiations

Unsolicited proposals usually originate from the private sector and are generally not requested by a Grantor. In unsolicited proposals the Developer uses its own resources to develop a project idea and then approaches the relevant governmental or Grantor for the required official approvals. Accessibility of Governments to receive unsolicited proposals could create incentives in the private sector to come forward with innovative proposals (Hodges & Dellacha, 2007). Additionally, in smaller municipalities where it may be too costly or difficult to arrange a competitive bidding process, direct negotiations increase the chance of private sector interest in infrastructure development projects. However, research confirms that a major disadvantage for Grantors of unsolicited proposals they do not guarantee the most effective or economical solution and as there is lack of competition are associated with lack of transparency. The granting of exclusive rights to private entities without the accountability of a transparent tendering process courts controversy and history suggests that such scenarios lend themselves easily to corruption (Hodges & Dellacha, 2007). Moreover, the competitive dialogue with the wider industry and financial sector which helps develop and refine the PPP Agreement is lost.

8.7 EUROPEAN UNION POSITION ON PPPs

This section describes specifically the approach that the European Union has taken with respect to PPPs, their definition and the issues surrounding their procurement and operation. The relevance of discussing EU in this book and its 'fit' in this chapter is that the EU has developed an approach flexible enough to incorporate countries with different legal traditions yet with a consistent approach to the manner in which each country uses the private sector to help deliver infrastructure services. Accordingly, despite the basic guidelines or distinctions on PPPs, EU law on public contracts and concessions do not prescribe whether or not a PPP arrangement should be used. A European Union Member Country, and thus by implication Accession Countries, are free to provide a public service themselves or to entrust it to a third party. Moreover, EU law allows a pragmatic approach to PPP whereby it recognises that under PPP arrangements, public partners are primarily national, regional or local authorities. There may also be public law bodies created specifically to fulfil certain State control interests. For the purposes of EU interpretation, the term 'contracting body' or 'contracting authorities' is used to designate all of these entities[12]. In this way, EU law may be a useful guide in developing countries seeking to create an enabling PPP environment and associated legislation. EU is not prescriptive about the PPP structure but ensures that in selecting the partner of a PPP, the general principles of the EC Treaty apply, as do the current directives on public contracts. These include that PPP Agreements must satisfy the requirements of transparency, equal treatment, proportionality and mutual recognition. Furthermore, all agreements in which a public authority awards a contract are subject to the general principles on freedom to establish and to provide services in Articles 43 to 49[13].

In addition, PPP Agreements that are considered public contracts are subject to the application of the public procurement directives. Certain special provisions apply to building concessions, but service concessions are not at present covered by secondary law. National laws apply on the implementation of PPPs which must comply with the provisions of the EC Treaty, which take precedence. Additionally, the Commission has introduced some degree of flexibility under procurement law. Notably, it defined, on the basis of the rules and principles derived from the Treaty and applicable secondary legislation, the outlines of the concept of concession in Community law and the obligations incumbent on the public authorities when selecting the economic operators to whom the concessions are to be granted[14].

In the absence of a uniform European legal framework for PPPs and in view of the heterogeneous nature of national laws the question arises, whether uniform provisions are needed to safeguard transparency and effective competition between market participants. The EU has decided against the creation of a separate legal regime for PPPs, but considers that there is a need for legislative initiatives in the areas of concessions for the provision of services, and institutionalised public-private partnerships (IPPPs). EU also considers that as a matter of principle, the law on public contracts needs to be applied whenever a private partner is to be selected. This is discussed in great detail in Annex 2.

Community secondary legislation establishes that any contract[15] for the execution of works or services between a contracting body and an operator is to be designated as a 'public works or public services contracts'. The concept of 'concession' is defined as the same type as a public contract except the concessionaire only has the right to exploit the construction or service and be paid for it. Thus, in relation to all PPP contracts the following should be considered:

- PPP Arrangements should emphasise fairness and transparency. Accordingly, transparency must be in evidence whenever public funds are involved, as well as ensuring that the PPP Arrangement allows for the right of elected representatives to inspect agreements and documents. Furthermore, the EU takes the view that, in the citizen's interest, transparent rules should be applied in awarding public contracts to ensure effective competition and protection from corruption.
- For those PPP Arrangements or contracts which have already taken place, and in order to avoid legal uncertainty, transitional periods should be applied for existing contracts that have been concluded in good faith in accordance with national law.
- PPP Arrangements should make adequate provisions for the protection of the citizen's rights to ensure secure safety, efficiency and quality standards.
- Member States are recommended to alleviate the task of the public sector by means of standard contracts and by stepping up the training of the decision-makers who have the task of selecting the private partners for PPPs.

Furthermore the EU has decided not to establish a European agency for PPPs, but welcomes other ways of sharing experience, such as the networking of national and regional PPP authorities.

PPPs are essentially a form of public contracting. Given the significant examples under case-law of the Court of Justice[16], the EU has decided against developing blanket regulations for PPPs. In any event, the Commission has taken the decision to go carefully in the legislative process as a matter of principle, because PPPs are often very complex agreements whose attractiveness can only survive in a climate of flexible regulation.

As stated earlier in this chapter, in legal terms there needs to be a clear statement that, wherever a private partner is being selected, transparent procedures have been used for this selection. Regardless of the type of PPP selected, transparency and fair competition between market participants should be used.

The EU considers two types of PPP Arrangements, Contractual PPPs and Institutional PPPs. In the case of PPPs that are categorised as public contracts the main issue is the question of the award procedure. With PPP structures, the competitive procedure to be used tends to be slightly more complex. In this regard, a new award procedure has appeared on the scene which combines elements of the tender procedure with those of the negotiated procedure and takes precedence over the negotiated procedure. It must be noted that difficulties do arise in defining precisely what is meant by 'particularly complex' contracts. In response to this, the Commission has clarified that a particularly complex project is that in which the public contractor is unable to specify objectively the technical means with which their needs and objectives can be fulfilled, and/or is objectively unable to specify the legal and/or financial conditions of a project.

Although PPP projects by their very nature need not be necessarily be more complex, each case must be assessed on its own merits. In any event, the Commission does allow for the competitive dialogue procedure to take preference in awarding of PPP contracts on principle because it combines the advantages of the restricted procedure with those of the negotiated procedure. Its flexible nature enables it to obtain optimum solutions under great competitive pressure. Although its disadvantages include its complexity and the extent of effort required from all involved, these are offset by its greater transparency compared with the negotiated procedure. This reduces risk of anti-competitive behaviour.

As mentioned earlier the concept of 'concession' is defined as a contract of the same type as a public contract except for the fact that the consideration for the works to be carried out or the services to be provided consists either solely in the right to exploit the construction or service, or in this right together with payment. The Commission takes the view that concessions should be of limited duration so that competitors are not excluded for an unduly length of time. At the same time it recognises that under these types of PPPs, competitive dialogue is also the appropriate procurement procedure, as its flexibility is as a rule suited to the complexity of concessions, without jeopardising the principles of transparency, equal treatment and proportionality.

Institutionalised PPPs (IPPPs) are joint industrial creations combining private and public partners. The objective of such a PPP structure is to provide or supply a service on behalf of the public, typically in the field of public welfare. IPPPs may be created by the formation of a mixed-economy enterprise or company, but may also arise when a private company assumes control of what was a public concern, that is, by changing the share ownership.

This raises the question whether the two processes, new formation or a change of ownership, are relevant from the point of view of procurement law. The Commission says that the private partner of such an enterprise must be selected in a transparent way and without discrimination for the tasks involved, irrespective of the actual form of the contract under the procurement directives or the EC Treaty. The selection of a private partner in accordance with objective criteria will safeguard fair competition.

Case law as understood in the Court of Justice in the Teckal and Stadt Halle cases (C-107/98 and C-26/03) set the principle that the participation of the awarding authority in a mixed-economy company does not justify exemption from procurement law principles. What is crucial is to award the contract to a body with independent legal personality. Exemption from procurement law is only recognised when the awarding authority exercises a control over the mixed-economy enterprise similar to that control which it exercises over its own departments, and when the enterprise essentially acts for the public body. The two criteria must be cumulatively fulfilled to ensure that it is equivalent to the internal entities of the contracting authority. The Stadt Halle case-law clarified the matter further. The Court decided that public contracting authorities may waive the procurement provisions to award a contract to an undertaking belonging to them only when they hold 100% of the undertaking's capital. In other words when there is not even a small private shareholding in the undertaking. Only then, says the Court, does the contracting authority exercise control as it would over its own departments.

In summary, the Commission considers it necessary, in view of the transparency requirement and the ban on discrimination, for procurement law to be applied when an IPPP is set up, in so far as the act of setting it up is combined with the assignment of a public contract, concession or other public tasks. Additionally, this same principle would apply when an IPPP is established by the sale of shares in a public undertaking.

8.8 SUMMARY AND CONCLUSIONS

In this chapter we have examined key issues to be addressed for selection and award of a PPP Agreement. Starting from a recognition that given the size and complexity of water PPP projects, the chosen procurement method has not only to meet local procurement standards but generally will need to be adapted to satisfy international norms and standards, to ensure effective involvement of financing institutions and developers. The overall aim is to establish an effective method of selecting a long term Developer who is financially, technically and operationally capable of developing and long term operation of an effective water and wastewater system under the PPP Agreement.

Important contractual issues related to the bidding procedure include the use of Bid Bonds which provide insurance that compensates the Grantor should the winning bidder decide to withdraw from the project. The bonds serve to act as a guarantee to the Grantor that the winning bidder will implement the project and has the means to fulfil the terms of the PPP Agreement. At the same time it will be important to make sure that the acceptance process procedures are clarified. Brook & Irwin (2003), suggest that the Grantor may require all bidders to sign a legally binding agreement which confirms their satisfaction with all aspects of the bidding process, that they will accept the outcome of the process and that they will not challenge the decision. This aims to limit bidders from seeking to re-open the bid evaluation process on the basis of complaints over procedures, evaluation criteria or lack of information. This is linked with a need for the Grantor to conduct the process with full transparency and treat all parties on an equal basis.

Upon selecting a winning bid, a long negotiation period between the Grantor and the winning bidder begins. During this period, there is risk of the selected bidder attempting to change parts of agreement. To mitigate this risk, the Grantor, prior to selection, may require the bidders to provide a signed PPP Agreement as part of the bidding package (Diaz, 2003; Dinar & Subramanian, 1997). In the event that a financial model is provided as part of the bidding documentation the following should be considered: (a) should a single model be specified as part of the bidding material, or should the model be created by the bidder[17]? What would the legal status of the model be and how would it be given that status?[18]; (b) Will the results of the model be binding, or merely provide a guide to the Expert Panel, which could be allowed to take other factors into account also? A binding model provides greater certainty, but runs the risk that it may not capture relationships between variables which in the event turn out to be important.

Other issues that contracting authorities/Grantors may come across include:

Dealing with Variant Bids: Contracting authorities will expect bidders to submit standard bids based on the bidding requirements. Additionally, bidders may submit variant bids with alternative proposals for technical or financial provisions. The Grantor has to decide whether variant bids will evaluated simultaneously with the standard bid or whether variant bids will only be considered after a preferred bidder has been selected. This system encourages innovation and can lead to the Grantor getting better value for money. Bidders must indicate the precise differences between their standard bids and variant bids and with explanations of the effects on costs and risk allocation. Variant bids also allow the Grantor to test the costs of different risk profiles between the authority and the potential Developer. Contracting authorities can hold meetings with the bidders to clarify or negotiate any issues emerging from the bids. All bidding teams must be dealt with in a fair and equitable manner.

Dealing with Unsustainable Bids: With complex and competitive negotiations, a mistake or deliberate misrepresentation of information is possible. Contracting authorities should develop clear and sustainable guidelines on how to deal with such scenarios. Deliberate 'under-bidding' or even a genuine mistake by bidders may assist his financial proposal and increase the attractiveness of his bid to the Grantor. However, should a bid be accepted based on an unsustainable project,it would mean that eventually the PPP Agreement would have to be renegotiated or dissolved and re-bid, both expensive in time and money for the Grantor. Experience shows that certain risk evaluation and mitigation techniques should be followed to ensure the sustainability of the bids and to avoid performance or financial issues. Such techniques include for example, so called use of 'shadow models' created by the Grantor to show what a standard and financially viable bid might look like. Real bids are compared to this shadow model giving greater information for the authority to query bids. It is also recommended that the Grantor create bidding procedure that permit discussions and clarifications, with the option to withdraw from discussions[19]. Requesting higher performance bonds could also protect the Grantor against the risk of Developer failures.

Every procurement process runs the risk of being challenged on the grounds of unfair competition, partisanship or questions over procedures. Experience in the field indicates a need to establish guidelines showing methods to deal with such eventualities[20]. These procedures need to determine unequivocally how complaints and appeals will be heard, on what grounds and how they are to be evaluated. The bid details should establish clearly the process to deal with a situation when the long term PPP Agreement ends, and should ensure that the Grantor has enough flexibility to be able to extend the PPP Agreement or to re-tender the whole agreement. This matter should be dealt with in the contractual agreements.

Furthermore, this chapter examines briefly the implications of PPPs in the European context. Finally, attention is drawn as to how best manage the procurement process and recommendations for early consultations with the market to assess suitability and 'market appetite'. Recommendations are suggested to use a consulting process to contract development at the bidding stage, as long as the right framework is established to engage with bidders directly and any alterations conducted fairly and transparently. The chapter ends with a review of possible ways to negotiate and recommends that in whatever process chosen, the grantor retains competitive tension as a means of achieving best value for money.

ENDNOTES

[1]Excessively costly and burdensome selection procedures may be a deterrent to bidders as well as being a burden on the Grantor, and therefore selection procedures should be designed efficiently to select a Developer within a short period of time, at a reasonable cost to both Grantor and bidder and with minimal administrative burdens.

[2]Time and money should be invested in proper procurement preparation. When badly conceived, tendering processes can lead to delays and wasted capital. An example of this is the failed bidding procedure for a lease type PPP Agreement in Dar es Salaam, Tanzania which was re-bid twice before Bidders felt confident that the procedure was well enough designed and had a reasonable risk allocation for them to spend money and resources in bidding for the project. The Grantor should ensure that, prior to beginning the process, he has conducted a thorough review of the project in detail, the potential risks, the design requirements and that the project will deliver value for money. Thorough project preparation will reduce the bidding costs for potential Developers, increasing the number of parties able to participate, the level of competition and thus reducing the cost of the project to the Grantor.

[3]Other advisers and consultants may be hired to assist the Grantor in the evaluation of proposals and drafting and negotiation of the project agreement. Financial, technical and legal advisors should generally be selected on a competitive basis, and should demonstrate that they have the skills relevant for the scope of the work.

[4]One approach is for the Grantor to calculate the comparable cost for the public sector itself to construct and run the project in order to ascertain a 'public sector comparator'. This comparator can be useful as a measurement standard for assessing bids and provides another method of ascertaining value for money.

[5]Many contracting authorities have opted for a process in which all the bidders bid on the same technical specifications or service requirements and the evaluation is based solely on the financial proposal.

[6]In discussions with Mr. David Stiggers, an experienced transaction advisor in the water and wastewater sector (February 2009), he strongly recommends that the Grantor only pre-qualifies firms with a track record and reputation at stake. Firms with experience will have observable PPP Agreement behavioural patterns and therefore evidence of 'low-balling' in their past can be factored into the bid evaluation.

[7]Discussion with three international business development managers for Severn Trent Water International (Simon Langton), Thames Water International (Iain Menzies), Suez Lyonnaise des Eaux (Charles Chaumin), Bechtel International Water (Dr. Geoff Thorpe) confirmed this statement. This was also evidenced by the retraction of Azurix from the bidding process of Tallinn, Estonia where five firms were pre-qualified.

[8]Compensation for Bid Costs: Bidding for Water and Wastewater PPP Agreements is expensive. It is our suggestion to retaining at least two bidders until final selection in order to maintain public bargaining power. Public support during final stages of the procurement process can offer good value for money by keeping competing consortiums involved and committed. However, structuring competition for unsuccessful bids can be difficult as actual bid costs are often unknowable. Bidder compensation based on a percentage of proven expenditures is one solution for compensation when parties can agree to qualify expenses and open information exchange. Detractors of this method include increased complexity and reduced private incentives to control costs. Alternatively, fixed sum 'honorariums' may offer a simpler option for compensating unsuccessful bidders. Setting the honorarium amount correctly for this latter method is an obvious challenge and will require expertise from similar procurement processes.

[9]A summary resulting from the evaluation of various tender documents in Ecuador, Tanzania, Estonia, Bulgaria, Chile, Argentina amongst others, confirms that generally these include details on: Design, planning and system management, examples of previous projects of a similar nature to the bid for the water and wastewater system in which the procedures proposed have been used by the bidder, Procurement and Subcontracting Strategy, Asset management, Civil Works, Operations and Maintenance, Health, Safety, Quality and Environmental Management, Project Program and Work Structure and a list of other Critical Events and Factors.

[10]A new process called 'Competitive Dialogue' is an innovative addition to the procurement options available to contracting authorities. Pioneered by the European Union, the process allows for dialogue between the Grantor and prequalified bidders to allow them to fine tune, specify and clarify their bids.

[11]More information on 'Competitive Dialogue' is available at www.4ps.gov.uk

[12]The term 'contracting body' or 'contracting authorities' is considered the same and falls within the meaning of Directives 92/50/EEC, 93/36/EEC, 93/37/EEC and 2004/18/EC and the contracting entities of the type 'public authorities' and 'public undertakings' within the meaning of Directives 93/38/EEC and 2004/17/EC.

[13]The rules on the internal market, including rules and principles governing public contracts and concessions, apply to any economic activity, that is, any activity which consists in providing services, goods, or carrying out works in a market, even if these services, goods or works are intended to provide a 'public service', as defined by a Member State.

[14]Interpretative Communication on concessions under Community law, OJ C 121, 29 April 2000.

[15]The contract has to be for a pecuniary interest that is, a contract where a consideration in the form of a fee.

[16]See for example Case C-324/98. Telaustria Judgment, the Court stated in this respect that '[the] obligation of transparency which is imposed on the contracting authority consists in ensuring, for the benefit of any potential tenderer, a degree of advertising sufficient to enable the services market to be opened up to competition and the impartiality of procurement procedures to be reviewed'. See also ruling of 30 May 2002, Case C-358/00, *Deutsche Bibliothek,* ECR. I-4685. These principles are also applicable to other State acts entrusting an economic service to a third party, as for example the contracts excluded from the scope of the Directives owing to the fact that they have a value below the threshold values laid down in the secondary legislation (Order of the Court of 3 December 2001, Case C-59/00, *Vestergaard*, ECR. I-9505), or so-called non-priority services.

[17]It must be noted that if the model is provided for by the Grantor there is a risk that the cost relations in the model may not match those that the bidders/Developer believe exist in the business. Conversely, if the bidder/Developer provides the model there is a risk that they could game the process, predicting where data is likely to change, and creating a model which overcompensates them for such changes. Experienced transaction advisers recommend that the Grantor develop its own model and as part of the bidding process request, bidders should provide detailed commentary on both the inputs to the model and also to its architecture.

[18]It is to be recommended that care be taken to ensure that the model cannot be altered after the contract is signed.

[19]This approach was used successfully in Tallinn, Estonia where bidders were allowed to comment on the bidding documentation. It seems that this was the first time such consultative approach was widely used in a long term Water and Wastewater PPP Agreement. The authors were involved as the lead transaction adviser for such project.

[20]A good example of this was the recently flawed PPP process on the island of St Lucia where a bidder had been selected (March, 2009) and the losing bidder has launched a legal contest on this decision.

Chapter 9

Dealing with existing PPPs and re-negotiation

"A deal is a prediction ... deal makers' predictions at a negotiating table inevitably have to confront the realities of change later on."

(Salacuse, 1991)

9.1 INTRODUCTION

The nature of contracts in the Water and Wastewater sector renders them susceptible to the impact of unforeseen or unpredictable events. With contract durations of 25 years and above, there is greater exposure of the arrangement to environmental, commercial and political risks. Such events may alter the economic balance of the relationship originally assumed by each party at the signing of the arrangement and may make operation of the PPP Agreement no longer commercially and financially viable. In most cases, neither party will wish to take the drastic step of terminating the contract or withdrawing[1]. Instead, experience and practice suggest that renegotiation and adaptation clauses should be included in contracts to help address such circumstances. The previous chapter mentioned clauses related to variations but this chapter argues that a 'structured renegotiation' process should be in-built in the PPP Agreement to allow both parties to maintain the contractual relationship with all the associated benefits whilst dealing with a difficult issue which would have otherwise triggered a termination right for one of the parties. Yet these provisions come loaded with consequences for the PPP Agreement and the business relationship between the two parties. It follows that there is a difficult balance to achieve within the PPP Agreement that allows the parties to deal with the issue at hand but does not open the contract for superfluous renegotiations. It should be emphasized that care should be taken from the outset of the design of a PPP Arrangement to establish general renegotiation clauses specifically covering a change of circumstances and its impact that is serious enough to trigger a renegotiation. This will increase the chance of the survival of PPP Agreements.

9.2 WHY SHOULD RENEGOTIATION AND ADAPTATION CLAUSES BE CONSIDERED AS PART OF THE PPP AGREEMENT

It seems that there are a number of common causes for renegotiation requests. For Governments this is usually when the fiscal regime (taxes, tariffs, subsidies) are affected by external events, altering the economic balance of the arrangement; when Developers obtain a higher return from their operations due to decreases in commodity prices; or due to pressure from the imposition of technical and environmental standards in existing contracts or enactment of said standards through legislation[2].

On the other hand, for Developers renegotiations may occur when, due to unforeseen events (such as improved knowledge of asset conditions etc.), the economic viability of the PPP Agreement is untenable through obligations imposed on the PPP Agreement (environmental obligations etc.) reducing the level of returns envisaged upon embarking into the partnership; or when government policies have a detrimental effect on the minimum rate of return through reductions on tariffs or demands to subsidize costs to allow greater access.

Other issues which may influence the need to specify the renegotiation clauses within a PPP Agreement may include, for example:

Government Intervention: Renegotiation may often be necessary due to the unexpected intervention of governments in contractual relationships such as new laws and so on. In some instances changes in law may not be considered as a Grantor's failure to honour his contractual obligations. Experience shows that the Anglo-American legal approach to contracts is preferred to court arbitrated changes to contracts[3].

Environmental Risks: Contracts in the Water and Wastewater sector are also likely to face environmental risks over the duration of the arrangements. Environmental pressure groups can lobby the government for the cancellation or suspension of a PPP Agreement that is viewed as threatening to the environment. Similarly, as mentioned above, the cost of having to comply with new environmental regulations and standards (water purity levels, treatment levels for wastewater) can also bring about unforeseen costs to the operation of a PPP Agreement. Events of this nature are not unusual in emerging countries which may not have developed their environmental regulations and who still take an accident by accident approach to developing them. Continuing the economic viability of the project in the face of these new environmental costs may require renegotiation.

Change in Government: A change in government does not always preclude a renegotiation nor by itself constitute a legal basis for renegotiation. However, experience has shown that any controversial contracts or PPP Agreements entered into by previous governments are often liable to be renegotiated by the new government which may cite allegations of impropriety in the original arrangement.

Against this backdrop, explicit renegotiation clauses may provide protection to both the Grantor and the Developer by guarding a state's sovereign right to change laws that may affect the PPP Agreement as well as providing protection to the Developer. A renegotiation clause is a contract provision that, upon the occurrence of a certain event or events, requires all parties engaged in the contract to renegotiate the terms of their agreements (Gotanda, 2003). Various legal arguments (Gotanda, 2003; Waelde & Kola, 2004; World Bank, 2004) recommend that contracts should not attempt to use stabilization mandates against governments changing laws that would disrupt the financial balance of the contract. Instead, Grantors should retain this right under the proviso that any changes will automatically give the Developer the right to renegotiate or adapt the contract in order to re-balance it.

Waelde and Kolo (2004), identify five main drawbacks with including renegotiation provisions in a contract:

- *They reduce contract stability*: Contracts and partnerships rely upon predictability and enforceability. Adaptation and renegotiation clauses can breed scepticism in the viability of the arrangement from the outset[4]. For example, US Courts system (other legal systems discussed later in this chapter) are notoriously reluctant to include compelled renegotiation and adaptation clauses as these are not used regularly in common law countries as they are in civil law countries.
- *They may raise the costs of transactions*: These provisions can create uncertainty of economic return which can either dissuade a Developer from entering into a PPP Agreement, or will force them to 'structure the investment in such a way as to increase returns to offset the risk created by the environment' (Salacuse, 2000)[5].
- *Difficulty in engaging a third-party to resolve disputes*: if a renegotiation clause is included and activated and each party is unable to resolve their issues satisfactorily, some legal arguments suggest that if no agreement can be reached, there is no dispute and no breach of contract (Gaillard & Savage, 1999)[6]. Without a dispute, an arbitral tribunal may not have jurisdiction and any decisions made may not be enforceable. Gotanda (2003) explores the situation when a contract is adapted on the request of an arbitral tribunal which may not necessarily constitute a disagreement. This is important as the recognition of a dispute is a prerequisite for arbitration under the ICSID Convention[7] and the UNCITRAL[8] Model Law which are the basis for many national arbitration laws[9]. The uncertainty over this issue often dissuades parties from inserting renegotiation provisions into the contract.
- *Lack of guidance for tribunal contract adaptation*: Without specific instructions on how the tribunal should go about adapting the PPP Agreement, the tribunal may be averse to involving itself in adapting a

contract. If tribunals proceed to adapt a PPP Agreement without specific guidelines, the final agreement may contain provisions which neither party would have agreed upon prior to engaging the tribunal and may produce a profoundly different arrangement than desired.

- *Events that trigger renegotiation may be within the control of government*: A major concern resides in whether it is sensible to allow renegotiation and adaptation clauses to be triggered by events within the control of the Grantor. If the contract is renegotiated favourably towards the Grantor, the government has gained an advantage and therefore profited from having the clause invoked. Obviously the best outcome for the Developer would be that renegotiation would place the arrangement back to the same economic balance which it enjoyed prior to renegotiation.

In contrast to the identified disadvantages, the provision of renegotiation clauses with attached clauses for third party arbitration to adapt the PPP Agreement in the event of unsuccessful renegotiation, has many advantages according to Gotanda (2003, pp. 69–71):

- *Stabilize the relationship between government and Developer*: Clauses allowing renegotiation can reduce the likelihood to terminate a relationship. Flexibility in a PPP Agreement which allows renegotiation of certain terms under certain circumstances can lessen the risk of disputes that can permanently sour a relationship.
- *Allowance for flexibility*: Any Developer, prior to committing to a long term contract, will need assurance of the stability of the investment regime, that the PPP Agreement will be adhered to and respected by the Grantor and the rules of the game will not be altered[10]. Furthermore, the Developer's sponsors (banks, insurance agencies and customers) will also require this assurance. However, the Grantor that owns the resources of the water and wastewater sector often enters into a long term arrangement for the management of these resources with imperfect data about the condition of the assets (See Chapter 11), resources available and the future price that it can command. It will therefore require flexibility in the PPP Agreement to deal with changes in circumstances[11].
- *Contract allowance to deal with unforeseen or 'force majeure' related events*: Provisions may prove more palatable to each party should the clauses be restricted to events that are beyond the control of either party. Provisions for 'force majeure' events are highly common in contracts and are often seen as facilitators for the contractual relationship rather than increasing uncertainty[12].
- *Maintain reputations*: The failure of a major PPP Agreement damages the reputation of the Grantor and the participating Developer. The political risk profile and investment climate perceptions of a country may be negatively impacted if there is a failure in a PPP Agreement. Similarly, the Developer

may face criticism from his peers and it could affect his ability to raise money in the investment community. Failed projects have serious negative repercussions and renegotiations may be a way to mitigate these repercussions. The possibility of tarnishing of a government or Developer's reputation and credibility can act as a strong incentive against violating the rules of the game.

9.3 LEGAL ISSUES RELATED TO RENEGOTIATIONS

Renegotiation of contracts including PPP Agreements opens up a whole series of legal debates on the sanctity of contracts versus flexibility of contracts. According to Waelde and Kolo (2004), this also reflects the tension between the binding nature of the legal instrument of a contract – designed to commit the parties for a long time whatever their changing position, relation and interests, and the practical and commercial difficulties to being fully and specifically committed to certain obligations and courses of action when the reasons underlying the contract in the first place have changed drastically. As a general principle of contract law, the concept of the sanctity of the contract is widely accepted by all modern legal systems and is upheld by international tribunals. But under no municipal or international system has the principle of sanctity been found to be absolute. Some believe that classic contract principles over discrete transactions are not applicable in long-term 'relational' contracts of long durations. Instead, a relational contract is a framework for co-operation between parties[13].

The provision of renegotiation has many advantages (Gotanda, 2003) as described above. Importantly, including such renegotiation clauses with associated clauses for third party arbitration to adapt the PPP Agreement in the event of unsuccessful renegotiation is key to being acceptable to both the Grantor and the Developer for these types of clauses. In other words, agreeing to include renegotiation clauses within the PPP Agreement does not dilute or preclude each party's right to arbitration.

Applicable Law – Common and Civil Law systems[14]

Renegotiations occur under the law applicable under which the original contract was signed. If each party chooses national law to govern the PPP Agreement, national law will determine the manner in which renegotiations may occur. Often in financial agreements, parties arrange to use of laws of another state, in which case, it will be the law of the other state which will determine the applicability of renegotiations. If international law only is chosen, certain difficulties may arise as international law governing contracts is not very well defined. If negotiations over adaptation requests are not settled and are referred to a contract's dispute resolution mechanism – an arbitrary tribunal – the tribunal arbiters will decide which law they consider applicable. In situations where parties, unable to agree on a system of law, choose a hybrid of national law, international law and generally recognized principles of law, arbitrators will attempt to calculate how each system treats the issue and try to identify any commonalities. Generally, arbitrators have a natural tendency to select principles which reflect international consensus and will reject principles of national law that counter generally accepted international practice. The lesson to be drawn from this is that both parties, when negotiating renegotiation clauses, should consider how a hypothetical tribunal would deal with a renegotiation and what the applicable law would be.

Common Law Systems[15], generally identified with English law, places particular focus on detailed legal drafting and regulation by contract and as a result display less flexibility for contract renegotiation. The sanctity of the contract is regarded as paramount. If renegotiations are to be considered, detailed provisions should be incorporated into the contract through escape/hardship/indexation and adaption clauses. Judges operating under common law systems generally choose not to intervene in contracts and uphold the validity of existing agreements. American courts tend to allow a narrow degree of flexibility, permitting contractors to withdraw from onerous contracts if changes directly affecting the contract's economic balance occur or if the changes were unexpected and if holding the contractor to the agreement could lead to damaging consequences for the contractor[16].

Civil Law systems are much more amenable to the renegotiation of long term contracts than Common law systems in the English tradition. Civil law systems view contracts as valid so long as the conditions that upheld the conclusion of a contract continued to exist (rebus sic stantibus- from the Latin: 'things standing thus' or 'things thus standing')[17]. While civil law systems do not provide simple exit strategies from a contract, lawyers can draw upon past cases and legal decisions to support their arguments for renegotiation and adaptation of contracts. The civil law systems in many countries are shaped by their national political and economic experiences[18] and recognize a restricted 'rebus sic stantibus' that allows for termination and renegotiation and empowers courts to enforce contractual adaptations[19].

Where a contract can still be operated, albeit under onerous conditions for one party due to severe changes in the assumptions underpinning the contract, civil and common law systems demonstrate a divided judgment on whether to permit renegotiation and whether adaptation is justifiable. Renegotiation is only likely to be agreed upon where there is a long term PPP Agreement between parties who have to continue performing under drastically changed and excessively onerous new conditions.

9.4 INTERNATIONAL BUSINESS PRACTICES FOR RENEGOTIATIONS

As a result of the grey areas between national and international law approaches to renegotiation and the specific mechanisms for contract adaptation, it has been seen that international investment business practices have developed templates for contracts that balance the binding nature of long-term contracts, with the necessary adjustment mechanisms required to maintain the economic equilibrium of a project throughout the PPP Agreement's duration. These templates have developed over the course of project finance transactions in the Water and Wastewater Sector, learning from the successes and perhaps most importantly from the failures. These templates seek to address the typical risks that Grantor-Developer contracts face – technical, commercial, regulatory, environmental and political risks. Furthermore, there are other techniques being developed (such as Comparative law described below) which can be used effectively to deal with variations and renegotiations. The international business community regards some form of PPP Agreement adaptation provision(s) as critical to ensuring the long-term health of the business relationship and the economic viability of the PPP Agreement. International investment/trade institutions and associations have recognized the need for clearly defined mechanisms for contract adaptation through the use of special rules provided by facilities such as the World Bank's Centre for the Settlement of Investment Disputes[20] or the Contract Adaptation Rules of the International Chamber of Commerce[21] or the UNCITRAL Conciliation Rules[22]. The ICC and UNCITRAL both advocate renegotiation proceedings should there be any drastic changes to the assumptions on which the PPP arrangement was originally signed.

The UNDROIT Principles of International Commercial Contracts provide guidance on conducting renegotiations and the conditions under which they can occur. The Principles allow for renegotiation in the event of a 'hardship' which is defined as:

> *The occurrence of events (that) fundamentally alter the equilibrium of the contract either because the cost of a party's performance has increased or because the value of the performance a party receives has diminished, and (a) the events occur or become known to the disadvantaged party after the conclusion of the contract; (b) the events could not reasonably have been taken into account by the disadvantaged party at the time of the conclusion of the contract; (c) the events are beyond the control of the disadvantaged party; and (d) the risk of events was not assumed by the disadvantaged party.*

In the event of a hardship, the disadvantaged party can invoke a renegotiation of the terms of the contract in order to modify it to reflect the altered parameters. If a successful renegotiation does not occur, the UNIDROIT principles call for the use of a third party to end the contract or adapt it to restore its balance. Both the government and the Developer can alter the contract to reflect unavoidable changed circumstances. Some courts have argued that the principle of fairness should be used to adapt the PPP Agreement. The UNIDROIT principles provide tribunals with guidance on this[23].

Other international commercial transactions have developed a similar concept of 'hardship' clauses which provide a party with the opportunity to request an adaptation of a contract's provisions if continued operation under those contract terms would be economically non-viable. Such hardships could include acts of god or economic (fiscal) problems. The main premise behind such provisions is that no party should be forced to continue operating under such circumstances, as this would compromise the long-term benefit of the contractual relationship[24].

Akin to the hardships clause, and as discussed in Chapter 9, 'force majeure' differs in that it may cause the termination or suspension of the contract. However, according to Yates (1999) when a 'force majeure' event occurs, the government and the Developer should rather 'seek to maintain the contractual relationship but to implement the term in a modified term, adapted to new circumstances' (Yates, 1990). Effective provisions for hardship and 'force majeure' clauses should not only stipulate a moral obligation for both parties to negotiate in good faith, but also to provide mechanisms for third party resolution of a dispute, should negotiations in good faith fail. Such clauses are recommended for long-term projects where unexpected events can mean that carrying out contractual obligations become onerous and economically non-viable.

Finally, *Comparative law* is a technique for comparing how major legal systems handle common subjects[25]. In the case of renegotiations, using a comparative law approach can assist each of the parties' negotiators with information on common triggers for that can lead to renegotiation across different legal systems[26]. When drawing up agreements, if neither party can agree on a particular legal system to govern the PPP Agreement, the parties may agree on certain common legal practices among major legal systems which can be used to govern the contract.

9.5 SUMMARY AND CONCLUSIONS

International commercial business practices in contracting, regard provisions for renegotiation and adaptation as an organic element in business contracts, particularly in long term arrangements. I fully concur with this principle. It seems that this principle stems from the main premise that parties to an agreement should not be forced to comply with unexpectedly tough terms in a PPP Agreement simply in the interests of maintaining the long-term business relationship[27]. Each party should be expected to absorb the short term costs of the unexpected event until renegotiation or a PPP Agreement adaptation has occurred. Furthermore, there are new techniques (such as Comparative law described) which can be used effectively to deal with variations and renegotiations.

It would also be advisable that renegotiation of an agreement should only take place in exceptional circumstances and that parties to a contract should bear their allocated risks, except in the event of unforeseen occurrences that affect the assumptions on which the agreement was made originally[28]. This is particularly important, since to allow a party to withdraw or not comply with contractual obligations creates uncertainty and can affect the long-term investment climate. Furthermore, legal assessment suggests that generally accepted legal principles do recognize (albeit it narrowly) the right of a party to seek renegotiation or termination of his contractual commitments if, through no fault of its own, his participation in a contract has become untenable as a result of disruption to the main assumptions that underpinned the original contractual agreement, In response to the grey area of renegotiation within the auspices of international law, international business has developed a proprietary set of rules to handle contractual renegotiation and adaptation[29]. Even in the absence of structured mechanisms, international business relies on general principles and use of mutual accommodation to reach a satisfactory renegotiation. Accordingly, renegotiation should be conducted in good-faith and with the knowledge that hopefully, each party will benefit from the re-balancing of the agreement. It is to be recommended that the party acceding to the other's request for renegotiations should receive some form of concession to recognize his flexibility so as to demonstrate that some kind of advantage from the adaptation of the contract was gained. Moreover, if renegotiation clauses are to be inserted, consideration must be given to the scope of events that trigger renegotiation – whether the events be unforeseen and beyond control or whether the applicable laws recognize the interference of an arbitrator or third party to adapt a contract should renegotiations fail; and to provide specific criteria to guide the arbitrator in acceptable adaptation procedures.

ENDNOTES

[1]Exercising a right to termination is not the issue. Experience shows that in most cases the Developer may have a right to termination but is loath to exercise that right as it would mean the end of the PPP Agreement. This was the case in Buenos Aires, Argentina where the clause linking the convertibility of the peso at parity with the dollar was breached by government's actions. The Developer clearly had the right to termination but did not wish to exercise such right. Similarly, my personal experience in Ecuador Guayaquil where Dr Mandri-Perrott acted as a mediator between the Grantor and the Developer Interagua, showed the reluctance of the Developer to exercise his termination rights. In the words of the Director for Regulatory Affairs of Interagua Mr Geoff Thorpe, 'We know we can exercise our right to termination, but we also know that would lead to our divorce. In that event we will only recover our investments in the courts and our Board [referring to the Board of Directors and shareholders of Interagua] will not get their money back in many years'.

[2]This was the case for example in Eastern Europe where allowances had to be made in contracts to ensure that there was compliance with EU standards such as the Water framework Directive.

[3]The critical factor demonstrated by decisions in the United States Supreme Court is that regulatory changes may not always be regarded as a failure to honour a contractual obligation. English courts (mainly related to England, Wales and Northern Ireland but not Scotland) also remain reluctant to allow parties of commercial agreements to alter contractual arrangements even in situations of unexpected changes in industry or sector law.

[4]The implication being that if a PPP Agreement needs a renegotiation clause from the beginning it would imply that such arrangement is incomplete or would require a renegotiation from the beginning.

[5]Transaction costs may increase as the cost of financing the PPP Agreement will increase as investors and or the Developer would seek assurances against uncertainty by demanding a higher return on its investment. Another consequence could be that the Grantor might lose foreign investment.

[6]Gotanda refers readers to Fouchard *et al.* (1999).

[7]The International Centre for Settlement of Investment Disputes (ICSID), an institution of the World Bank group based in Washington, D.C., was established in 1966 pursuant to the Convention on the Settlement of Investment Disputes between States and Nationals of Other States (the ICSID Convention or Washington Convention). As of May 2005, 155 countries had signed the ICSID Convention. ICSID Regulations and Rules have been amended and these revised rules are in effect from April 10, 2006.

[8]The United Nations Commission on International Trade Law (UNCITRAL) was established by the United Nations General Assembly in its Resolution 2205 (XXI) of 17 December 1966 'to promote the progressive harmonization and unification of international trade law'. It is the Commission that formulates and regulates international trade in cooperation with the World Trade Organisation.

[9]If national laws do not recognize a tribunal to adapt the terms of a PPP Agreement in dispute, the tribunal will not have the jurisdiction to adapt the agreement/contract. If the tribunal asserts its jurisdiction without authorization, the contract may be unenforceable under the New York convention.

[10]By rules we mean, issues to do with rights, titles, currency regulations, labour and environmental laws.

[11]Take the example of a Grantor that is unsure and unable to obtain information about the condition of the underground water assets. It then prices the cost of refurbishments and repairs into the subsidy and tariff levels in the PPP agreement with a Developer. If the Developer discovers that the underground assets do not require the level of repairs predicted in the PPP Agreement, the Grantor may feel that the tariff and subsidy levels are overly generous and may demand a readjustment of the PPP Agreement. On the other hand, the reverse may be true and the Developer may require a greater subsidy.

[12]It must be noted however that this may not erase the issue of whether a tribunal's adaptation of a contract is viewed, as discussed above, as a dispute.

[13]A number of French, Spanish and British operators were interviewed and it was interesting to note that generally the French companies (Vivendi, Sour and Suez) all regarded long term PPP Agreements as contracts were the relationship between the parties was crucial – thus 'relational' contracts. This may be related to the fact that their original water contracts were completed in Napoleonic times.

[14]Consultations and discussions were had with the international legal firm of Allen and Overy, specifically with Mssrs Troy Edwards, Nigel Pritchard and John Scriven (various consultations between February 2008 and May 2009), with regard to renegotiations and

the implications of the various legal systems with respect to this. Some of these views are included in this chapter based on what is considered international best practice.

[15]For a more detailed description on the general differences between Civil and Common law legal systems see Chapter 5, Legal framework.

[16]During the research for this book it was evident that United States law remains very strict and narrow with regards to renegotiation.

[17]The concept of *rebus sic stantibus* stipulates that, where there has been a fundamental change of circumstances, a party may withdraw from or terminate the agreement in question. In public international law, clausula *rebus sic stantibus* is the legal doctrine allowing for treaties to become inapplicable because of a fundamental change of circumstances. It is essentially an 'escape clause' that makes an exception to the general rule of *pacta sunt servanda* (promises must be kept). Clausula *rebus sic stantibus* only relates to changed circumstances that were never contemplated by the parties. Because the doctrine poses a risk to the security of treaties as its scope is relatively unconfined, it requires strict regulations as to the conditions in which it may be invoked. Further details can be found in Vienna Convention on the Law of Treaties (Vienna, 23 May 1969. Entered into force on 27 January 1980. United Nations, Treaty Series, vol. 1155, p. 331)

[18]For example, Germany allows the termination of judicial adaptation of PPP Agreements and other contracts in the event of severe changes to the main parameters governing the economic balance of a contract which has made the contract valueless for the affected party. French law allows contractual escape should the foundation of a PPP Agreement be affected by an unforeseen event which severely disrupts the contract's benefit allocation that is that the economic equilibrium has been breached to such an extent that it is no longer economically beneficial to one of the parties to continue with such contractual arrangement.

[19]It must be noted that restrictions do apply in many countries which uphold the '*rebus sic stantibus'* principle and it is not easy for parties to escape their contractual obligations.

[20]http://icsid.worldbank.org/ICSID

[21]http://www.iccwbo.org/

[22]http://www.uncitral.org/

[23]*Article* 6.2(3), provides for 'In (adapting the PPP agreement) the court will seek to make a fair distribution of the losses between the parties. This may or may not, depending on the nature of the hardship, involve price adaptation. However, if it does, the adaptation will not necessarily reflect in full the loss entailed by the change in circumstances, since the court will for instance, have to consider the extent to which one of the parties has taken a risk and the extent to which the party entitled to receive performance may still benefit from that performance'.

[24]Experience from the oil, gas and mineral sectors show that it is feasible to have PPP agreements with formal renegotiation clauses that allow each party to invoke negotiations to restore the economic balance of a contract. Additionally, stabilization clauses are increasingly common. These clauses are invoked when government interventions such as change in law, change in taxation regime and so on affect the

balance of the PPP agreement. Furthermore, research done for this book suggests that certain petroleum contracts contain highly specific renegotiation and contract adaptation mechanisms that deal with specific issues (conclusion reached after discussions with Mr Christopher Sheldon Principal Energy, Oil and Mining Expert of the International Finance Corporation and World Bank June 2011, as well as review of some gas contracts in Latin America (details withheld for confidentiality reasons). Furthermore, discussions were also had with Allen and Overy Mr Troy Edwards. It is the authors recommendation that such practice can be replicated in the water and wastewater sector, should either party have particular concerns over a specific risk.

[25]Wikipedia defines Comparative law as an academic study of separate legal systems, each one analysed in its constitutive elements; how they differ in the different legal systems, and how their elements combine into a system. Several disciplines have developed as separate branches of comparative law, including comparative constitutional law, comparative administrative law, comparative civil law (in the sense of the law of torts, contracts and obligations), comparative commercial law (in the sense of business organisations and trade), and comparative criminal law. Studies of these specific areas may be viewed as micro- or macro-comparative legal analysis, that is detailed comparisons of two countries, or broad-ranging studies of several countries. Comparative civil law studies, for instance, show how the law of private relations is organised, interpreted and used in different systems or countries.

[26]Generally recognized legal practices may include a given approach to compensation, use of independent (experts) in the determination of tariff reviews and so on.

[27]Fiscal volatility and political upheavals are not as unexpected as they once were. The international business community has developed means by which to address these issues and so therefore should governments. In this respect international business practice is to place more importance on maintaining the relationship is more important than holding a party hostage to the original contract terms.

[28]Allowing post hoc renegotiations every time there is a change in circumstances could reduce substantially the use of the PPP agreement as a risk allocation tool. In other words the PPP agreement sets out the overall risk allocation that the parties are prepared to absorb and the allowance of renegotiation events should be carefully circumscribed.

[29]Experience from negotiated agreements from different countries and private independent consultants can assist governments and Developers in designing appropriate clauses.

Part III

Developing Sustainable PPP Arrangements

Chapter 10

Improving key contract provisions

10.1 INTRODUCTION

This chapter considers some of the major issues in formulating the contractual arrangements for a Water and Wastewater PPP scheme. By their nature PPP arrangements are complex, and each arrangement has its own set of interlinking contracts and agreements needed to establish allocation of risks and responsibilities, as well as to deal with financial obligations and cash flow. Depending on the risk allocation of each type of contract, as discussed in Chapter 3, Section 3.3 each contract will allocate risk and responsibilities differently between the parties. Arguably, the most complex of these contractual arrangements is a *long term concession* type contract. For this reason, a PPP concession structure is used to illustrate some of the main contractual issues to be considered. Some of the associated contracts that would typically be used by the Grantor are also shown as a means of sharing the risk among the project participants.

This chapter is designed to outline the key issues to be considered when designing and applying a new water PPP Agreement. The majority of the new water PPP projects reviewed have involved major project investment, mainly through Private investment commitment. It should be noted that many of the points covered here (i.e. those not specifically related to investment) will also apply to the other non-investment contract forms, such as a management contract. The provisions in this section will also apply to contracts which include delegated management to the private sector but with no (or little private investment) such as lease/affirmage type contracts. This chapter also assesses the basic need for subsidiary agreements to the main PPP Agreement, specifically for an appreciation of the potential complexity of a complete contract structure and interlinking of the various contracts, land lease, water abstraction license (as appropriate), direct agreements of lenders and technical annexes involved.

For ease of reference we have divided the recommended provisions in the following way:

- Provisions during Development, mainly aimed at indicating the role to be established of the Grantor after signature of agreement and prior to service being effective.
- Provisions related to Construction and Assets: this, whilst not the main focus of this type of PPP Agreement, details the General Construction requirement conditions and the relationship between Contractors and Sub-contractors. It also includes valuation and calculation of terminal payments of assets at the end of the PPP Agreement period.
- Provisions related to Service Obligations: this is the main section and great attention is paid to making recommendations as to what has worked and what could be improved in the future.
- Provisions Applying to Financing to include protections for lenders through drawdown schedules, construction milestones and reserve accounts and so on. including security rights to ensure lenders are protected from unsecured or junior creditor actions.

10.2 CONTRACT FORM AND RISK ALLOCATION

Contracts are important tools, regardless of the PPP type chosen or, the amount of government support, or the Developer is privately owned, or by the Grantor or another public body. Furthermore, contracts are required whether or not the project is subject to a competitive procurement process. They can define a wide variety of objectives: economic, financial, social, and environmental, not only on behalf of the Grantor but also of other local, regional, or national stakeholders. They help to give clarity and an objective basis for the development of the Water and Wastewater sector. Consequently, the use of PPP Agreements is often promoted by external regional, national or supra-national agencies to enshrine policy objectives and to implement specific policy actions.

Contracts have been used to change the relationship between Authorities and local publicly-owned service providers, either by replacing these providers with new Developers or to distance Authorities from the local publicly-owned service providers and to make the providers clearly accountable for quality-of-service.

Each PPP contractual structure has to be tailored to suit the specific PPP scheme. We will use a typical long term Concession Contract to illustrate some of the key issues to be considered when establishing a PPP arrangement. In this type of contract, the Public Authority contracts an outside organization to construct, rehabilitate, operate and, simultaneously provide services to specified quality-of-service standards at a pre-defined tariff. Consequently, the Developer must, provide the required inventory of fixed and moveable assets from internal resources or via external financing. The Developer will also retain all revenues and will absorb either all, or a contractually agreed portion of, traffic/revenue risks as well as the risks associated with construction. Risk-sharing may also extend into areas of regulatory risk.

Any contract allocates obligations and risks to the parties and consequently sets out the financial linkages between the parties. These arrangements are complex. One way to demonstrate this complexity (and a useful tool in the analysis or development of a full contract arrangement) is to show the linkage of the different parties in two ways:

(1) The contractual instruments linking the parties.
(2) The cash flows (or potential cash flows) linking the parties.

The following diagram illustrates the necessary contractual and financial linkages between the parties to a PPP Agreement. When designing a PPP arrangement it is useful to start by defining key linkages in this way. This approach can be used to establish agreement on broad principles of the contract structure, before moving on to detailed contractual design. The first diagram establishes all the main contractual linkages, each of which requires its own contract document or agreement, and in the second diagram we note some of the main cash flows which are also dealt with contractually.

Contractual linkages

The diagram Figure 10.1 shows the typical contractual relationships within a PPP Agreement. The Grantor grants a right to a Developer to operate, maintain, construct, rehabilitate and charge for Water and Wastewater services. The terms of the PPP Agreement detail the rights and obligations of the parties. The PPP Agreement allows the Grantor to allocate project risk to the Developer and to define the risk sharing among the project participants. The PPP Agreement may also set out the legal and tax regimes applicable to the project, including the Developer's environmental obligations.

In defining the PPP Agreement it will be necessary for the Grantor to be clear on a number of matters including, in particular:

- The overall risk allocation that is which party assumes, or is responsible for managing associated risks. Annex 4 provides a risk allocation matrix which could be used as an aid to determine the optimum risk allocation.
- How is it anticipated that the Developer will raise the revenues necessary to fund a significant capital expenditure and, as a related issue, how much and what type of financial support will be available from the Grantor.

For this reason, as part of the PPP agreement, it is not uncommon for the Grantor to issue subsidiary documents defining and describing in some detail the other contractual relationships considered crucial for the integrity of the PPP Agreement.

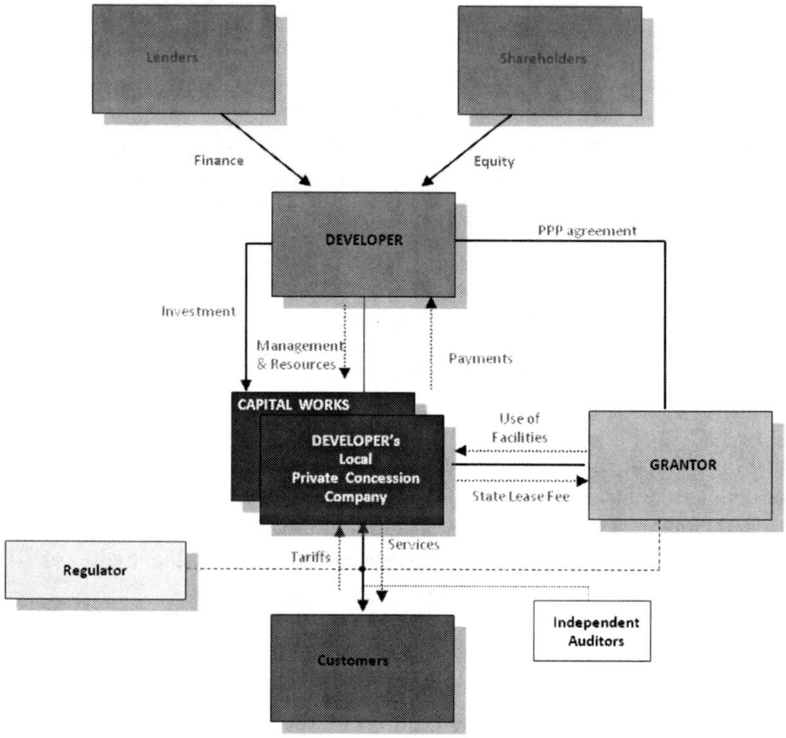

Figure 10.1 Contractual linkages.

The Grantor will also seek to ensure that the Developer performs against a minimum set of standards and overall requirements. The Grantor's requirements will cover issues such as coverage, quantity and quality of service provision. Below is a summary of the main issues the PPP Agreement will seek to cover which include (adapted from Delmon, 2009):

- *Completion date*: The Grantor's need for the infrastructure in question is generally immediate (often as much for political as practical reasons).
- *Performance of the project*: The Grantor's requirements will cover issues such as input consumption, efficiency of operation, maintenance needs and costs, life-cycle, health, safety, environmental, quality/quantity of the output/service generated and cost of operation.
- *Maintenance regime*: The Grantor will want to ensure that the maintenance and rehabilitation regime implemented is adequate and sufficient. This is even more important late in the project as the incentive for the Developer to invest funds during the final phase of the concession period may be diminished, owing to the imminent transfer of the project to the Grantor.

- *Construction and operation*: The Grantor will want to ensure that the Developer's construction and operation activities meet certain minimum standards, both those imposed by law and those specified by the Grantor to ensure the quality of the services provided and the protection of the public.
- *Government guarantees*: The government may provide guarantees for public sector bodies taking part in the project whose credit risk is otherwise insufficient.
- *Exclusivity*: The Grantor will need to provide the necessary assurances that the developer can abstract, discharge and have the authority to serve the customer base. As previously noted, the Water and Wastewater sector tends to be a natural monopoly and therefore these conditions will need to be confirmed in the PPP Agreement.
- *Know-how transfer*: The Grantor may want to maximize the interaction between the Developer and local partners or the Grantor's personnel in order to ensure the proper transfer of know-how[1].
- *Government interference*: Notwithstanding the regulatory provisions and authority under the PPP Agreement, the Grantor may guarantee to protect the Developer from a specific sub-set of political risk or that the host government will not, for example, act against the interests of lenders, shareholders, the Developer, the performance of the Developer's obligations or the project itself.
- *Concession fees*: The Developer may be required to pay concession fees for the privilege of obtaining a concession and to offset Grantor costs, payable before commencement and possibly periodically during the concession period.
- *Restrictions on share transfers*: The Grantor may want to place restrictions on the transfer or change of Developer shareholding. The Grantor may want to disallow any transfer (direct or indirect) until a certain point in time after completion of construction (a lock-up period)[2], a right of approval over the identity of any transferee, and/or to maintain some guarantee from the original shareholders.
- *Grantor step-in/continuous operation*: The Grantor may want the right to continue operation of the project when it terminates the concession agreement, sometimes referred to as the 'right to continuous operation', to ensure continuous delivery of services.
- *Hand-back*: At the end of the concession period, the Grantor will either put the project out for re-tender or it will require the Developer to transfer the project assets to the Grantor or to a replacement Developer.

Financial links

When allocating risks and responsibilities in any contractual arrangement, it necessarily involves some flow of money. A diagram similar to the previous one, is presented Figure 10.2 and illustrates financial links between the parties. Having a diagram like this is an extremely useful tool to obtain an agreement on key financial issues. This is crucial for lenders and shareholders who need assurance that their rights and obligations are protected within the PPP Agreement.

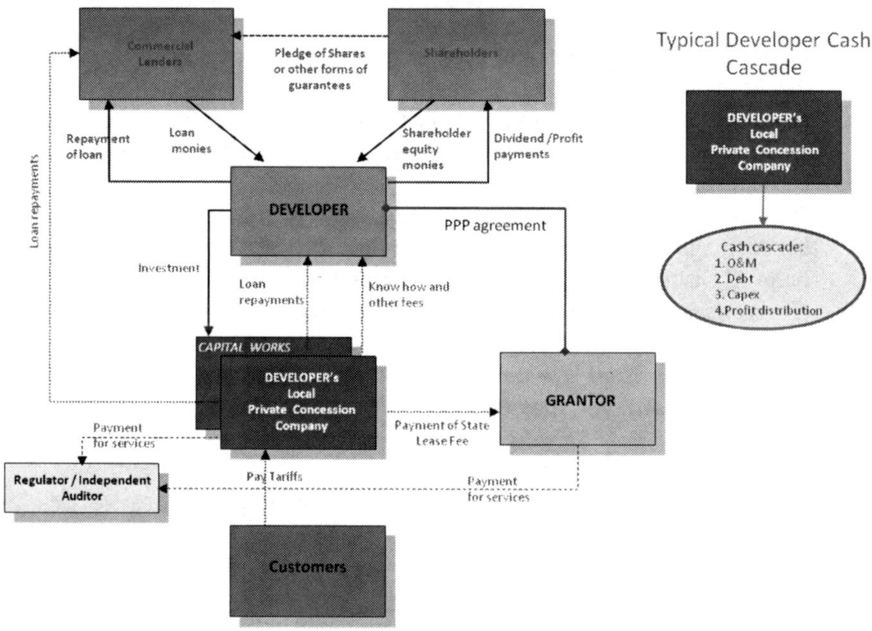

Figure 10.2 Typical financial linkages.

For complex contractual structures, such as for PPP contractual arrangements, the Term Sheet approach may help the Grantor ensure that key issues are systematically and comprehensively established and agreed upon, before detailed contract drafting is carried out. As a preparatory step, the Grantor or his advisors prepare a Term Sheet for each contract or agreement, briefly summarizing the key issues to be covered in the document. The main advantages of this approach are that issues and contractual principles are sorted out at an early stage, and the detailed drafting of the PPP Agreement and other related arrangements are carried out with more certainty, thus reducing the need for later detailed revision. This approach is also beneficial for establishing supporting documents, such as the operational and technical annexes.

Analysis of existing concession contracts has suggested a draft 'Heads of Agreement' as shown in Table 10.1. This table also shows which of the provisions in the 'Heads of Agreement' discussed later in this report.

Table 10.1 Typical heads of agreement in a typical PPP agreement*.

Section in heads of agreement	Where discussed in this chapter
1. Definitions	Not discussed
2. General Provisions	Section 10.3
3. Conditions precedent	Section 10.3
4. Grant of concession	Partially discussed in Section 10.3
5. Duration of concession	Partially discussed in Section 10.3
6. Transitional arrangements	Partially discussed in Section 10.3
7. Representations and warranties	Not discussed
8. Service obligations	Section 10.3
9. Other obligations	
a. Labour issues	Section 10.11
b. Taxes	Not discussed
c. Concession fee	Not discussed
d. Asset use and transfer	Section 10.10
e. Disconnections	Not discussed
f. Insurance	Not discussed
g. Compliance with laws	Section 10.4
h. Performance bond	Section 10.4
i. Procurement	Not discussed
j. Minimum equity	See section 10.2 and chapter on funding and finance
k. Developer debt	See chapter on funding and finance
l. Existing projects	Not discussed
m. Technical submission	Partially discussed under Section 10.3
n. New assets	Section 10.8
10. Tariffs and indexation	Section 10.6
11. Periodic and extraordinary tariff adjustments	Section 10.7
12. Reporting requirements	Section 10.5
13. Monitoring and enforcement	Section 10.5
14. Variations	Section 10.10
15. Termination provisions	Partially covered in Chapter 12
16. Dispute resolution	Section 10.9
17. Contract management mechanisms	Not covered Section 10.10

*A number of international Water and Wastewater operators were contacted to discuss the issue of making performance information available to the public. It is interesting to note that there was consensus that information pertaining to performance of the PPP Agreement should be made available to the public. There was an element of discomfort for full disclosure of financial information. The argument being that Developers did not want to 'open their books' to other operations as this was considered sensitive commercial information that could be used by competitors or even Grantor's for purposes other than that of monitoring.

10.3 GENERAL PROVISIONS

Under the general provisions of the PPP Agreement the status of a Water and Wastewater PPP project and the legal basis on which it will be established must be clearly defined[3]. The contract starts by stating the details of the parties to the contract, including:

- Rights and obligations of each party to the PPP Agreement that is the Developer and the Grantor.
- Term of the Agreement sets out the date that the PPP Agreement becomes effective, and defines the term of the contract (e.g. 30 years) with a mechanism to agree on any change in duration.
- Start of the PPP Agreement or commencement date in order to distinguish the time (if any) from the signing of the PPP Agreement to the moment in which the services or actions detailed under the PPP Agreement will start in earnest that is that the PPP Agreement becomes effective[4].
- Transfer of Rights and Obligations, Authorized Body and Mandatory Requirements to the Developer. Although the Developer will generally be required to work within local laws, norms and standards, this section details the specific covenants made by the Developer on his status and ability to carry out the contract, and all necessary legal and contractual arrangements carried out and necessary approvals obtained for him to enter the contract.
- Compliance by the Developer and any subcontractors with the lawful demands by public authorities.

Under a concession type of PPP scheme, the assumption is that the design, construction, integration, testing, commissioning, operation, maintenance and ultimate performance of any asset developed by the Developer should be the Developer's responsibility. Accordingly, as a general principle, if the design meets the Grantor's requirements it should be the Developer's risk.

The Grantor's role after the signature of the PPP Agreement and prior to the services being effective should be clearly specified to include at least:

- Review, and as appropriate, comment on the Developer's plans for achieving the level of service outputs and, as appropriate, any specific capital program,
- Defines right of access to and use of the water resources, land and so on and Grantor support to the Developer in obtaining necessary permits and licenses.

Conditions precedent, effective date and financing agreements

Conditions Precedent (CP) are in essence a summary of the issues that need to be completed prior to the PPP Agreement becoming effective on the Effective Date. The Effective date is when the CPs are met. Part of those CPs would be the signature of the financing arrangements. These are the Developer's (including any associated companies) financing agreements for the project. CPs are often checklists of tasks or documents needed and depending on the legal regime of the PPP Agreement they will be part of the agreement itself.

Parent company guarantees and minimum equity requirements

It has been usual to require parent companies to guarantee the performance of the company entering the PPP Agreement, or that the Developer has at least a specified minimum level of equity. Minimum equity, performance bonds and (limited sum) parent company guarantees are varying tools to achieve the same objective. It is good practice to ensure that the concession company has enough equity or parent company guarantees to remain viable even if things do not go as expected (Ehrhardt & Irwin, 2004). The Grantor should determine the total sum it expects to be placed at risk, and then allow the Developer flexibility to use whichever of these mechanisms, so long as the total amount at risk is at least equal to the specified minimum.

Provisions related to construction and assets

Conditions related to the construction should stipulate such issues as conditions precedent to commence construction, the timetable and construction plan to which the Developer is committed (including specific milestones as appropriate), as well as the General Construction requirements and the Contractor and Sub-Contractors. It would be necessary to detail, as appropriate, the nominated sub-contractors.

Generally under along term PPP Agreement the spirit of the contract should allow flexibility to the Developer to design and implement his own construction program to meet the pre-agreed levels of service. Notwithstanding this, it is important that the PPP Agreement should include a procedure for:

- The Developer to inform the Grantor of his capital investment program regularly and any other relevant information.
- The Grantor should retain a certain amount of flexibility to make minor modifications that do not have an impact on the Developer's proposed cost or service.
- The maximum allowable time for any further reasonable comments made by the Grantor (if any).

It must be noted however, that the Grantor should be wary of taking on too great an overseeing role. The Developer should be granted the freedom to manage the project without interference from the Grantor. Issues related to compliance with Health and Safety standards including documentation and procedural requirements to be followed should be included.

Acceptance of the infrastructure constructed: Notably, the PPP Agreement needs to specify who is responsible for the test satisfaction assessment. As a general rule it is advisable that this be done by a joint assessment between the Grantor and the Developer. Better still, and in order to minimize any potential disputes, the test could be undertaken by an independent third party or Expert Panel[5].

Early Works Agreements: In most cases, *Early Works Agreements* are considered bad practice for the following reasons: They may generate questions over whether the agreements have been procured in accordance with procurement laws and regulations:

(a) Under ordinary circumstances the contracting authorities would not make any payments to the Developer or sub-contractors prior to the completion of the relevant facilities and commencement.
(b) Under ordinary circumstances, contracting authorities should not be obligated to make any payments prior to financial close.
(c) Reaching agreements on early works contracts can act as a distraction from the negotiations on the main contract and delay contract closing.
(d) They can lead to project integration issues.
(e) Early works agreements can alter the balance of risk on the procurement by altering the bargaining position of the Grantor.

In cases where early works agreements are required, the following rules should be applied (United Kingdom HM Treasury, 2004):

(a) The enabling works should be planned in advance and be incorporated as part of the procurement strategy.
(b) A decision should be made as to who should perform the works – whether the Developer should do so or whether the Grantor should commission a third party to do them.
(c) The works program should deliver time-savings and provide value for money.
(d) Only essential and necessary works should form part of the enabling works. However, this might include surveys, access roads or other pre-construction enabling works not specifically related to the Developer's project plans but would be of general use to the Grantor whether a contract is signed or not.
(e) The enabling works should not affect the risk allocation of post contract work.
(f) The costs of the works should be controlled and the Grantor should designate funds for this purpose.

10.4 PROVISIONS RELATED TO SERVICE STANDARDS AND OTHER OBLIGATIONS

The PPP Agreement needs to specify the actual services expected to be delivered to the Grantor under the PPP Agreement. Some contracts have failed in this basic premise as there is often difficulty in reaching a consensus as to the baseline against which to measure service improvements (see Chapter 11 on 'imperfect data problems'). Furthermore, the Service Standards themselves may not be well specified. For example, important dimensions of service may be omitted, or targets may not be disaggregated by geographic area, even though needs and services may vary widely between areas. Finally, in some instances, there have not always been clear and sensible links between the Developer's remuneration and his service standard achievements.

In drafting the provision for service standards, it is generally accepted that the contract provide standards covering all aspects of service relevant to customers and potential customers. These include, for water:

- Coverage of households or other potential customers.
- Water quality.
- Water pressure and reliability of pressure.
- Customer service response times.

For Wastewater Service, key standards include:

- Coverage of households and other potential customers.
- Service quality and reliability (frequency of sewage backups and overflows or frequency of collection of sewage from holding tanks).
- Customer service response times.
- Treatment level and outflow quality to the environment from Treatment plants.

In some instances it may be advisable to specify such standards for certain geographic areas, not necessarily for the PPP scheme as a whole. This can guard against particular areas being 'left behind', and ensure that the benefits of expanded and improved service are widely felt. It also allows for differential standards to be set based on the characteristics of different areas. For example, service standards (and tariffs) may be lower in low income or hard to serve areas. In all cases, customers' willingness to pay for the service will be taken into account when setting the service standards[6]. It is recommended that the PPP Agreement should allow for:

- A definition of each service standard indicator, including water quality, water pressure, continuity of service, coverage, water billed as a percentage of water produced and customer service levels.
- The level of service to be attained, for each customer category and area.
- The date by which the standard is to be met – the opening value of the service standard should be at the level currently achieved by the utility, with a realistic timetable for improvements.
- How compliance is to be measured and monitored.

- Events which excuse compliance.
- Sanctions in the event of non-compliance.

It is to be recommended that the PPP Agreement specifies the services to be provided but not the technology to be used. This basic principle of specifying the services is presented in Box 10.1 giving details on the Bucharest contract. The Bucharest contract provides an example of good drafting of service standard provisions[7]. Poirier (2004) argues that this is especially important for services in low income areas, where innovative technologies may be required. For example, the contract may provide service to be extended to a specific number of households in certain areas with at least 40 litres of water per day for at least 12 hours a day, no more than 50 meters away from the house.

BOX 10.1 LINKING TARIFF ADJUSTMENTS TO SERVICE IMPROVEMENTS – THE CASE OF BUCHAREST

The Bucharest concession used an innovative approach to provide incentives for achieving service targets. The contract had a specific schedule of service targets that the Developer was expected to achieve at different points during the term of the concession. The Developer was given direct monetary incentives in the form of tariff increases and penalties linked to achieving these targets.

Level of services indicators include: potable water quality, pressure, continuity, coverage, billing, and other customer service indicators. Targets for these indicators were set for years 2, 3, 5, 10 and 15 of the contract. Each target was assigned a weight; for example, achieving the potable water service target in year 5 was worth 23% of the targets for that year. The Developer was required to indicate in its offer the tariff increases (i.e. difference between bid tariffs for two consecutive target years) that it would expect to receive for achieving the service targets. Upon achieving the service targets for a particular year, or a portion of them, the Developer is granted the corresponding tariff increase, or the corresponding portion depending on the targets achieved and the weights assigned to them.

In the event the Developer failed to achieve a level of service target by the deadline set in the contract, its corresponding level of service tariff adjustment would be delayed until the target is met. If the target is not met within 3 months after a failure notification by the technical regulator, the Developer would be required to pay US$1.5 million for each target that it failed to achieve. The contract makes a distinction between limited and general failures. The Developer is required to monitor its performance against service level targets, and to inform the technical regulator when a penalty is due. The Developer would be subject to additional penalties if it fails to comply with this provision.

Source: Operators Roundtable, World Bank (2004), Concession Contract between The City of Bucharest and APA Nova Bucuresti S.A,Privatization of Bucharest Water and Sanitation Services, Corporate Finance Services, IFC, Amy 2000.

10.5 PROVISIONS APPLYING TO IMPLEMENTATION, MONITORING AND ENFORCEMENT

PPP Agreements are fundamentally contracts which essentially set out service obligations. As discussed earlier the premise is that contractual obligations need to focus on achieving service levels, customer focus and based on monitoring outputs but not necessarily inputs. Starting from that philosophical perspective, it is nevertheless essential to design the PPP Agreement in such a way so as to provide mechanisms to give the Developer the incentive to actually fulfil his contractual obligations and that there are sufficient disincentives if he fails to fulfil his obligations. The following sub-sections discuss problems that have arisen in each of these areas and recommendations for improved contractual approaches.

Monitoring and contract monitoring units

PPP Agreements generally specify service standards and other targets which the Developer is expected to meet and will also define who is responsible for monitoring such obligations. Monitoring and enforcement of obligations within the PPP Agreement should be directly linked to the institutional capacity of the given country where it takes place. It is usual to charge a government agency (as in Port Vila, Vanuatu or Tallinn, Estonia) or a regulatory body (as in Manila, Sofia or Bucharest) with monitoring the service standards. However care must be taken by the Grantor in designing such arrangements to ensure that the entities given the task of monitoring and enforcement are capable and motivated to do so.

Experience seems to suggest that, in some instances, there is insufficient clarity on procedures to audit reports received from the Developer, or that the Grantor agency may not be competent or motivated enough to carry out the monitoring, or that monitoring is not transparent, leaving the public uncertain and suspicious as to the Developer's performance. In this respect, improved contractual drafting is to be recommended and should stipulate:

- Clear definition of service standards and tariff rules in the PPP Agreement,
- Specify the Grantor agency responsible for receiving information from the Developer, detailing the format and content of each report, the standard to be monitored, and the Developer's report on these standards at regular and specified intervals[8],
- Provide the Grantor agency with some audit process to check the accuracy of the Developer's performance reports against accepted norms,
- Allow the Grantor agency to take specific enforcement actions when performance falls short of standards,
- Make the standards and the Developer's actual performance available to the public at large[9].

A sound recommendation is to contract an independent contract monitoring unit. This is discussed in the next section.

Contract Monitoring Units can be an effective way of monitoring contractual obligations, as in the case of Sofia, Bulgaria. One of the main recommendations in this chapter is to contract out performance monitoring to an independent, reputable, international firm. Under this model, the Developer's regular service performance reporting would be submitted to the contracted monitoring unit. The unit would check the accuracy of the report. It would then report the extent to which the Developer met the standards to the government and recommend enforcement action where it appears standards are not being met. The reports of the monitoring unit would be public. It is also recommended that customer groups are involved in the monitoring performance. For example, this may include creating a group of customer representatives, and providing for it in the PPP Agreement so they have a mandate to check service levels compared to standards, as an additional feature in the monitoring agency's monitoring process. It may also be a good idea to provide an explicit monitoring of the contracting authorities performance against its obligations.

Penalties and grantor ability to intervene for persistent minor breaches

Given the long term nature of Water and Wastewater PPP Agreements, it is prudent to consider situations where the Developer makes persistent breaches but that by themselves do not justify termination of the Agreement. The introduction of a *persistent minor clause* introduces a specific mechanism for the Grantor to remedy this. It provides for deductions or financial penalties, and in the event these are insufficient, consideration should be given to the use of warning notices with pre-defined resolution procedures where the ultimate sanction would be a right to early termination[10]. Cleary, such arrangements need to go hand-in-hand with specific penalties if the service standards are not met.

It is accepted good practice to specify penalties for all breaches of the contract, with the penalty proportionate to the harm caused by the breach[11]. Care must be taken to ensure that the inclusion of such penalties is permissible. For example, in common law countries it may not be possible to include penalties within the PPP Agreement itself. Penalties may have to be introduced as liquidated damages, for example, or included in specific regulations. It is advisable to cap the level of penalties to the Developer as a result of any specific incident or over any given period[12] and allow a procedure for expedited resolution in the event of disagreement over the application of a penalty[13]. The cap should be set at a level which is high enough to provide the Developer with a strong incentive to perform, but low enough to protect the Developer from financial distress caused by penalties (Operators Roundtable, World Bank 2004). Additionally, and in the interest of equitable treatment of parties to the PPP Agreement – sanctions also need to be levied against non-performance of contracting authority[14]. The form of such penalties may be a direct penalties or set-off rights as discussed below.

Performance bonds and set-off rights

Research suggests that the use of Performance bonds is a well used technique for PPP schemes in infrastructure development projects[15]. The reason is due to the 'limited recourse' nature of project finance (as previously discussed in Chapter 7). The Water and Wastewater sector is no exception. Many contracts have included performance bonds posted by the Developer.

A performance bond is an amount of money lodged with a neutral third party, such as a bank or financial institution, which the Grantor may claim in the event of breach of pre-specified obligations of the Developer detailed in the PPP Agreement. The bond ensures the Developer faces a penalty for non-performance, and that a sanction can be imposed. Without a performance bond, the public authority might need to sue the Developer for non-performance, which can be costly and time-consuming, and even if the court finds in the public authority's favour, it might be difficult to actually make the Developer pay.

It is good practice to include performance bonds in contracts. However, problems have arisen – for example in Guayaquil (see Box 10.2) – that conditions under which the bond may be called are not well defined, creating risk. A related problem has been 'first-demand' bonds – those which are callable without the public authority having to demonstrate that the Developer has defaulted. The idea of 'first-demand' bonds is to protect the public authority by avoiding delays and litigation when the public authority believes the Developer has defaulted. However, they have, in some cases, unbalanced relationships by making a substantial sanction available to a public authority, which it can exercise at will.

BOX 10.2 COVERING THE RISK OF WRONGFUL CALL OF PERFORMANCE BONDS – GUAYAQUIL

The Guayaquil concession contract required the concessionaire to issue a US$10 million performance bond in favour of the Grantor to cover his performance obligations. The bond could be called to cover payments due to the contracting authority to cover damages to third parties, unpaid penalties or early termination, as well as unpaid renewal of insurance policies. Beyond these traditional trigger events, the bond also covered the obligation of the Developer to make the capital investments to the amount offered in his bid. The Developer felt that this was an unduly harsh requirement, and that there was significant risk o the contracting authority wrongfully calling this event. To protect itself against this risk, the Developer requested a counter-guarantee from MIGA to cover a wrongful call of the performance bond. This was the first guarantee that MIGA provided to a water sector transaction. The US$18 million guarantee covered risks of expropriation, war and civil disturbance, and notably, wrongful call of the Developer's performance bond.

Source: Interview with Interagua (Dr Geoffrey Thorpe) and MIGA, and Price Waterhouse Coopers Securities LLC, IFI Risk Mitigation Instruments and Direct Sub-Sovereign Lending, Final Report, 2004.

Emerging options to improve practice include:

- Avoiding 'first demand' bonds, and instead including a provision that the bond may be called when the independent monitoring unit certifies that the relevant breach has occurred.
- Making lodging of performance bonds reciprocal. In many cases the Developer is required to post a bond to protect the government from risky behaviour, but not vice versa. When the Developer is required to lodge a bond to ensure he meets his obligations, the government should also lodge a similar bond to cover its obligations.
- Using set-off rights which would allow the Developer to adjust cash payments unilaterally, any service obligations and any investment roll-out if the Grantor breaches the PPP Agreement in ways which increase the Developer's cost or reduce its revenue. Similarly, the government should be able to withhold payment of any fees payable to the Developer to set-off any money owed by the Developer. If either party is unhappy with such an adjustment, the dispute resolution process (outlined below) should be followed.

The proposal involves practical measures and pragmatic solutions at the core of any PPP Agreement that is that although a party has a right to something does not necessarily allow it to exercise that right, and that if one party is in breach it should not excuse the other party from performing its obligations[16]. By avoiding performance bonds on demand, by making performance bonds reciprocal (i.e. both Grantor and Developer are obliged to post a bond) and by allowing 'rights to set-off' in the contract, there will be (a) less likelihood for unilateral actions by the Grantor, (b) an automatic compensatory mechanism that allows the Developer to unilaterally adjust cash payments, service obligations and any investment roll-out if the Grantor breaches the PPP Agreement in ways which increase the Developer's cost or reduce his revenue.

Enforcement of customer payments

A Developer cannot be financially viable if it cannot collect payment for customers. This means it needs a way to sanction non-payers. Disconnection is the most effective tool. It is accepted practice that all the normal sanctions against customers who do not pay should be available to the Developer, including the ability to disconnect after a specified period. Developers may be required to report on the number of disconnections, but should not be required to report in advance of disconnection, or to seek approval to disconnect. However, this needs to be balanced against ensuring that there are mechanisms to protect consumers. These may include for example, a requirement that in cases where a customer disputes the bill, he has the right to take the issue to an independent body, during which time service may not be disconnected until the dispute has been settled, or that the Developer is obliged to offer reasonable terms for settlement of payment in arrears. Another possible suggestion is that the Grantor set up a fund whereby he has the discretion to pay water bills on behalf of families in genuine hardship, and can work with them to provide budgeting and other assistance. Such a fund and agency would recognize the role of the government in social assistance, allowing the Developer to focus on commercial performance.

10.6 PROVISIONS APPLYING TO FINANCING

Financing – developer obligations

Financing is provided by the lenders and the shareholders. Financing will include up-front fees, development costs, design, construction and ultimately operation. Lenders will advance funding progressively with payments usually linked to milestones and verified by an independent expert acting for the lenders and possibly the Grantor. During this first phase, the lenders will insist on a careful balance of equity and debt funding and may require recourse over and above project assets, for the shareholders or some other guarantor. The lending agreements will need to indicate protections for the lenders, such as:

- Drawdown schedule and the conditions precedent that must be satisfied before each drawdown, in particular related to completion of construction milestones and aggregate paid-up equity.
- Repayment schedule.
- Funding and control of reserve accounts, where the project company must set aside money for contingencies, in particular to cover a number of months of debt service in the event of revenue shortfall, periodic major maintenance expenses and annual costs like insurance and taxes.
- Default events, such as failure to satisfy ratios (DSCR, LLCR, debt: equity, etc.), late payment, project contract defaults, unsolicited changes in management or project contracts.
- The right for lenders to stop disbursements to shareholders, to control voting rights and other project company discretions ('reserved discretions') and seize funds in the event that things are not going as well as the lenders would like (e.g. when default events arise or might arise).

Security, step-in rights and direct agreements

Security rights (over different project assets and in favour of creditors) are both 'offensive' and 'defensive': offensive to the extent the lenders can enforce the security to dispose of assets and repay debt when the project fails; defensive to the extent that senior security can protect the lenders from actions of unsecured or junior creditors. If comprehensive security rights are not available, the lenders may seek to use ring-fencing covenants in an effort to restrict other liabilities, or, security over project company shares to allow lenders to take control of the company or the creation of a special golden share that provides the lenders with control in the event of default. Security rights may also allow the lenders to take over the project rather than just sell the project assets, since the value of the project lies in its operation and not in completed assets (Delmon, 2009).

The lenders and the Grantor may enter into direct agreements with the project participants indicating step in rights, notice requirements, cure periods and other issues intended to maintain the continuity of the project when the project company defaults and/or falls away. A project may not require direct agreements where provisions can be included in the relevant project document or when some other solution is available.

Exchange rate issues

From the contracts analysed it would appear that there is no real best practice for drafting exchange rate risk provisions. Generally this is dealt with by a combination of indexation, review, financing and enforcement provisions. As explained in the risk chapter, exchange rate risk can have a significant effect on the ability to attract private finance. Accordingly, there are a number of possible contractual mechanisms which may help to manage foreign exchange rate risk and include[17]:

- *Mandated local currency financing* can eliminate the mismatch between a project's revenues denominated in local currency and its debt denominated in foreign currency. The disadvantage is that the amount and duration of local finance will be poor (or not available) in many markets. Another factor is that even if local finance is available, tenors are usually too short making it impractical or too costly. Nevertheless, Gray and Irwin (2003) suggest that, while reducing the reliance on foreign currency, debt may initially lower the amount of finance available, in the long-run more robust long-term investments could be beneficial.
- *Mandated hedging* to help mitigate exchange rate risk by requiring the Developer to take out compulsory hedge contracts (as discussed in Chapter 7). In this way, a Developer could purchase the foreign currency requirements for debt servicing at a fixed forward exchange rate. Like local currency financing, this approach is hard to implement due to lack of depth and tenor in local currency hedge markets[18]. However, there may be cases where local currency financing and hedging instruments will available, and in such cases it will usually be a good idea to use these instruments as far as possible.
- *IFI local currency swaps*: The Asian Development Bank has initiated a new instrument in which it swaps a given amount in US dollars for the country's local currency (Bestani & Sagar, 2004). The IFI then on-lends the local

currency to financial institutions in that country, which in-turn lend to borrowers for investment projects. This provides the benefits of local currency financing. The difficulty, however, is that the government of the country, by making the local currency swap, takes on a large portion of the exchange rate risk.

- *Liquidity backstopping facility*: Another approach is to use a devaluation liquidity backstopping facility, as proposed in the Camdessus report (2003), which could be provided by an international agency with an excellent credit rating, which would pay foreign lenders the Developer's debt servicing due to devaluation of the local currency. This requires an 'affordable exchange rate' to be established, on which the facility would intervene to repay foreign debt. Repayments made by the facility then become a loan from the facility to the government. This loan is repaid over a long time period of time through an additional surcharge added to tariffs. Further details are given in Box 10.3. However, since 2003, it seems that there has not been another project – other than the Tietê power project in Brazil – where a similar structure has been used. Given the current financial crisis it is not clear if this proposed structure could be commercially feasible for financial institutions to take the risk involved in backing such a facility.

BOX 10.3 LIQUIDITY BACKSTOPPING FACILITY IN TIETÊ POWER PROJECT IN BRAZIL

A similar type of facility was used for the Tietê power project in Brazil. In this case, the operator raised finance for the project by issuing U.S. dollar bonds. Tariffs from the project are indexed to local inflation but not to changes in the exchange rate. If the operator is unable to meet its debt repayments, it may draw on a liquidity facility provided by the U.S. Overseas Private Investment Corporation. In the long-run, this is repaid from cash that would otherwise have been paid to shareholders.

Source: Gray and Irwin (2003).

10.7 PROVISIONS RELATED TO 'FORCE MAJEURE' & TERMINATION

The intention of each party in an arrangement is to work towards fulfilling the terms and completing the contract until its expiry date. Provisions for early termination of the PPP Agreement should be dealt with within the arrangement. The Developer's lenders will require precise details in the agreement of the compensation payable if the arrangement is terminated early[19]. The key consideration for contracting authorities is to ensure the Developer is motivated to achieve optimum performance and should not be encouraged to deliver substandard performance because the financial penalties are not sufficient enough to motivate a remedy for the problem. A compromise should be found that meets both expectations. The cause of termination will determine the level of compensation to be granted. Additionally, a compromise will be needed to define persistent breaches and the list of grounds for termination as well as a definition of 'Force Majeure'. The Grantor should be careful not to make the termination triggers too stringent as they are unlikely to be accepted by the Developer or his lenders.

It seems that the main causes for early termination include[20]:

- *Grantor Default*: The Developer should be granted the right to terminate the arrangement when the Grantor or the government conducts itself in a such a way to compromise the contractual relationship and makes service provision so impractical or impossible[21]. This should be a last resort and provisions should be made to allow the contracting authorities or government to rectify its actions before triggering a termination. Compensation should be calculated on the basis of compensating the Developer and his financiers fully that is no better or worse off than if the arrangement were to continue as expected.
- *Developer Default*: The Grantor may seek to terminate the agreement due to inadequate service provision. A balance must be found between the effect of early termination due to inadequate service provision and the Developer's and financiers' interest in ensuring termination only occurs after severe defaults and after all possible attempts at rectification have been made and all other options have been exhausted. To counteract this eventuality, the Grantor could incentivize the Developer to deal with persistent performance breaches. It is recommended to include a performance points system in the PPP agreement regarding all defaults, in which there is a right to terminate the contract if a certain threshold of points is reached. Where such a points system is not applicable, the Grantor should retain a right to terminate the contract for persistent breach of unrectified defaults. Warning procedures should be used before termination procedures begin.
- *Corrupt Gifts and Fraud*: Termination can be triggered by instances of fraud or corrupt acts perpetrated by the Developer, sub-contractors or the Grantor. Careful consideration must be given to each party's interests in handling issues of corruption. It is perfectly understandable for the Grantor to wish to distance himself from a corrupt agreement and any form of corporate corruption should lead to immediate termination. However, it may be suggested that the Grantor not proceed immediately to terminating the agreement if the fraudulent act has been perpetrated by a sub-contractor or by an employee acting on his own, in which case the Developer should be given sufficient time to impose sanctions on the relevant party, terminate the relationship and locate alternative service providers.

- *Voluntary Termination by Grantor*: Certain events may occur that make the agreement between the Grantor and the Developer untenable, forcing it to voluntarily terminate the agreement. Such events could include policy changes which make service provision redundant. In these instances, the Developer should receive a termination payment to ensure it remains in the position it would have been in had the PPP Agreement been fulfilled.
- *Authority Break Points*: The Grantor may also structure the agreement to grant an option to terminate the agreement on specified dates for a specified compensation amount for equity and junior debt. The authority break points and the degree of flexibility for the Grantor will depend upon the extent to which the specified compensation amounts are less than the alternative amounts payable under voluntary termination by the authority circumstances.
- *Termination for Breach of Refinancing Provisions*: The agreement between the Grantor and the Developer should cover the possibility of termination due to breaches of refinancing provisions. A balance must be struck between the Grantor's need to give incentives to the Developer to be open and honest in regard to its refinancing and the Senior Lender's fear of their loans being threatened due to the Developer's actions.

'Force majeure'

It is essential that the PPP Agreement should have a detailed list of 'Force Majeure' events that may trigger termination and the respective rights of each party. Unfortunately, definitions of *force majeure*, and their consequences, are not always clearly defined or well thought out. Different people have different interpretations of what *force majeure* means, so it is important for the PPP Agreement to be clear on the matter. Generally it is accepted that *force majeure* refers to something which neither contractual party can control or mitigate, and which affects the performance of the services substantially. Basic principles suggests that if such an event arises and neither party can reach an agreement within a specified period (generally 6 months), each party has the right to terminate the contract. The Developer is entitled to compensation payments but the Grantor can attempt to prevent the termination request by continuing to pay the Developer as if the event had not occurred. The Grantor should define the time period for these payments, after which it will review the situation.

Despite what would appear to be a fairly un-contentious issue, Water and Wastewater contracts appear to have run into difficulties with the definition of *force majeure* (FM) events. For example, the Grantor may want to include events which increase the cost of providing the service (such as a currency depreciation), or if events which are foreseeable but uncontrollable should be included in the definition, or whether it should be limited to unforeseeable events.

A considered recommendation is to limit the FM provision to events which make performance of the service physically impossible[22]. However, other provisions are to be recommended such as Termination for Breach of Refinancing Provisions, Emergency Adjustment, Grantor or Developer Default provisions to manage the financial consequences of unforeseen and uncontrollable events which have a major impact on the cost of service. Where events are foreseeable, explicit contractual provisions should be included to deal with them. If any event is not explicitly included in the contract, and is beyond the ability of either party to control or mitigate, it should be considered within the definition of 'force majeure'.

Step-in rights[23] and remedial action

At any time during the term of the contract, the Grantor should reserve the right to suspend the Developer's rights under the PPP arrangement for specific circumstances: these generally include events of war and long periods of unavailability of service. Where it does step-in, the Grantor should also comply with performance criteria and be responsible for proper maintenance and operation of the system. The Developer will require compensation for any damage caused during a step-in period. If the step-in continues for a long period of time, the Developer will have the right to terminate the agreement. At the same time, the financiers of the project will undertake their own assessment of the project and will attempt to rectify any breaches which have led to termination. Financiers will not want to take on significant risk before they have had the opportunity to assess the project and assess whether they can put together a substitute plan which could take over the project and rescue it. If the financiers confirm that a breach has been rectified, control over the project may be returned to the Developer. However, if the financiers decide that a permanent replacement of the Developer is required, they would then seek to find a suitable substitute to take over the rights and obligations of the Project. The Grantor should retain the right to approve the suitable substitute. The Substitute should be granted a 'clean sheet' in terms of pre-substitution termination defaults.

Handover provisions for assets which transfer to the authority

The methods for dealing with the transfer of assets will need to be determined in the PPP Agreement. Key areas to be determined are:

- The condition of the assets to be transferred, repairs required, costs and how they are paid for.
- Any design life requirement after the Expiry Date.

- Inspection tests prior to handover.
- Provision for warranties, contracts and other rights.
- Resolution of any disputes related to any of the above.

Employees are also regarded as assets and provisions should be made in the agreement as to how they are to be dealt with in transferring to another Developer or to the Grantor.

Valuation of terminal payments on expiry where residual value risk has been transferred

The Grantor has two main options to calculate asset payments at the end of the PPP agreement period:

- Calculate the market value of assets in their existing use.
- Refer to the amount bid by the Developer during negotiation of the original contract and its index during the life of the PPP Agreement.

The most valid basis for determining terminal payments is the market value of the assets. But in a scenario of an extraordinary increase in the market value of critical assets (without which service availability would be compromised), a payment cap should be set to guard against excessive payments.

Calculation and payment of early termination liabilities

The principles applying to the calculation and payment of early termination payments are specified:

- *Compensation Payment Amounts and Changes to Financing Agreements*: when any of the above termination triggers are activated Compensation payments are calculated in whole or part in reference to the level of the Developer's senior debt at the time of the termination payment. The Grantor must ensure that senior debt outstanding at any time is not inflated and therefore will not increase the Grantor's liability.
- *Set-Off on Termination*: The Grantor should be entitled to place any outstanding liabilities of the Developer against the amount it pays in compensation if the Developer defaults. Where termination is triggered for reasons other than Developer default, the Grantor is entitled to place the Developer's outstanding liabilities against all amounts paid in compensation, except sums paid to compensate senior debt.
- *Method of Payment*: When an incoming Developer pays market value on Developer default termination, or the agreement is terminated due to Grantor default, the Grantor should pay the Developer a lump sum. Unless affordability constraints hinder this option, payment in instalments should be avoided as interest will accrue on the amount of compensation.

Transition on handover

(a) Handover of Water and Wastewater system at end of PPP Agreement or early termination.
(b) Procedures, documentation, inspection, bonds. Surviving rights and remedies after handover.
(c) Assignment of Assets & Stock.
(d) Retention of Assets by Contractor on Termination:
In situations where the Developer retains assets after termination, the value of the assets should be deducted from any compensation payments made. Where assets have no alternative use, their value will be approximately zero and therefore no contractual options are required. Assets with residual value must be carefully considered, as the commercial incentives could be significant. For example, if residual values exceed original returns, they could distort the effect of any compensation agreement and reduce the Developer's incentives to perform.
(e) Information handover.
(f) Position of personnel.

Treatment of assets on expiry of service period

Two distinct types of contract approach deal with the way assets are handled upon service period expiry (HM Treasury, 2007):

(a) Contracts which represent value for money for the Grantor to take over the assets on expiry. Includes assets which have no other feasible alternative use and is only of value to the particular public sector entity.
(b) Contracts in which the residual value of the Assets is best transferred to the Developer. Such assets have alternative uses and are not required by the public sector in the long term. 'Residual value' refers to the market value of the assets associated with the contract at the time of the expiry of the Agreement. This is classified as a risk as the residual value of the assets is unknown at the time of signing the contract. Estimations of the value of assets should be made and incorporated into the financing structure of the Agreement.

Key allocation questions to be considered are:

(a) Which party retains the assets on termination?
(b) Do these assets have alternative uses?
(c) How does this affect the termination payment (if any) payable by the Grantor?

The UK Government's PFI Unit recommends that the contracting authorities long term objectives will be best served by requiring either automatic transfer or reversion of the assets to itself on expiry of the agreement or at the very least, an option to purchase the assets at nominal cost. This is the case where legal constraints prevent any practical alternative option or when assets have a useful economic life if retained by the Grantor and conversion of the assets for other uses may be costly. The Grantor may also require the asset in order continue providing service.

The Grantor should protect himself by not reducing his available options just before the termination of a contract. These options include:

(a) Taking possession of any assets at no cost.
(b) Retendering the service provision with the outgoing Developer with any assets made available to the Grantor at no cost.
(c) Removing any assets.

Where the Grantor retains the assets at no cost, provisions should be made to ensure the Developer's obligation to deliver the assets in a serviceable condition. This does not apply if the assets have reached the end of their useful economic life. Importantly, the Grantor should ensure that for the use of assets that have reached a prescribed useful life but which are still in effect being used, specific operational procedures are developed to allow for this, and ultimately ensure an effective transition back to the Grantor.

Preserving the conditions of the assets on expiry

In the past, terminal payments related to the value of the assets at the end of the agreement, were used as an incentive mechanism for the Developer to maintain high standards of service throughout the contract period. However, this system is flawed in that it confuses payments for services and payments for asset transfer as the assets may be well maintained, yet the quality of service may be low.

Instead, the availability payment should be the main vehicle for providing incentives to the Developer to maintain standards throughout the life of the Agreement. If at the end of the contract there will be a re-tendering of the service, the Developer has further incentive to continue to meet the Grantor's service requirements.

An alternative means of encouraging the Developer to maintain service standards when there is no alternative use for the asset, would be to structure the Agreement in such a way as to grant the Grantor the option of a secondary contract with the Developer. The Grantor could exercise this secondary contract option at the end of the Expiry Date after holding an open competition with other bidders. Under this system, the Developer would have the incentive to maintain standards through to the Expiry Date without the need for terminal payment. The major difficulty with this approach is how to calculate the price for the secondary contract during negotiations for the first contract.

10.8 VARIOUS STANDARD CONTRACTUAL ISSUES

Transparency provisions

Transparency in the PPP Agreement is particularly important in relation to tariff indexation and adjustment procedures, to help protect against corruption and increase legitimacy in the eyes of the customers and other affected parties. It seems that most contracts do not provide full disclosure and transparency in tariff-setting[24]. The process usually involves only the Developer and the government or a regulator, without any opportunity for public scrutiny and input. Nevertheless, there are a number of 'good practice' options related to customer service standards and tariffs which allow the public a right to ask questions and be heard in periodic and other kinds of adjustments[25]. In other cases it has be known that Consumer representative groups have been involved in monitoring performance against service standards. Other important provisions to increase transparency might include for example, making a Developer's determination of its estimated tariff increase publicly available, public hearings for any tariff increases and the proceedings and findings of such tariff reviews public. And in the event of dispute between the Grantor and the Developer, making the result of such proceedings publicly available[26].

Intellectual property rights

In long term PPP Agreements it will be necessary to specify explicitly any Intellectual Property Rights (IPR) resulting from the Agreement. In the event of expiry, early termination or authority step-in, agreement should be reached to ensure the Grantor has the right to use the IPR required to continue providing service. This is particularly important in the case of the Water and Wastewater Sector in which part of the contractual obligations the Developer may be required to develop a customer data base or commercial/billing system[27]. In such circumstances the Grantor should be permitted immediate access to the IPR and any information required to operate the system. New service providers will also require access to the IPR prior to expiry dates in order to familiarize themselves with the service.

Labour issues

Although in the main PPP Agreement this may not be very detailed, this is a crucial point in terms of liabilities, pension commitments and so on and other associated issues for a successful PPP Agreement. Usually the existing labour force in municipal water and wastewater services is a major source of potential efficiency gain. There maybe major implications for the sustainability of the PPP Agreement in the type of contract for staffing and the manner with which it is dealt. It is recommended that if major labour issues are involved in part of the PPP Agreement, the Developer should explicitly clarify the way in which staff issues are handled. For example, this may include the level of compensation payable, pension contributions, the retrenchment numbers allowable over a period of time, outreach and post employment obligations and so on.

10.9 SUMMARY AND CONCLUSIONS

PPP Agreements in the Water and Wastewater Sector are fundamentally contracts whose essence is to state service obligations in terms of quality, quantity, management, operations and investments. As has been discussed earlier, it is to be emphasized that the contractual obligations need to focus on achievement of levels of service, customer focus and be based on monitoring outputs but not necessarily inputs. Taking that philosophical perspective as a starting point, it is nevertheless essential to design a PPP Agreement in such a way so as to provide mechanisms to give the Developer incentives to undertake his contractual obligations with sufficient sanctions if he fails to fulfil its obligations.

One of the main recommendations in this chapter is to contract an independent, reputable, international firm to monitor performance. A further recommendation is that monitoring and enforcement of PPP Agreement can be improved by:

- Defining service standards and tariff rules clearly in the PPP Agreement.
- Specifying the Grantor agency responsible for receiving information from the Developer, detailing the format and content of each report, stating how each standard is to be monitored, and requiring the Developer to report on these standards at regular and specified intervals[28].
- Providing for the Grantor agency to check the accuracy of the reports through some audit process and requiring it to validate Developer reported performance against standards.
- Permitting the Grantor agency to take specific enforcement actions when performance falls short of standards.
- Making the standards and actual performance of the Developer available to the public at large[29].

Furthermore, in this chapter we recognise that although a party may have a right to something, it does not mean that it necessarily has to exercise that right or in fact wishes to exercise that right. Similarly if one party is in breach of the PPP Agreement, it should not excuse the other party from performing its obligations. Accordingly, one recommendation is to avoid on-demand performance bonds, but instead make performance bonds reciprocal (i.e. both the Grantor and the Developer are obliged to post a bond). This concept together with allowing 'rights of set-off' in the PPP Agreement, will ensure that there is less likelihood for unilateral actions by the Grantor. Furthermore, if an automatic compensatory mechanism is introduced into the PPP Agreement allowing the Developer to unilaterally adjust cash payments, service obligations and any investment roll-out if the Grantor breaches the PPP agreement in ways which increase the

Developer's cost or reduce its revenue, should allow for a pragmatic and efficient mechanism to deal with contract variations.

This chapter argues the importance of allowing variations that take into account pass through and other cost items that are not in the direct control of the Developer. Additionally, certain costs associated with the PPP Agreement financing structure that need to be included to provide for unforeseen changes. All these issues require resolving with pragmatic solutions which include mechanisms that permit these changes in the PPP Agreement for its ultimate long term survival and sustainability. In this respect structure recommendations are made to allow periodic changes in the PPP Agreement. Several options for introducing periodic adjustments are proposed. These include (a) UK system of price-cap regulation, (b) US-style rate of return regulation, (c) Chilean style Rate of Return over the life of the PPP Agreement, (d) Rate of Return Bands and Stabilization funds, and (e) fixed NPV of revenue with a variable length of PPP Agreement. All are discussed briefly. The applicability of any or part of these methods will depend on the actual conditions on the ground, the type of PPP Agreement and the local legal and legislative environment where the contract will take place.

ENDNOTES

[1]For further discussion of the transfer of technology and know-how, see UNIDO (1996) at 75–90.

[2]There are a number of PPP Agreements that include a lock-up period. For example in England and Wales the 'undertaker/operator licenses' had a provision for a golden share that granted the government a veto right on any change in shareholding structure. Mendoza Argentina, Tallinn Estonia and Sofia Bulgaria all included in their respective PPP Agreements restrictions on changes in shareholding for the Developer within the first five years of operation.

[3]Depending on the legal tradition of a country, in some instances the subject of the Water PPP contract is described in a specific general or sectorial law or regulation.

[4]Conditions to be met by both parties before the contract becomes effective or external conditions to be met are usually detailed as Effective Date Undertakings or as Conditions Precedent to the effectiveness of the contract and are usually detailed in an annex.

[5]It is important to note that the Grantor should not permit third parties appointed by the lenders to act as their third party, as the adviser appointed by the lenders will not have the same overall objectives as the Grantor. The Grantor should appoint a fully independent third party acceptable to both he and the Operator.

[6]In some instances this may imply that the PPP Agreement specifies service levels lower than European Union, United States standards, or World Health Organization guidelines.

[7]Although the concept of level of service attainment tariff adjustments is an innovative one, and could provide direct and strong incentives to the Developer, this did not work well in practice in the PPP scheme in Bucharest. The winning bidder offered a declining bid tariff schedule, therefore removing most of the incentive to achieve level of service targets. This

shows the importance of a careful design and consultation on such provisions. To work well and be publicly acceptable, the increased tariffs need to be planned and expressed as being designed to ensure the sustainability of the service improvements.

[8]For example, in the Manila concession contracts the Developers must report quarterly on service performance indicators such as coverage, quality standards, and customer service data.

[9]During the course of writing this paper a number of international water and wastewater operators were contacted and the issue of making performance information available to the public was discussed. It was interesting to note that all agreed that all performance information pertaining to a PPP Agreement should be made available to the public. There was an element of discomfort for full disclosure of financial information. The argument being that Developers did not want to 'open their books' to other operations as this was considered sensitive commercial information that could be used by competitors or even Grantors for other purposes than that of monitoring. Having worked for a private international water and wastewater operator I have heard this same argument being made by senior management at the time.

[10]It must noted that this is not a remedy normally used in this sector. However, there are positive experiences of this in other infrastructure sector for example public transport services in the UK and the UK's Private Finance Initiative procedures (HM Treasury, 2008).

[11]A good example is the provisions of the Port Vila, Vanuatu concession contract. This involves a monetary penalty for the Developer in any instance where he fails to fulfil his obligations. The monetary penalty for any failure is then calculated by taking the prevailing volumetric tariff multiplied by the appropriate multiplier. For a particular failure, for example, interrupted supply or insufficient water pressure, the pre-specified number of cubic meters is the multiplier. As service standards will be clearly defined in the contract, it was envisioned that this approach would enable penalties to be easily determined when standards are breached.

[12]The Guayaquil, Ecuador PPP Agreement specifies that any breach of the 'Reglamento de Servicios' (operational rules) which governs the relationship and interface between the Developer and the customers is liable to a fine of US$5000 per event. The PPP Agreement is silent on the manner in which such sanction is be imposed. So for example, not responding to a customer call could be considered a customer complaint by the Regulator, and a fine could be imposed. There are no provisions defining a customer complaint nor time allowances for the Developer to respond to such complaint. This demonstrates a failure in the design of the PPP Agreement for a clarifying clause stating a 'reasonable' time in which the Developer may be fined US $5000 for 'failures to perform its obligations under the terms of the concession'. If it is not specified it can have a cumulative effect that could potentially lead to the Developer being faced with multi-million dollar fines imposed on it for not responding to customer calls. In that case it would be because there was no cap associated with such provision or a mechanism within the PPP Agreement to allow the Developer to disagree with such a penalty application. The result would be that the Developer had a contractual

obligation to pay the fine or else he would be in breach of his contractual obligations. His only legal recourse would be to avail himself of the arbitration provisions in the PPP Agreement, a procedure he would loath to initiate as it would put him in direct 'conflict' with the Grantor.

[13]In a dispute as to whether or not a service standard has been breached, the penalty should not be applied until a final legal determination on the matter has been reached. Although a provision like this may be advisable, care must be exercised to avoid nullifying the effect of the penalty provision. This means that the provisions of the PPP Agreement should not allow the Developer to shield himself and delay the application of a given penalty unduly. until there is a legal determination on the matter. That is why it is recommended to include a procedure(s) in the PPP Agreement that allows for an expeditious resolution on this. If the parties cannot agree then they should resort to other dispute resolution procedures contained within the PPP agreement.

[14]Arguably a provision like this might have helped in the case of Guayaquil, where the Regulator ECAPAG had too much discretion. This is a personal opinion based on my assessment of the situation when I was employed as a mediator for a dispute between the Interagua and ECAPAG between 2004 and 2005. Maybe delete …

[15]Performance bonds have been used extensively in the Water and Wastewater Sector. For example, the contracts mentioned here Guayaquil, Manila, Sofia, Estonia, Trinidad and Tobago, Port Vila amongst others, all had performance bonds linked to their respective PPP Agreement.

[16]For example, in Manila the government did not build a new water source it had promised, but the Developers were still legally required to provide the level of coverage and service they had committed to. In some African contracts for example, the lease contract in Dar es Salaam, Tanzania, government agencies have not paid their bills, leaving the Developer in a difficult financial position, but the Developer is still expected to deliver the contracted levels of service. In Buenos Aires the Grantor did not allow tariffs to increase as the contract required, as the peso depreciated, but the Developer was still contractually bound to make investments and expand service, even though government action meant it did not have the money to do so.

[17]Interview with Ms Tomoko Matsukawa, Senior Investment Officer, Finance Economics and Urban, Sustainable Development Vice-presidency, World Bank (April 2009).

[18]Matsukawa *et al.* (2003, p.10) argues that in lower-middle and low income countries it is unlikely that forward foreign exchange transactions could be arranged at an affordable cost with sufficient tenor to serve as the basis for financing an infrastructure project. As a practical matter, forward foreign exchange transactions have not been used to finance infrastructure projects in developing or developed countries.

[19]In some instances the maximum liability payable under a termination event is limited or uncertain in law leading to the project not being considered by lenders as bankable. In this case lenders are likely to require assurances for partial or full compensation in the event of early termination and in some cases even for termination due to the Operators' breach.

[20]For more detailed discussion on termination triggers and drafting of appropriate texts, readers are advised to consult the H.M.'s Treasury's 'Standardisation of PFI Contracts Vol. 4 document.

[21]For example, if the Operator requires visa for its foreign personnel to be in country and these visas are not granted by the Government (or the grantor) this would make it impossible for the Operator to meet its contractual obligations.

[22]It is generally good practice to list all events which might constitute *force majeure* and state explicitly whether or not they are included in the definition. Earthquakes, floods, fires and military or terrorist action are generally included in such definition. Furthermore, the Developer may be required to maintain insurance against some or all of these risks. Other factors like industrial unrest may or may not be included, depending on the parties' view of how much control the Developer has over such matters.

[23]A step-in right is when the Grantor can exercise his right under the PPP Agreement to take action.

[24]This has caused suspicion and reduced legitimacy in several long term PPP agreements such as Buenos Aires, Manila and Port Vila.

[25]The precedent for this is in the United States and the tariff/rate cases and which has been successfully used in places such as Barbados.

[26]Incorporating these provisions will enable customers to check for themselves that the Developer is not trying to put through an unwarranted tariff increase. They also ensure that the customers, being those directly affected by tariff increases, are able to make objections or general submissions on any proposed increases. An example of such transparency provisions is provided by the US rate of return regulation process, which allows public participation throughout the process. The New Jersey regulatory system does this.

[27]For example, the concession contract in Tucuman Argentina, the Developer (a French multinational water company) exited the contract in acrimonious terms with the Grantor and decided it would remove the commercial and billing systems of the company.

[28]For example, in the Manila concession contracts the Developers must report quarterly on service performance indicators such as coverage, quality standards, and customer service data.

[29]For those items where it says 'Not discussed' it means that the particular topic does not include a specific provision that has been included in this section. However, it does not imply that this issue has not been discussed throughout this paper. For example, the category called Employee Matters is a major risk issue in developing PPP Agreements, in the associated liabilities in terms of staff, (over-staffing is typical in Water and Wastewater utilities that are publically owned), pension liabilities and other associated contingent liabilities crucial in the development of a PPP scheme. This topic is discussed in earlier chapters.

Chapter 11

Maintaining the economic and financial equilibrium of the PPP Agreement

11.1 INTRODUCTION

PPP Agreements typically tend to be for long periods of time. It is inevitable that change will occur over long periods and that not all can be anticipated at the contract design stage. Accordingly, adequate flexibility needs to be considered in the design and implementation of the PPP Agreement. At the same time, it is important that this flexibility does not open the door to full re-negotiation of the basic terms of the contract. Adequate balance needs to be struck between price, long-term flexibility and certainty of whole-life costs.

This chapter starts with a discussion on the importance of allowing variations within the PPP Agreement that do not jeopardise the legal standing of the PPP Agreement (i.e. they do not materially or adversely change the conditions of the PPP Agreement from the time it was originally bid). However, certain costs associated with the financing structure of the PPP Agreement need to be included to provide for changes that cannot reasonably be foreseen by the Developer either at the time of bidding or during the implementation of the PPP Agreement. These issues require pragmatic solutions which imply that specific mechanisms should be included in the PPP Agreement to allow for these changes. The ultimate sustainability of PPP arrangements depends on the flexibility of the agreement. Furthermore, PPP Agreements must allow for periodic changes and variations. There are several options for introducing periodic adjustments. These include (a) UK system of price-cap regulation, (b) US-style rate of return regulation, (c) Chilean style Rate of Return over the life of the PPP agreement, (d) Rate of Return Bands and Stabilization funds, (e) fixed Net Present Value (NPV) of revenue with a variable length of PPP Agreement, and (f) and hardship clause. These are discussed in this chapter, and specific recommendations made for sample clauses and principles that may be included in PPP Agreements.

One of the major recommendations in this chapter is the use of a Tariff indexation formula to protect the Developer from costs which are beyond his control. The recommendation made is an indexation formula which is flexible enough to protect against certain occurrences such as general inflation, changes in prices of particular inputs (e.g. electricity and chemicals increase by more than the average rate of inflation), prices of inputs such as fuel, personnel, imported goods and import taxes imposed on the Developer. Notably, the manner in which different parameters are used within the formula may lead to (a) a hybrid system between a traditional price and revenue cap which reflects the cost structure of the business, (b) a price cap system in which the utility over-recovers compared to costs if demand is above expectations, and under-recovers compared to costs if demand is below forecast, or (c) a revenue cap, in which the utility under-recovers compared to cost if demand is above expectations, and over-recovers compared to costs if demand is below forecast.

11.2 DEALING WITH THE LONG TERM NATURE OF PPP AGREEMENTS: PROVISIONS FOR PERIODIC, EXTRAORDINARY AND EMERGENCY VARIATIONS

Under any PPP structure, it is important that the Grantor reserves the right to request variations in the design, PPP Agreement and the financial model. To maintain an appropriate balance between the parties' rights and obligations, variations should be permitted with the consent of both parties but should not be imposed on one partner by the other. If the variation is carried out, the pricing should be agreed as part of an initial agreement and reflected in sub-contracts, rather than agreed after construction. This is likely to lead to a better and more efficient process and diminish possible areas of disagreement.

Indexation alone cannot always ensure tariffs remain in line with costs. An evaluation of the contracts studied reveals that adjustments also often need to be made for changes in the original assumptions[1]. Research by the World Bank in consultation with major international water Developers (Operators Round Table, 2004), as well as Berger (2003), summarizes some of the major problems encountered as:

- The long duration of Water and Wastewater/PPP Agreements makes them vulnerable to unforeseen political or economic influences.
- The adjustment provisions have not been adequate to preserve the financial equilibrium of the contract[2].
- The way in which adjustments should be applied has not been clear. Vague terms like 'financial balance of the contract' have been used, and processes have not been well-defined.
- The Developer generally has sunk significant costs at the outset of the contract and is therefore dependent on the payoff from the PPP Agreement based on its initially specified length and under its initial assumptions.

- Governments may push for renegotiations to the contract to take advantage of higher realized revenues leaving the Developer in a vulnerable position due to the significant initial costs sunk in the project.
- These problems have led to renegotiation of contracts. Renegotiation involves stepping outside the contractual framework, and therefore risking the benefits of competitive award of contracts. Renegotiations have also been frustrating and time-consuming for all parties.

In addition to indexation and the other changes described above, changes in law including any adoption, modification, or repeal may occur at any time after a PPP Agreement has been signed and become effective (the Effective date). Accordingly, change of law provisions included in the PPP Agreement should deal the responsibilities for the costs arising from such changes of law and how such costs should be funded. Normally, the Grantor is in a position where it can control any consequences of changes in law. However, in the case of Water and Wastewater Projects, there are instances where the Grantor may not have control over changes in law. In which case, careful consideration should be given to the possible mechanisms that can be included within the PPP Agreement in which the Developer may prove to be more adept at managing the effects of changes of law and minimizing their impact. For example, it may be possible to pass on the costs of changes in law to the customers of the PPP scheme through an increase in tariffs. However, whenever the Grantor bears some of the risk of changes in law, the Developer should be obliged to keep any associated costs to the minimum[3]. It will be important that the PPP Agreement allow for costs to be recovered, but it should not allow the Developer to recover costs that it would have normally had to incur.

These can be broadly classified as periodic, extraordinary and emergency adjustments (HM Treasury, 2007).

Periodic Adjustments[4] are applied at regular intervals to bring tariffs back in line with costs. These essentially correct errors in the indexation mechanism. Research finds that the main objective of these adjustments is to allow a reasonable rate of return for an efficient Developer.

Extraordinary Adjustments[5] alter tariffs in response to specific, unexpected but manageable changes in cost resulting for example from changes in laws; new environmental standards; or demand movements above a specific level. The main objective of these adjustments is to restore the financial position that the Developer would have been in had the event not occurred.

Emergency Adjustments[6] are designed to keep the contract working in the face of unexpected changes. These changes are considered so significant that they would either bankrupt, or have a major negative financial effect on the Developer, or lead to a renegotiation of the terms of the PPP Agreement. The main objective of this adjustment would be to ensure continuity of the service and enable the Developer to return to normality after a reasonable period of time. In the following section we discuss the different forms of periodic adjustments and make some recommendations on how these three types of adjustments may be built into the PPP Agreement for greater sustainability.

11.3 PERIODIC VARIATIONS IN PPP AGREEMENTS

UK system of price-cap regulation

Due to the historic evolution of the private sector in the Water and Wastewater sector in which early examples have been the privatization of the sector in England and Wales in the late 1980's, there has been a tendency to try to emulate the UK style of regulation[7]. This basically consists of an assumption that tariffs should be reset at regular intervals – for example every five years – to bring them in line with the costs of an efficient Developer. In principle, no changes to the tariff are permitted within these pre-defined periods, although tariffs may be indexed to inflation and certain other defined cost-drivers. Furthermore, certain adjustments are permitted for pre-defined extra-ordinary events. The implications of such a system is that the Developer may lose money as no changes are permitted in tariffs if there are costs increases or demand is lower than expected (i.e. the Developer assumes full risk of cost variations), the flip side of this system is that the Developer may make profits which are higher than forecast, if costs are low or sales volumes high. This approach has now become the standard, and was adopted with some modifications in Buenos Aires, Manila, Vanuatu and elsewhere[8].

US-style rate of return regulation

Under this system, a Developer would be allowed to set tariffs at a level to recover his costs and give it a reasonable return on prudently incurred capital expenditure[9]. If as a result of increases in costs or additional investment, returns fall below a reasonable level, the Developer may apply to an independent regulator or the Grantor to allow an increase in tariffs. Conversely, if the Developer is deemed to be making excess profits, the Grantor may reduce such tariffs or customers may petition the regulator for a tariff reduction[10]. Rate of return regulation is often criticized for providing incentives for over-investment and so called 'gold-plating', and the inadequate incentives for cost control[11]. Incentives for efficiency are in-built so Developers will try to keep costs down and avoid tariff reviews (rate case) as these are costly and disruptive. At a rate case, efficiency levels are scrutinized, and inefficient costs may not be recovered through the tariff alone.

It should be noted that the US approach functions well since it has a comprehensive legal structure that underpins the Grantor's behaviour and also provides incentives to the Developer to invest. One of its major strengths is that the Supreme Court has ruled that a regulator setting tariffs at a level which would not allow an efficient Developer a reasonable return on capital would not be allowable. This provides significant investor protection as it allows the Developer to earn a reasonable return on capital invested, and to adjust tariffs when this objective is threatened. As discussed in Chapter 7, it may not be possible to provide this protection within the legal framework of the country and should therefore be explicitly covered in the PPP Agreement itself.

Chilean model: Fixed NPV of revenue, variable length concession

The Chilean toll road concessions specify the Net Present Value (NPV) of revenue which the Developer needs to receive over the life of its PPP Agreement as part of the bidding variable (GET PPP, 2009). During the life of the contract, if revenues are lower than expected – for example, as a result of lower than expected traffic – the term of the PPP Agreement will be extended to allow the Developer to recover the required total NPV of revenue. This approach has the advantage of giving the Developer a strong incentive to control costs, while providing some protection against traffic demand risk (Engel *et al.* 2000). Applying this concept to a Water PPP Agreement may be a challenge given the typically 'lumpy' investments required in the sector. A typical PPP Agreement requires large investments throughout the life of the contract (unlike in a road where most of the costs are incurred initially and are controllable). The Developer will want to ensure that its NPV of revenues is at least equal to the NPV of costs over the life of the PPP Agreement. However, given that investments would change over the life of the contract, extensions may be required every given period, making extension to contracts difficult to manage and agree. A better approach may be to adjust tariffs to achieve a particular rate of return over the life of the PPP Agreement. Alternatively, it may be possible in some cases, to extend the life of the PPP Agreement, rather than increase tariffs, if returns fall below the target band.

Rate of return bands and stabilization funds

Lam (2002) describes a similar rate of return concept used in Hong Kong under the Schemes of Control for telecommunications and electricity. The regulatory regime set a permissible range for returns on equity. If returns rose above the band, the excess profits were put in a stabilization fund. The stabilization fund could then be drawn on to finance new capital expenditure, or set aside to be used to top-up profits in years in which returns fell below the bottom of the target return band. Only if returns were persistently above or below the target band would be tariffs be adjusted.

A similar structure could be used for long term PPP Agreements as it would provide stability of returns for the Developer and stability of tariffs for customers. Furthermore, such a structure encourages investment, especially with the ability of using the stabilization fund to finance new assets. It also contains efficiency incentives, since higher profits can be saved for use in bad years, or used to expand investment.

Rate of return over the life of the contract

In the case of Bucharest the rate of return concept has been adapted allowing tariffs to be adjusted whenever the Internal Rate of Return (IRR) of the contract, projected over the contract life, is outside a predefined band. This adjustment potentially allows the Developer to benefit from tariff increases if the net income falls below the amount required for servicing contracted long-term debt obligations. Furthermore, it allows customers the potential benefit of tariff decreases if the Developer IRR increases beyond the level of a comparable business. The economic regulator supervises the rebasing process, and uses an Expert Panel (water engineer, regulatory economist, and financial analyst) to provide an opinion on the rebasing petition of the Developer[12].

Hardship clauses

Some PPP arrangements have developed a concept of 'hardship' clauses which provide a party with the opportunity to request an adaptation of a contract's provisions if continued operation under those terms would no longer be economically feasible. Such hardships could include 'acts of god' or economic (fiscal) problems. The main premise behind such provisions is that under such circumstances, no party would be forced to continue operating, since this would compromise the long-term benefit of the contractual relationship.

11.4 RECOMMENDATIONS FOR TARIFF INDEXATION AND COST PASS-THROUGHS

Water projects, as many PPP projects rely on the availability of cash flows. These cash flows derive fundamentally from the tariffs that the Developer is allowed to charge for the services he provides. As discussed in the risk chapter, the rules associated with how tariffs are set and the variations permitted are fundamental to the sustainability of the PPP Agreement. Depending on the type of PPP Agreement, opening tariffs will be specified either in the initial bids or in the contract, and indexation provides a mechanism for keeping those tariffs in line with costs and allowing the Developer to pass on cost increases to customers through its tariffs.

Essentially any tariff adjustment mechanism will have to allow for (a) cost-pass throughs and (b) tariff indexation.

Cost pass-throughs effectively allow changes in the costs of certain inputs that are immediately passed on to the customers through the tariff. Cost pass-throughs are generally allowed for major cost items over which the Developer has no control. These include, for example, changes in taxation regime, price of bulk water arrangements and so on. However, one must be careful not to allow for costs for which a normal and diligent Developer should be responsible. For example, if a Grantor is considering the cost of electricity to be a cost pass-through, it must do so in a way that does not relieve the Developer from seeking reductions in operating costs that he can adequately manage through reduced pumping, or using low night-time electricity tariffs. In Guayaquil there were major problems[13] (see Box 11.1) which demonstrate the importance of permitting automatic increases in the tariff in accordance with a prescribed indexation mechanism, while providing some check that the increase is in line with the contract[14].

BOX 11.1 AUTOMATIC INDEXATION PROVISIONS BLOCKED BY REGULATOR – GUAYAQUIL

In early 2001 the Water and Wastewater Utility of Guayaquil (ECAPAG) entered a 30-year concession contract with Interagua. Shortly after the contract was signed, Interagua failed to raise project finance debt as a result of a perceived increase in the regulatory risk. Regulatory risk was seen by financiers as more imminent after inflation-based tariff adjustments were not approved by ECAPAG.

According to the contract, tariff indexations are automatic and applied every quarter. However, ECAPAG, which acts as both the counterpart to the contract and regulator, had a different interpretation of this contract provision, and disallowed the first tariff indexation adjustment. As a result, the current tariff is 11% lagged in relation to increases in underlying costs. After a long negotiation process, both parties recently reached an agreement whereby 50% of the tariff lag will be recovered over a period of two years. The end of this 2-year period coincides with the end of the first 5-year period of the contract when tariffs are to be reset. It was also agreed that Interagua would be granted, during the upcoming tariff reset, a compensation for the lost revenues, in exchange for reducing energy costs. Both parties also agreed to a revision of the wastewater expansion program over the first 5 years, essentially to compensate the Developer for his inability to raise finance due to the circumstances of the dispute.

Source: Concession contract between Interagua and ECAPAG and interview with Dr Geoff Thorpe, Director of regulatory affairs Interagua.

Tariff indexation formulas also seek to protect the Developer from costs which are beyond his control and indexation formulas seek to anticipate certain changes in the cost of providing a service. Some of the contracts evaluated suggest that there are two main problems with this provision: (a) problems with approval of adjustments, and (b) problems with using indexation to offset the effects of foreign exchange devaluation. Foreign exchange problems are dealt with in general terms in the chapter related to Funding and Finance. Detailed recommendations are provided in the following section.

The indexation formula needs to be flexible enough to protect against certain issues such as general inflation, changes in prices of particular inputs (e.g. electricity and chemicals increase by more than the average rate of inflation), prices of inputs such as fuel, personnel, imported goods and import taxes.

Based on research of various PPP Agreements, a possible tariff indexation provision could be developed using the following general formula. The use of its different components will lead to different outcomes which are discussed below. Additionally, a brief discussion of how the given parameter is to be used in the formula is provided in the 'notes' section, as well as a recommendation of the minimum value to be applied to each parameter.

The basic principle behind the design of this formula is to emulate as much as possible the price cap and benchmarking behaviour and other forms of regulation, depending on the use of the different variables within the formula, as discussed in the previous sections of this chapter.

Accordingly, the recommended formula is:

$$R_t = R_0\{\alpha_1 \mathrm{RPI} + \alpha_2 \Delta \mathrm{Ex.Rate}[\gamma \mathrm{RPI_F} + (1 - \gamma)] + \alpha_3 \mathrm{Elec} + \alpha_4 \mathrm{Labour})\}\beta(\Delta Q) + (K - X + Y)$$

where:

R_t, R_0 = Developer revenue in the current period (t) and the reference period (0).

RPI = Retail Price Index (or any other consumer index which is subject to good independent statistical verification). It is recommended that this be set at one (1) plus the percentage change in the retail price index in the home country between time (t) and the reference period (0).

Ex.Rate = One (1) plus the percentage change in the exchange rate index of the home country[15] currency against a relevant international currency between time (t) and the reference period (0).
Note: The relevant international currency will be the currency in which the majority of imported inputs are priced (these inputs may include chemicals or equipment), or the currency in which the majority of foreign financing is denominated. If these currencies are different, the formula would need to be adjusted to reflect this. (See also recommendations on hedging provisions and currency swaps which are directly applicable to the manner in which the ExRate delta is to be used in the formula).

$\mathrm{RPI_F}$ = One (1) plus the percentage change in a price index tracking foreign input prices, in the relevant international currency between time (t) and the reference period (0).
Note: This is important as some inputs of the Developer will be subject to variations since they are not produced locally, for example, chemicals and equipment prices. It is recommended

that RPI_F could be a general foreign inflation term, or where possible specific price indices if available.

Elec = One (1) plus the percentage change in a price index of electricity prices in the home country between time (t) and the reference period (0).

Labour = One (1) plus the percentage change in a relevant wage index in the home country between time (t) and the reference period (0). *Note:* The wage index is usually a good reference value that is available in most countries. If this is not available, an RPI proxy could be considered.

α_i = Weighting of the following term in the indexation formula with $\sum_i \alpha_i = 1$

Note: The applicability of α_i will depend very much on the cost allocation of the given PPP Agreement. For example, if the intention is for the Developer to bear the risk of changes in certain input prices (e.g. cost of electricity, chemicals etc.) and exchange rates between reviews, then the relevant α_i would be set at zero, or somewhere between zero and the impact of that factor on total costs. Alternatively, in a situation where this risk is not to be borne directly by the Developer, that is a formula which tracks input costs closely, each α_i should be set equal to the share of the cost represented by the indexation factor as a proportion of total costs. For example, if electricity accounts for 20% of total costs, α_3 would be set equal to 0.20.

It must be noted however, that regardless of how much cost risk is to be borne by the Developer, the sum of the α_i should *always* equal one. For example, if the electricity indexation factor was set at zero, the weight given to the general RPI index should be increased to offset this.

γ = Weighting of foreign operating and maintenance costs in the share of total foreign costs may be set to 'one', if the intention is to index exchange rate risk on imports but not on foreign currency financing. It is important to make a clear distinction between foreign O&M costs and the financing cost incurred by the Developer.

ΔQ = One (1) plus the percentage change in relevant volumes of output (for example, water supplied or waste water treated) between time (t) and the reference period (0).

β = *Note:* The manner in which the β is dealt with in this formula will create the effect of the UK, US type of regulatory systems. For example, setting β to one creates a price cap, in which the utility over-recovers compared to cost if demand is above expectations, and under-recovers compared to costs if demand is below forecast. If β is set as the volume weighting factor determined by the percentage of short-run variable costs of the firm in the total costs of the firm, it creates a hybrid between a price and revenue cap which reflects the cost structure of the business. Finally, Setting β to zero creates a revenue cap, in which the utility under-recovers compared to cost if demand is above expectations, and over-recovers compared to costs if demand is below forecast[16].

K = Real tariff adjustment factor to reflect increases in tariffs to cover new investment.

X = Real tariff adjustment factor to reflect reductions in tariffs as a result of efficiency gains.

Y = Real tariff adjustment factor to reflect increases in tariffs to adjust for initial tariffs below costs.

The basic principle behind this provision is a very general formula which can be adapted as needed. For example, setting $\alpha_2 = \alpha_3 = \alpha_4 = 0$, and $\beta = \Delta Q = 1$, yields the classic UK RPI + K structure. An alternative approach is to change tariffs only when input cost changes exceed some trigger. In this case, the Bucharest contract provides an example of the drafting:

Ordinary Tariff Adjustment [i.e. index-driven changes] will be implemented no more frequently than on a monthly basis and only if:

$$\left| \frac{CPI_{t-1} - CPI_{t-1}}{CPI_{t-1}} \right| > 5\%$$

Some contracts, such as the one in Port Vila, Vanuatu do not have any specific provision in the contract for the Grantor to check that the formula has been applied correctly. In Manila, the contract does allow for the application of the formula to be checked by a regulator[17].

However, it is recommended that a similar provision be drafted that would allow an automatic increase but taking into account specific details about:

- Who is responsible for applying the indexation formula, who checks and how disputes can be resolved if the parties do not agree.
- Making the results of such calculations available to the general public to enhance transparency and Grantor accountability.
- Procedure for recovering any over charge that may have resulted from a wrong application of the formula and in the event that the Developer does not comply, the dispute resolution provisions will apply.

The use of an independent expert panel or an independent third party as suggested in the earlier section on contract monitoring unit.

11.5 CONCLUSIONS

Research reveals[18] that periodic adjustments and extraordinary adjustment provisions are accepted good practice (Green *et al.* 1999). Emergency adjustments are an emerging option as demonstrated by their use in the UK Water Licenses[19] (the 'Shipwreck Clause') and are implicit in the US rate of return regulatory approach as well as French civil jurisprudence. Such clauses are designed to avoid the need for renegotiation by providing a structured way of dealing with events which fundamentally alter the financial conditions under which the services are to be performed.

Specifically to be recommended are the following points in their respective clauses[20]:

(1) For the *Periodic Adjustments* the Developer submit its estimate of the revenue (and the tariff increase required) required by an efficient Developer[21] . An expert panel reviews the Developer's proposed increase and makes a recommendation, followed by a decision to be made by the two parties or (in the absence of agreement) arbitration provided that the proceedings are made public.

(2) Also for *Adjustments* linked to traditional indexation, utilise the *Tariff indexation* formula suggested, which protects the Developer from costs beyond his control and which anticipate certain changes in the cost of providing the service. The application of the formula could lead to different outcomes depending on the use of the different variables within the formula for example price cap, benchmarking or a hybrid between both. The different parameters of the formula would be enshrined in the PPP Agreement and their possible application. As per the Period Adjustment, an expert panel reviews the Developer's proposed increase and makes a recommendation, followed by a decision to be made by the two parties or (in the absence of agreement) arbitration. Similarly the proceedings are to be made public.

(3) For *Extraordinary Adjustments:*

- The mechanism would follow a similar process to the Periodic Adjustments. The key difference would be in the approach taken to determine the required tariff increase and a specific requirement by the Developer to demonstrate to Grantor that the event has actually occurred and that it was uncontrollable and/or unforeseeable[22]. As the objective would now be to restore the financial position to its original position, the Developer would submit its required tariff increase to achieve this restoration. The remaining process would be the same as above.
- The Developer would have the right to suggest possible solutions to permit it to continue operating without having to renegotiate the contract. For example, the Developer may suggest that tariffs be

increased or that the Developer fee be reduced, or that the term of the concession be extended.

(4) For *emergency adjustments* (that deal with unforeseen events not contemplated in the original contract) the mechanism would be broadly similar to that detailed above but would no longer be focused on tariff changes and achieving a reasonable rate of return for an efficient Developer, but to enable normality to be restored over a reasonable period of time. The expert panel would consider changes in tariffs and service standards, changes in subsidies, financial restructurings, and changes in the duration of the contract, with the objective of ensuring continuity of service and the balancing of the interests of customers, government, Developers and investors. The remaining adjustment mechanism would work as described above[23].

ENDNOTES

[1]A good example of this is demonstrated by the following quote from the regulator's guide to the Manila Water Concessions 'However detailed and skilful the concessionaires have been in their original forecast, the one thing that can be guaranteed is that their forecasts will be wrong, one way or the other. Without a mechanism for making general adjustments to the level of Rates, there would be every likelihood of cumulatively declining financial performance, leading to bankruptcy, or cumulatively financial performance, which could threaten the foundations of the Concession arrangements from a political perspective.' (NERA, 1997, p.III-4-1)

[2]For example, economic shocks in Manila, Buenos Aires, Jakarta and elsewhere overwhelmed the ability of the contractual mechanisms to keep costs and revenue reasonably in balance rendering the PPP Agreement unsustainable.

[3]This same principle is applied under PFI type projects in the United Kingdom.

[4]Manila is a good example where tariffs are adjusted at five-yearly intervals to allow a rate of return for an efficient Developer.

[5]Bucharest: tariffs can be adjusted for events such as changes in laws. The amount of the adjustment would be the tariff change required to offset the change in cash flow resulting from the event.

[6]UK Shipwreck clause (as contained in the England and Wales Water Licences): tariffs can be adjusted if an event occurs that results in an unavoidable change in cash flows by more than 20% turnover.

[7]Price caps were introduced as an alternative to the hitherto prevailing system of rate of return regulation. This is discussed later on.

[8]The key benefit of the price cap is that by separating tariffs from a Developer's costs, three is a direct profit incentive to reduce costs. However, this benefit comes at the cost of increased risk. Because tariffs and costs diverge, utilities can make catastrophic losses as happened to Maynilad in Manila, Philippines. In the case of the United Kingdom, profits were found to be too high in the first ten years of privatization and thus found to be politically

unsustainable. This caused such a backlash from customers that the new Labour Government of Tony Blair in introduced a tax to profits (as occurred in the UK initially) in the period between price reviews.

[9]This system relies on the concept of 'reasonable' cost as would normally be accepted by a reasonable person in undertaking his business.

[10]The regulatory system for water in the US State of New Jersey uses such system. Rate of return regulation has been successful for electricity and telecommunications in several developing countries, including Barbados.

[11]The term 'gold plating' refers to when a Developer would undertake an investment without it necessarily being needed. For example, for a given asset such as a pipe with a given life span the Developer may decide to replace such pipe prior to the expiry of its useful asset life. Notwithstanding this fact, typically in developing economies, the problem is usually getting enough investment, rather than there being too much. Using this system the United States has been successful in providing privately financed water and waste water systems to many communities – even those very remote and small ones. The structure in the UK by contrast emphasizes efficiency over investment. Arguably, developing countries may do better to emulate the US approach with its emphasis on investment in assets.

[12]This approach gives the Developer certainty through fixing the overall project's minimum and maximum project IRRs. It may be argued that this may reduce efficiency incentives compared to standard periodic reviews or the variable length concession approach. However, this structure does promote investment and reduces financing costs. Provided the IRR band is wide enough, and taking into account that Developers have a built-in incentive to avoid tariff increases in order to limit political risk, it can still be reasonably efficient.

[13]This is an example of where an automatic tariff indexation provisions in a contract is not sufficient. In this case tariffs should have been raised with inflation-based increases. This highlights the importance introducing a mechanism to guarantee and provide security to strict compliance with contract provisions. The Developer argues that in this case, although the PPP agreement included specific provisions for automatically adjusting tariffs to fluctuations in the underlying costs, the regulator took into account political considerations, rather than applying the contract in a neutral way.

[14]Index-driven tariff changes are usually done at prescribed regular intervals. For example in the United Kingdom PFI contracts are indexed every six months. However, there is a trade-off between frequent reviews, which may limit the Developer's risk and provide for smoother tariff changes during periods of rapidly increasing prices, and the administrative complexity of frequently changing tariffs. Clearly, the right solution will depend on the macro economic conditions where the PPP agreement is being implemented. So for example in countries/circumstances where input prices may change rapidly, it is advisable to allow for monthly indexations.

[15]The 'home country' is considered to be the country in which the PPP Agreement is being implemented.

[16]One important caveat in this explanation is that the references to under- and over-recovery of costs for the Developer refer to short term impacts. Long term impacts will depend on

how actual customer demand affects the capital expenditure required, and how that capital expenditure is treated in reviews.

[17]The Manila contract specifies: 'Prior to October 31 of each Prior Year, the Concessionaire shall submit to the Regulatory Office its proposed Standard Rates for the following Charging Year, along with sufficient supporting information to enable the Regulatory Office to verify that the proposed Standard Rates is consistent with the Rates Adjustment Limit'. Upon examination of this provision it would appear that there is a major problem with this clause as it does not give any detail about the procedure and redress that the Regulatory Office may have in the event there are disagreements with the adjustment.

[18]As demonstrated for example by the periodic adjustment clauses of Bucharest, Buenos Aires, Tallinn Estonia, and for the extraordinary adjustments the case of the two concession contracts in Manila.

[19]It must be noted that although these are referred to as the UK water licences they only actually refer and are applicable to England and Wales, as the rest of the United Kingdom ie Scotland and Northern Ireland operate under different schemes without private sector involvement.

[20]The above procedure should be clearly established prior to the commencement of the PPP agreement.

[21]This procedure involves adjusting costs up or down through benchmarking, market testing or expert analysis to a reasonable level that an efficient Developer could achieve; forecast average demand over the period until the next periodic tariff review; determine the value of the Developer's assets including the Developer's proposed investment (if approved) incorporated at historic cost into the capital base; determine return on capital required by Developer (including debt-equity ratio, premiums etc) or use pre-determined rate of return, and calculate the required tariff increase by calculating the tariff required to cover fixed and variable operating costs, plus a return on capital.

[22]It is to be recommend that a financial trigger point, such as NPV of the cumulative impact equal to 20% of annual revenues, is used to determine when an event is major enough to constitute an 'emergency'.

[23]Work by Gotanda (2003) points to a similar mechanism for renegotiation and adaptation clauses, provided such clauses are limited to unforeseen events that cannot be triggered by either of the parties. He argues that that such clauses could be beneficial to investors, when included with a provision allowing a third party to adapt the contract in the event of a failure by the parties to reach agreement, as provided by the international arbitrator in the mechanism above. A similar view is presented in Berger (2003).

Chapter 12
Improving PPP structures

12.1 INTRODUCTION

The contractual recommendations mentioned in Chapter 10 discussed the conventional methods of private sector entry into the Water and Wastewater sector. As a result of issues identified in conventional models and discussed in previous chapters of this book, two prospective new models have been identified for private sector participation in the Water and Wastewater sector. These new models will be analyzed in this chapter in relation to the issues affecting conventional models and detailed recommendations will be made.

Previous chapters have each attempted to identify some of the critical issues and mistakes made by policymakers and Developers when pursuing reforms in their Water and Wastewater Sectors through the introduction of private sector participation. Often, these issues contend with the inability of a Developer to recover his costs or fund operations and maintenance (O&M) or expansion plans because tariffs are set too low or subsidy levels are insufficient. Chapter 6 has explored the 'conventional' models of PPPs in the Water and Wastewater Sector and have shown how critical it is to develop a PPP structure that allocates responsibilities correctly and provides lenders and Developers a suitable environment in terms of stability and predictability for investments. Furthermore, Chapter 9 on Key Contract Provisions demonstrates the need for a pragmatic approach in contract development, the need for variation provisions and how to maintain the economic sustainability of the PPP arrangement.

At the heart of any PPP arrangement should be a balance between expanding access to water and improving quality of service balanced with concerns over possible tariff raises, job cuts and dealing with suspicions of whether access and efficiency will be truly improved under private management.

During the writing of this book, it has been noticed that Government strategies to involve the private sector have not always been successful or have met with queries or resistance from various stakeholders. At the same time, a variety of PPP type contracts have been developed to ease the worries for Developers averse to risk. For example, medium-term management and lease/affermage contracts have been implemented as a compromise to ease the financial and foreign exchange risks of long-term PPP Agreements (usually concession type contracts). However, under these types of contracts, the Developer bears substantial commercial risk as he has to generate an operating surplus sufficient to cover his operation and maintenance costs. It seems that problems arise when issues fall into the grey area of either operational or investment issues. If these issues cannot be dealt with under the PPP Agreement terms, there is likely to be conflict between the parties leading to a failure in the PPP Agreement. As the decisions on operation and investment are shared between the Grantor and the Developer, neither side will have the incentive to implement the cost of improvements the system may require[1]. Concession contracts place responsibility for capital and O&M costs onto the Developer eliminating this as a potential area of conflict. Integrating operating and capital expenditure decision-making occurs most effectively through a concession type PPP Arrangement. However, long term PPP Arrangements following the traditional 'concession type' structure are not always appropriate for Grantors or in areas of high regulatory and political risk. Figure 12.1 shows a simplified representation of a standard concession type arrangement. Of note is that monies from lenders and sponsors (typically a mixture between debt and equity) is used by the Developer to meet the capital expenditure requirements of the PPP Agreement. The Developer's local company uses tariff income to service obligations, represented in the diagramme as the cash cascade. This cash cascade has a follows a specific hierarchy depending on the needs of the business, namely paying for operations and maintenance costs, servicing debt, investing in capital expenditure programmes (sometimes referred to as capex), and ultimately distributing profit to the project's sponsors.

For this reason, included here are some specific recommendations to develop new models for public private participation in the Water and Wastewater Sector. The approach is to tackle the issue of cost recovery and financing sources, as these appear to be at the heart of most issues with private sector contract operations. It should be made clear that cost recovery and financing issues are not linked and must be solved independently[2]. Suggestions outlined in this chapter are a first step in identifying areas in which new models can improve existing systems and if further subsidization is required.

12.2 NEW MODEL: OUTPUT-BASED LONG TERM PPP AGREEMENT

Under this structure the use of an output based arrangement is incorporated into the structure. Output based aid or OBA[3], is a financing strategy whereby public funds can be used to support the delivery of basic service in which policy concerns would justify public funding to complement or replace user fees. Usually payment is only made once a pre-agreed output has been reached. In the proposed model, the Grantor enters into a PPP Agreement with the Developer. The Developer commits to performance targets specified in the PPP Agreement and receives financing for capital expenditure from commercial capital markets. Tariff revenues provide the Developer's source of cash flow. These are used to cover O&M expenditure, debt repayment, capex financing and any remainder is retained as profit. Tariff revenues are supplemented by a Subsidy Fund which is to pay on an output basis (Box 12.1).

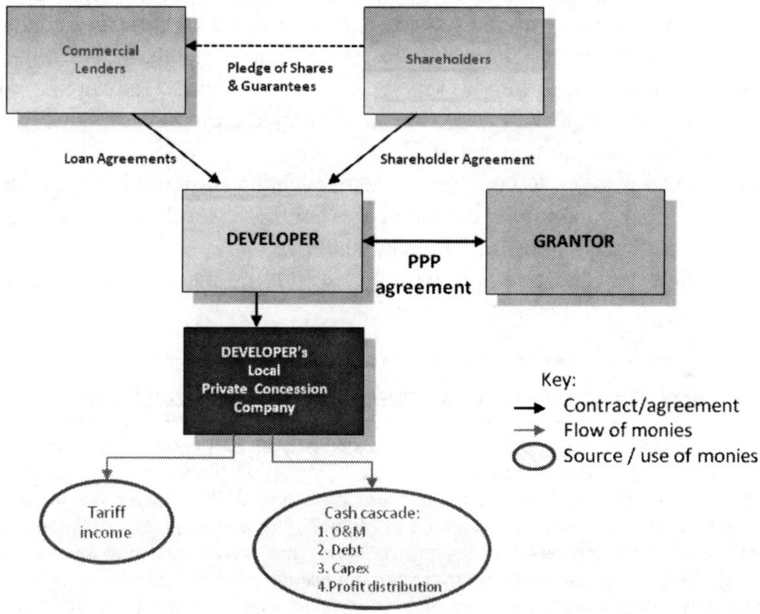

Figure 12.1 Simplified representation of a Standard PPP Agreement (Concession type).

The Subsidy Fund component is financed by an IFI and the subsidy is based on a predefined output to give incentive to the Developer to provide that output. The subsidy is paid into a fund which would be administered as a neutral escrow account, backed by an independent auditor to confirm that the service has been provided. The debt incurred by the subsidy fund would be the responsibility of the Grantor.

The main advantage of introducing OBA subsidies is in reducing the burden on the Developer of recovering all its costs (O&M expenditure, debt repayment, capex financing and profit) through the tariff itself and connections charges[4]. Instead, tariffs can remain low and the Developer is able to obtain full cost recovery and in some instances, a return on investment. The predefined outputs which are linked to incentives could include:

- The provision of access to service defined as the delivery of working connections demonstrated through a paid water bill. In this case, the total or partial cost of a connection is paid by the Subsidy Fund based on each new connection. An additional advantage is the increased accountability that comes with this system as customers can monitor the performance of the Developer on the basis of provision of connections.
- An incentive to improve and extend billing capacity and reduce unbilled water by using the subsidy to pay a proportion of the customer's bill to the Developer. In this way the subsidy is paid at the same time as a customer is billed, providing an incentive the Developer to increase the amount of water billed and improve service delivery. Another advantage of this mechanism is that the customer can identify the amount of subsidy on his bill thus improving transparency.
- Assurance is given to commercial lenders cautious about the regulatory and political risks to debt repayment, by linking the subsidy to outputs but paying it directly to the commercial lenders.
- A wastewater incentive is to base the subsidy on the volume of wastewater treated.

BOX 12.1 MOZAMBIQUE WATER: OBA AND THE WORLD BANK

Access to water in urban areas, currently estimated at 37 percent, has significantly improved through two consecutive national water projects funded by a loan from the World Bank of around US$126 m between 1999 and 2007. They have increased access to water in the five major cities of Maputo, Beira, Nampula, Quelimane, and Pemba, strengthened sector institutions and the regulatory framework, and successfully piloted the delegated management framework. This framework separates assets from operations, contracts out operations management to private sector operators, and is overseen by a regulatory body. The urban water strategy not only provides increased coverage, it also aims to secure full cost recovery through tariffs for larger cities – but this will not include connection costs.

The OBA component, Water Private Sector Contracts ($ 6 million funded by the Global Partnership on Output Based Aid (GPOBA), is targeted for poor households (with expenditures equal to or below US$1 per day) who cannot afford to pay connection costs which currently are totally funded by users and are not recovered through the tariff. Connection fees range from about US$160 per connection to well over US$240 per connection. Whilst most middle and high income households, industries, and commercial establishments can afford such connection charges, it is often very difficult for low income households. The output-based subsidy is funded by GPOBA and expected to encourage uptake by poor households, and as such is a critical component to achieve the desired outcomes of the IDA-funded scheme.

Source: Interview with Mr Lars Johannes, 12 May 2009, GPOBA, a multi donor trust fund administered by the World Bank.

It is generally recommended that the OBA Concession model be utilized in situations where Developers are unable to set tariffs to achieve full cost recovery for political or social reasons and in cases that can be comfortably managed in which the political and regulatory risks are not so great as to threaten the project. As can be seen in Figure 12.2, the use of the Independent Verification Agent is crucial to (a) ensure that the pre-defined output has been achieved, (b) provide certainty of revenue to the Developer and thus enhance the bankability of the scheme overall[5], and (c) avoids dispute between the Grantor and Developer through an external Independent Verification Agent. Nevertheless, as suggested earlier it may be necessary to provide for contract adaptations and renegotiations to take into account unforeseen changes in the initial assumptions underpinning the PPP Agreement; tariff indexations and periodic and emergency reviews.

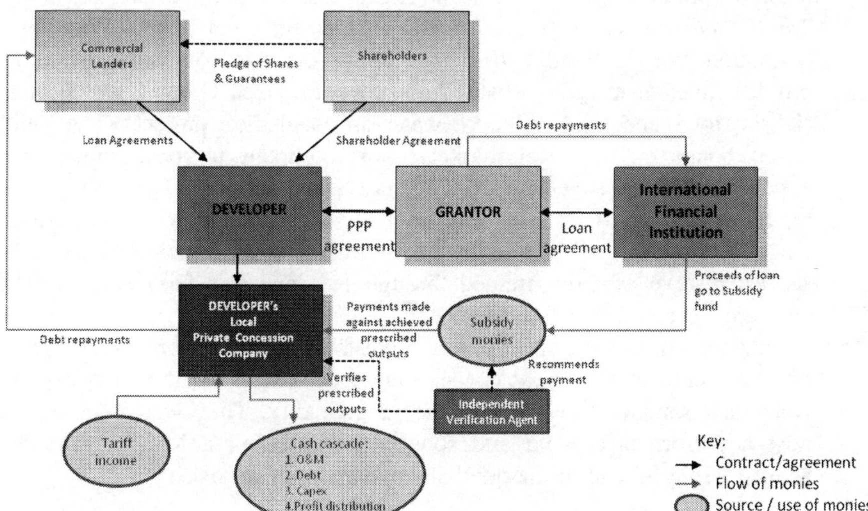

Figure 12.2 Output-Based long term PPP Agreement.

Through the use of an Independent Verification Agent to monitor the Developer's performance, this model provides a greater certainty against Grantor interference and ensures performance indicators are being met and appropriate tariff indexation practices are occurring[6]. In addition, the model will include specially constituted work group made up of independent specialists – an *Expert Panel* to advise negotiation proceedings between the government and Developer on issues related to tariff reviews, contract adjustments and renegotiations. Also, as a last resort, an *Independent Arbitrator* should be introduced in order to avoid entrenched positions of the Grantor or Developer and should an unresolved dispute fail to be settled by the Performance Auditor or the Panel of Experts.

12.3 NEW MODEL: PPP TRUST STRUCTURE

The PPP Trust Structure builds upon the successful application of similar structures in the cases of Tlalnepantla, Mexico, Tamil Nadu, India and Guayaquil, Ecuador which created reliable financing structures for the Water and Wastewater Sector operations and investments that utilized private sector expertise without equity requirements.

- *Tlalnepantla* – a legal Trust structure was used to issue debt that received a partial guarantee by the IFC ($3 m of $9.6 m issued) and backed by a letter of guarantee by a project finance specialist bank ($5.3 m from Dexia). The Trust used the debt proceeds to make a loan to the Municipality of Tlalnepantla and its municipal water utility to finance water conservation projects. The debt is repaid by user fees and local taxes. Debt holders receive assurance through Dexia and IFC guarantees in the event of default.
- *Tamil Nadu* – In Tamil Nadu, USAID enabled the creation of a Water and Sanitation Pooled Fund (WSPF) to allow the state of Tamil Nadu to provide investment funds to small and medium sized Urban Local Bodies (ULBs) to finance and refinance water and sanitation projects. The Fund raised bonds on the local market to extend loans to local utilities for expansion and improvement of WSS assets and services. The local utility repayment is based on user fees as well as state and federal transfers. USAID provides a 50% standby guarantee on bond principal if the debt service reserve account (funded through fees and transfers) is exhausted after any 90 day period.
- *Guayaquil* – In Guayaquil, a private investor/operator, International Water Services, entered a 30 year concession agreement to upgrade and expand water and sanitation services in the municipality. They were required to post a performance bond and sought and received a MIGA guarantee against wrongful call of the bond and against political risks.

This model builds upon the successes of the above mentioned projects but because of the legal structures and transaction costs is more appropriate for middle income countries and for medium and larger sized water utility companies (yet the model could be adapted for poorer countries and smaller utilities). The key structural elements are as follows:

- Create a financial and contractual structure that is ring-fenced and insured against political manipulation and partially guaranteed against default by using a Trust structure (created as a new entity or legal construct) which is then used to raise financing for the utility, and also to repay debt which is backed by guarantees. (Debt can be raised on local and foreign capital markets).
- Debt repayment is achieved through a financing structure based on a combination of user fees (tariffs), state transfers, local taxes, donor grants[7] and loans and OBA schemes.

- Private sector is engaged by allowing them to bid on the design, building, operating, maintenance, rehabilitation and finance arrangement.

In this arrangement, a utility receives user fees through tariffs and in some situations, through local taxes, state/federal transfers and sometimes donor grants/loans.

Figure 12.3 shows the basic parameters of this structure, whereby a Trust[8] is created which receives funds from local, state and federal government as well as tariffs. The Trust is positioned between the utility and local government. (This is the principle structure used in Tlalnepantla and Tamil Nadu). Depending on the national context, in some instances federal funds are mandated through the legislative framework. For example, in Mexico, the Federal Government is obliged to make available monies every year for the Water and Wastewater sector[9]. These funds can be easily pledged to the Trust as lenders are absolutely certain that those funds will be made available. Other main features include:

- Cash flows into the Trust can be pledged against debt raised by the Trust. The Trust helps mobilize local capital as it can obtain loans from local banks and bonds which are backstopped by a third party guarantee. Cash flows from IFIs to Trust can also be used to pledge against foreign denominated debt.
- Debt is used to finance capital expansion and rehabilitation of facilities. The debt is repaid through flows into the Trust. The private sector is not encouraged to provide equity in the project.
- Ownership of the utility can be retained by the municipality but the rights to operate, maintain, rehabilitate and expand are ceded to the private Operator/Developer through the contract and debt covenants[10].
- The Trust is likely to find it easier to borrow money than the Local government, given the conflicting demands on its unpledged revenue streams. The structure also helps to create a degree of transparency by separating the Local Government, local politics and the financing of the utility.
- The Trust can be structured in two ways:
  - o A legal construct established for the benefit of the debt-holders, or
  - o An actual institution with responsibilities for financial management and oversight of the utility. The Trust could also act as an investment management vehicle for other utilities and could assist in the identification, preparation, packaging, contracting and structuring of arrangements.
- Funds in excess of the required debt service to the Trust can be ploughed back to prepay debt or used for reinvestment. If the Trust is created as a legal entity, it will unlikely that it could reinvest the funds. If an investment Trust entity is created, it would be allowed to operate as an investment manager of the funds.

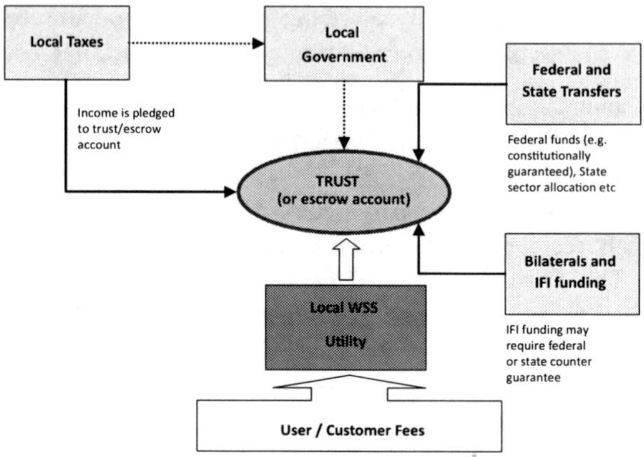

Figure 12.3 Trust Structure.

Simply establishing a Trust structure will not be sufficient to dispel private sector concerns about the risks of the utility's default payment obligations. The Trust fund creates a ring-fence around the utility's revenue flows but lenders will require greater security for debt to be repaid. This could be achieved through a third-party guarantee. Research reveals that the guarantee structures for the Tlalnepantla and Tamil Nadu Trust with partial credit guarantees (that pay out only a percentage of the outstanding principle in the event of default) played a critical role in their successful implementation. An alternative system would be to provide a guarantee that mimics mono-line insurance products to assure bond holders of complete principal repayment in the event of default. Each country's utility will require different forms of guarantees to reflect the varied nature in cash flows into the Trust (local or foreign denominated debt, IFI grants/loans).

The way to encourage bidding for this kind of structure would be to invite the private sector to tender for the ability to provide design, build, operate, maintain, rehabilitate and/or provide financial arrangement services. However, the private sector is not required to provide equity in the project. This something that Developers claim suits their business models and desired risk exposures. Instead private sector Developers compete to provide a package of services specified in bid documents and a Trust structure enables their participation and management expertise in a system designed to provide sufficient incentives without equity participation. Operators can also arrange and structure financing using a Trust structure which allows them to raise finance against public sector cash flows pledged through the Trust structure and backed by appropriate guarantees. To ensure a winning Developer's commitment to the project, the bidder would be required to post a performance bond to assure the debt-holders that the proposed work will be fulfilled as specified. The Trust could enlist the services of an independent evaluator to assess the performance of the selected Developer against a set of criteria specified in the contract.

The main incentive to meet the performance criteria of the private sector is a share in the savings arrangement. This mechanism works in the following manner: if the cash flow after debt service from the utility exceeds a projected rate of return for the investor, then this excess cash flow is divided between the Operator and the Trust. The risk allocation of the Trust Structure would be as follows:

- *Construction Risk* – private Developer backed by performance bond.
- *Tariff Collection Risk* – private Developer within contract-set parameters.
- *O&M Performance Risk* – private Developer within contract-set parameters.
- *Debt Repayment Risk* – Guarantors (first call in public sector funds collected from the local and national government) and Creditors for non-guaranteed portions including tariff based financing.
- *Regulatory and Contractual Risk* – risk of default by government covered by IFI guarantee, default by private Operator/Developer covered by performance bond.

The benefits of this proposed structure is that it enables long-term private sector participation in the Water and Wastewater Sector on a competitive, least cost, highest value basis. It recognises that the existing condition has insufficient tariff levels to cover capital and O&M costs but provides capital investment and improved O&M efficiency. In addition, this structure reduces cost of accessing finance since no return on equity is required and pledges from public sector flows guaranteed by a third party should reduce structural risk perceived by lenders.

Furthermore, this structure recognizes the decreased private sector appetite for equity investments in the water sector. It might also assist in local capital market development if the Trust chooses to borrow from local markets using financing in local currency. A Trust structure could be potentially converted into an equity investment vehicle provided it is granted a stable institutional platform and is adequately staffed.

12.4 CONCLUSIONS

The two proposed schemes are good steps towards developing integrated long term PPP solutions. Whilst they are not totally 'new' structures, they do focus on particular features, which if used, could enhance the overall bankability and sustainability of long term PPP Arrangements in the Water and Wastewater Sector.

Under an Output Based long term PPP Agreement, the Developer commits to performance targets specified in the PPP Agreement and receives financing for capital expenditure from commercial capital markets. Tariff revenues supplemented by funds payable by a Subsidy Fund against prescribed results and on an output basis and would provide the Developer's source of cash flow. These are used to cover O&M expenditure, debt repayment, capex financing and any excess would be profit. The main advantage of introducing OBA subsidies is to reduce the Developer's burden to recover all his costs.

On the other hand, the Trust structure recognises that equity is expensive and increases the overall perceived risk profile of a long term PPP Agreement. By using a trust structure, there are no equity requirements but other financial resources can be leveraged effectively through providing revenue security in pledges from public sector flows guaranteed by a third party.

It must be noted that none of these structures would work by themselves. The PPP Agreements developed should also take into account some of the recommendations on how best to deal with imperfect data, termination by convenience for both parties and other key recommendations made in Part III of this book.

ENDNOTES

[1]An example to illustrate this point would be, if underground assets are determined to be in poor conditions, a new network may need to be constructed. Reducing non-revenue water by fixing leaking pipes could also be a solution. The Grantor may argue, under the terms of the lease/affirmation agreement, that repairing pipes is the optimal solution and would classify as an O&M issue, and therefore the responsibility of the leasee. The leasee might argue the opposite, preferring the construction of a new network, which would require capital expenditure and thus fall under the responsibility of the Grantor.

[2]Cost recovery refers to the ability of an Operator/Developer to recoup its operational and capital costs. This is normally achieved through subsidies and tariffs. Financing refers to sources for capital expenditures required by the Operator/Developer to fund spending over and above the amounts received in revenues and subsidies.

[3]What is OBA? Output Based Aid (OBA) is a mechanism that ties the disbursement of public funding to achieving clearly specified results directly supporting the delivery of basic services. Basic services include improved water supply, electricity delivery, health care and education, communications services (ICT), and roads. In the case of OBA, 'outputs' are defined as close to the desired outcome or impact as is contractually feasible. For example, an output might be the installation of a functioning household connection to the water network. In some cases, an 'output' might also include a specified period of water delivery demonstrated through billing and collection records. 'Subsidies' are defined as public funding used to fill the gap between the total cost of providing a service to a user and the user fees charged for that service, justified by the need to improve basic living conditions or the existence of positive externalities. For more information on OBA schemes and the World Bank, please visit www.GPOBA.org.

[4]Another interesting example of the applicability of OBA schemes can be found in the project for the expansion of Water Services in low income areas of Jakarta, Indonesia. The objective of the project is to increase piped-water access to poor urban and slum households in Jakarta through the incumbent operator, PT Pam Lyonnaise Jaya (PALYJA). PALYJA, majority owned by international water management group Suez, has a 25 year water supply

concession contract for western Jakarta and has been operational since 1997. The project uses output based connection subsidies to connect low income households located within larger areas that are already served – the project is not focused on green-field areas, but on areas in the proximity of a secondary main. The project is able to provide services to poor urban households that would not normally be served due to their inability to afford the upfront connection charge. The OBA subsidy is able to transfer the performance risk successfully to PALYJA by paying 75% of subsidy upon independent verification of the connection. The remaining 25% is paid after three months of satisfactory service delivery. Construction began in mid-April 2008. A total of 3324 household connections have been made as of Feb 2009.

[5]As mentioned previously, significant political and regulatory risks act as deterrents for Developers to engage in long-term investments. This model provides certainty of funding Grantor obligations through the IFI loan. It also provides a guarantee to commercial lenders that payments will be made through the subsidy fund.

[6]In this regard, an independent regulator may not be required since the Independent Verification Agent will provide the ongoing monitoring to ensure the Developer is fulfilling its contractual obligations and responsibilities. The Expert Panel with an Independent Arbitrator should deal with tariff review processes if certain issues are not resolved between the parties, as described below.

[7]Donor participation may be in the form of donor grants (i.e. there is no financial obligation to repay such monies) or through soft or preferential loans in which case, depending on the IFI monies would need to be guaranteed either by the Federal or State government. For example, in the case of the World Bank, funding needs to be guaranteed by the federal entity which then 'on-lends' to the state entity. Other financial institutions such as the European Bank for Reconstruction and Development are able to lend directly to sub national entities such as state government which are guaranteed at a sub national level.

[8]A Trust is a financial structure which is ring-fenced and insured against political manipulation. In some instances this is similar to an 'escrow account' in which funds are only used against specific pre-defined criteria.

[9]Mexico and India also have other federal programmes which provide federal funds to the Water Sector every year based on the level of poverty in each state or region. These programmes also provide for a 'revenue stream' which could be potentially pledged into the proposed trust structure.

[10]It must be noted however that legal requirements may make it necessary to transfer the rights and responsibilities of the municipality for the utility to the Trust to permit the integration of the financial documentation with the legal documentation.

Chapter 13
Dealing with imperfect data

13.1 INTRODUCTION

Traditionally water and wastewater assets are underground and their condition unknown. Good management would require the utility to have an accurate understanding of the exact location and condition of its distribution and sewerage network. However, in some instances detailed asset conditions are not known. For example, a utility may have sections of the network that were developed at different times and consequently in different materials such as iron, steel, concrete etc. Each material (and depending on the condition of the soil surrounding it) will show different degradation. Accordingly, the rehabilitation or renewal of the network will depend on the type of material of the network, its current condition and the service[1] given. Under a long term PPP scheme, one of the major risks is how the risk of underground assets are transferred to the Developer and in what condition. Given the unavailability of accurate data during the bid stage and the impact that this imperfect data may have on the PPP Agreement, we shall look at ways in which mechanisms can be introduced to handle improvements in data at the bidding and the PPP Agreement stage. The accuracy and reliability of this data will be crucial to improving the operational efficiency of the partnership between the Developer and the Grantor during the PPP Agreement.

In the context of PPP schemes, and in view of developing specific recommendations to deal with imperfect data, 'data' has been defined as all quantifiable and qualitative facts to be gathered at the bid stage which may have a significant impact on the operational efficiency and on the financial balance of the contractual service as defined in the PPP Agreement. Accordingly, this section provides:

- A mechanism to identify and categorize major data items at the bid stage.
- Specific suggestions on how best to define priority data in the early years of a PPP Agreement.

- Recommendations on possible procedures for dealing with data during the life of a PPP Agreement.

The importance of dealing with imperfect data is twofold: firstly, if the PPP Agreement operation is executed efficiently and if the service is financially balanced, then reasonable targets (social, political and contractual) can be achieved and the service will be self-sustainable; secondly, given that PPP schemes are partnerships played out within the framework of the PPP Agreement, expectations must be managed and achieved for both the Grantor and the Developer.

13.2 IDENTIFICATION AND CATEGORISATION OF DATA

Data can be categorised on the basis of its increasing relevance to the service obligations detailed in the PPP Agreement:

(1) Data related to the status of the respective country/city, which may include demographic, economic/development level, social, geographic/environmental and political and cultural.
(2) Background data related to the enabling environment of the service/contract including economic and legal/regulatory issues.
(3) Data related to the service to be delivered under the PPP scheme including physical and financial data.

Sets of data ascribed under this categorisation are easy since they correspond to qualitative numeric measures. However, other necessary data includes items not so easy to consider, for example:

- Political and cultural issues (level of social pressure/political requirements for improved service, level of corruption), which may rely on subjective judgment.
- Legal and regulatory issues including availability of relevant information about the regulatory mechanisms, mechanism for co-ordination with the Grantor and quality of its enforcement at the bid stage.

Data resulting from estimations (including 'engineering judgement') and choices (targets and stage in a bid procedure not flexible enough to accommodate more realistic adjustments and over which the Developer may not have control.

Therefore, any information to be gathered for the bid will have a certain impact and will depend on its accuracy, hence an imperfect set of data will be defined as unreliable information if it is:

- Incomplete.
- Inaccurate.
- Needs to be assessed periodically because of instability over time.
- Difficult to measure.

- Is otherwise unreliable due to one or more of the following factors:
  - o Based on old studies and badly maintained records.
  - o Based on questionable or lacking studies or surveys requiring minimum technical or institutional capacity.
  - o Based on faulty estimations and inherit error in basic data.

As a starting point to deal with imperfect data it must be realized that it is rare to find 100% accurate data sets at the bid stage. Therefore, it is important to distinguish between baseline data and derived data calculated from baseline data. Derived data sets may come from different sources and can be manipulated in many different ways in order to show plausibly different results, with varying degrees of accuracy. The danger in a bid, is that the same base data may lead to different interpretations or calculations and consequently resulting in differences of opinion and expectations between the Grantor and potential Developers. This is particularly relevant for coverage targets calculation, service perimeters or levels of investments required. Bidding competitions can be won or lost depending on the accuracy or, more often than not, the inaccuracy of the data upon which bids are based. Much of the bidding criteria is decided prior to bid initiation and used by the government to set the performance baseline against which bids will be judged, with no input from prospective bidders.

Furthermore, the way in which data is dealt with may be equally as important as the inherent errors in the data itself. This and the optimism bias that inherent in setting targets based on imperfect data is important to understand and will assist both the government and potential Developers at the contract negotiation stage. Data particularly prone to inaccuracy includes:

- *Revenue and costs data*: Revenue collection performance, scope of investments required, billing ratio, price of bulk water.
- *Condition of assets*: Metering rate, unaccounted for water, condition and performance of assets, distribution network mapping, inventory of meters.
- *Population data*: Number of population, number of households and number of persons per household.
- *Customer base*: Number of households served, service coverage ratio and consumption patterns.
- *Data on water resources*: Availability and quality of ground water resources,
- *Targets*: Water quality and service targets.
- *Procedures for investment decisions*: Procedure of coordination with conceding/regulatory authority concerning decisions on investment planning.
- *Political and economic background*: Political stability, inflation and exchange rate.

The availability of accurate data is important for the Developer at the bid stage. The way the data is handled is also critical for the Grantor and his advisors when they come to define the targets and parameters of the PPP Agreement (targets, flexibility, and adjustments during the contract). Bias introduced at the data handling stage is equally as important as the quality of the data itself.

In addition, data relating to political stability, inflation and exchange rates and regulatory processes for adjustments are particularly difficult to predict over time and should probably not be analyzed in the same way as data on service parameters.

13.3 DIFFERENTIATING BETWEEN DATA TO BE COLLECTED AT BID STAGE AND POST CONTRACT AWARD

Given the constraints identified above, it is recommended that a PPP scheme focus on a limited, but targeted list of important data items at the bidding stage. In this respect it will be essential that the data to be collected should be clearly defined and the person in the best position to collect it. In most instances the Grantor will spend significant time and effort in gathering data which may not be of sufficient quality for the Developer to use or that may need to be fully warranted within the PPP Agreement, thereby shifting the risk back to the Grantor. This may not be an acceptable position for the public sector. Moreover, if the Grantor is responsible for collecting the relevant data, it may be subject to political or social pressures. For these reasons it may be more advisable to develop a mechanism whereby the private sector is responsible for data acquisition and its quality. The Developer is usually more experienced in this field than the Grantor. Usually they have a more informed judgement on data reliability issues, and are better equipped to assess and collect data related to operation and service parameters, level of service, underground infrastructure, and in some cases, water resources management. The important issue then would be to develop a mechanism to validate the accuracy and reliability of such data.

Table 13.1 Data that can be reliably determined at bid stage.

Summary of data that can be determined at bid stage		
Value of bill payments received	Value of bulk water sold	Tariff structure
Volumes of bulk water sold	Production capacity	Perimeter
Asset register (above ground)	Power availability	Pipe network mapping
Number of staff and labour costs	Cost of production per m^3	Affordability and accessibility
Debt service (historical)	Surface water resources assessment (availability and quality)	Current water quality standards

Source: Babtie (2004).

Table 13.2 Data that can only be determined reliably after several years of operation.

Summary of data that can only be determined after 5 years of operation		
Unaccounted for Water – physical and commercial losses	Accurate population figures and socio-economic data	Scope of investments required – detailed investment plans for below ground assets
Physical losses as percentage of total water produced	Consumption patterns/ water demand	Level of service and service targets
Service coverage	Metering rate	Service coverage and coverage targets
Safe yield and reliability of Water resources	Resources limitations	Assessment of groundwater resources

Source: Babtie (2004).

After consulting with major international water developers commissioned by the World Bank in 2004 (Babtie) the type of data that could be reliably obtained at bid stage and post contract award stage are summarised in Tables 13.1 and 13.2:

Likewise, the following are typical data best left to the Developer to establish post contract award.

Accordingly, in order to develop a sustainable PPP scheme the following main data would be required:

- *Financial data on costs and revenues*: The most important data that Developers need to establish prior to bid stage is basic financial data which is generally reliant on reliable information difficult to obtain, such as consumer base, commercial losses and so on. It is recommended that this data set be obtained through a Transaction advisor who knows the information. Where data is hard to obtain, it should be acknowledged that estimates require post contract revision.
- *Availability and quality of water resources and accessibility of bulk water*: Water resource issues are rarely assessed and it can be difficult to obtain information on this. Availability and quality of surface water can be measured through yields estimated through monitoring rainfall and river gauge stations and sampling and quality tests. Specialist consultants should be contracted to conduct an evaluation prior to the bid.
- *Level of service parameters*: Data related to levels of service often encompass the most politically and socially sensitive issues and are often the most inaccurate. They are often the most quoted data sets behind any drivers for reform (e.g. losses, service coverage, availability and affordability) and are extremely susceptible to biased reporting for political reasons for example, often overrepresented levels of service and population served. Quality and

service levels data are not hard to obtain yet unaccounted water data generally requires years to obtain. It is recommended that Developers with experience or independent specialist consultants be used to gather this data and bidders must be given access to it at the due diligence stage. It should be clarified that the Operator will refine the data gathered at this stage during the initial years of the contract.

- *Infrastructure asset base*: Infrastructure assets, particularly those below ground are particularly susceptible to bias and data inaccuracies. Obtaining reliable data on below ground infrastructure is difficult and it is recommended to leave this evaluation to the Operator during the initial years of the contract so that operational data can be used to determine areas requiring repair or relaying. Transaction Advisors can conduct the evaluation of above ground assets during due diligence performed by potential Operators. A monitoring regime may be established between the Transaction Advisor and the client for items such as power availability and resource availability to ascertain the performance picture.
- *Staff*: Staff costs are fairly simple to calculate and are important as they usually form a large part of the utility's operational budget. This analysis should be conducted by the Transaction Advisor. Performance evaluation should be undertaken by the Operator post contract award.
- *Customer base*: Local governments rarely maintain accurate and reliable records of households served. The customer base data is difficult and time consuming to assess. As such, it should be investigated by pre-bid consultants but cannot be meaningfully explored during due diligence proceedings.
- *Population*: Population data includes number of persons, growth rate, number of households, social distribution and income. Data on these items are usually inaccurate and rely upon periodic census data usually rife with inconsistencies. It may not be possible to assess this data pre-bid and Babtie (2004) recommend that the Operator gather this data in post contract award.

In conclusion, there are no 100% accurate data sets at the bid stage. Therefore it is better that the Grantor focus his efforts on a limited set of data centred on the utility operation. Furthermore, there is a need for clear and reasonable objectives within the PPP Agreement within manageable horizons coupled with a periodic review and revision of data baseline. A suggested response to this is detailed in the section below.

13.4 RECOMMENDATIONS ON DEALING WITH IMPERFECT DATA

The inaccuracies and issues identified in the data used in agreements can be handled through the provision in the design of the PPP Agreement for flexibility and adjustment mechanisms. However, given that data inaccuracies can have such a great effect on the outcome of the PPP scheme it is recommended that (a) the Grantor acknowledge that baseline data is inaccurate to some degree and will require periodic re-evaluation; (b) potential Developers should come to an agreement on clear and realistic targets and schedule revisions and target adjustments with details on the impact of these adjustments on their contract obligations; and (c) when evaluating Developer performance, performance indicators based on difficult to obtain data in the first years of operation should be de-linked from the remuneration system in the formative years of the PPP Agreement.

The following Figure 13.1 represents the way in which this process could be achieved:

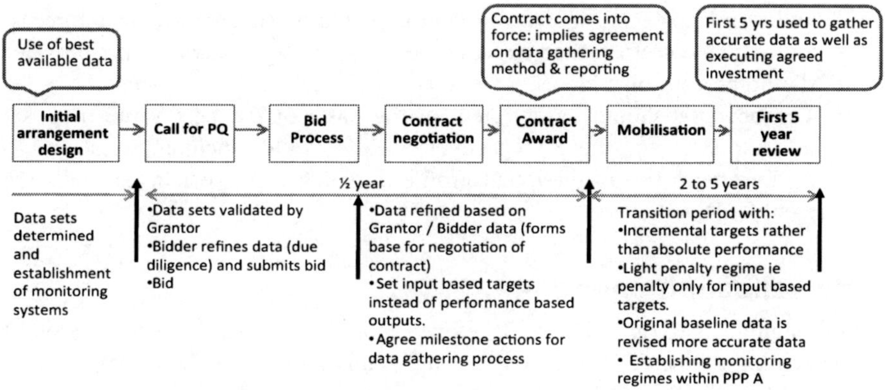

Figure 13.1 Representation on how to deal with Imperfect Data from initial bid preparation to bid submission and contract award. *Source*: Based on Babtie (2004).

In summary the proposed approach are detailed as follows.

Data gathering leading up to the bid period

When the bid procedure is launched, the Grantor must ensure that it collects all relevant data needed for the Bidder/Developer to present his bid. During this period the Grantor should act as Transaction Adviser who will collect all relevant information. The type of information to collect is detailed in the Figure 13.1. The Bidders will present their bids based on the initial data provided by the Grantor. Such information presented by the Bidders is likely to be an improvement on the Grantor's base data.

The long term PPP Agreement will be developed depending on the availability and quality of data the basic structure. Splitting the long term PPP Agreement into two phases is to be recommended: an initial period of 2–5 years followed by the remaining period of PPP Agreement contract. The contractual implications of this are discussed in the section below.

The bidding procedure should be as follows:

- Grantor should agree on a set of specific key indicators to set reasonable aspirational targets for the duration of a long-term contract based on a starting point derived from best available data at the bid stage.
- Targets will reflect the reform and performance improvements that the Grantor is seeking in the sector. Long-term incremental targets can be set at the beginning of the PPP Agreement or at the end of a pre-defined transition period. The Grantor and the Developer should agree clearly at the negotiation stage on the adequate timeframes to set the targets.
- For each indicator, the Grantor should be able to forecast the trend that the indicator is expected to evolve over the period of the contract. This forecast indicates the performance target at the bid stage, against which the operator should prepare his bid.
- The operator should be selected on the basis of the best value proposal in response to these objectives, and should include a short to medium term fixed investment profile in which early year investments are identified[2].

Data validation period

Before signing the contract, the Developer should agree to a revision period at the beginning of the PPP Agreement. During this period, the Developer will be obliged contractually to obtain accurate system data based on his experience of operating a utility. The original baseline data will then be revised using more accurate data. The following guidelines should be considered during this transition period:

- The time-frame must be identified clearly in the contract (usually between 2–5 years)[3]. The time-frame will reflect the varying amounts of time required to obtain complex data.
- Input based targets should be set, instead of performance based outputs[4]. The targets should be specified in physical terms and could be linked to incentives (e.g. number of km of pipes laid)[5].
- Milestone actions which will assist in the data gathering process are also recommended (e.g. aerial mapping photographs, house count surveys, zone metering monitoring etc.).
- The target during the transition period should be incremental increases within allotted periods rather than any absolute performance requirements.
- During the transition period, the penalty regime should be more lenient to allow for stability in the overall PPP scheme in which there should be no

target based penalties enforced which affect tariffs or investments. Instead maintained penalties should be focused on input-based targets.

Data acquisition between the bid stage and first major review period of contract

During the initial years of the contract, more accurate data on the utility will be obtained and reported to the Grantor when required. The data will usually lead to the adjustment of targets and contract parameters and therefore it is critical that procedures on adjusting PPP Agreement terms are clearly delineated in the contract. There are two possible options to agree on the new values of more accurate data:

(a) The use of a regulatory body or relevant institution.
(b) In situations where regulatory capacity and experience is limited, an independent expert may be used.

Provisions should be made in the PPP Agreement on how the expert is to be selected, how the expert will be paid and his respective terms of reference. It is also critical that the manner in which the new data is to be handled should be stipulated in the PPP Agreement. Varying methods of calculation and evaluation should be agreed upon and listed in the contract.

Contract adjustments based on revised baseline data should be agreed upon at the negotiation stage. After several years of improved data collection post contract award, the more reliable data would establish the true position of the performance indicator, and the actual improved profile would be compared to the predicted position using the original baseline data. Of course, these values could be compared against the original base data to assess the real 'degree of improvement'. The precise parameters of the revisions will be agreed upon during the step indicated above.

Key features of the initial period include:

- The arrangement is akin to an enhanced lease/affermage arrangement with considerable responsibilities devolved to the Developer.
- The Developer is responsible for the operation, maintenance and working capital from his own account.
- Minimum performance standards should be specified:
  - A limited number of service standards.
  - Investment milestones.
  - Network outputs delivered on time: asset evaluations, improved collection and billing systems, and so on.
  - A great deal of importance would be placed on data-gathering during the initial period in order to provide the raw data and analysis that will underpin the price review to take place at the end of the initial period.
- The Developer given a further incentive to improve operation and maintenance costs.

- Reduce risks to both parties and simplify relationships, by indicating clearly each party's responsibilities in the contract.

The second step is to define through contract provisions how the revised data will impact the PPP Agreement through revised targets, service parameters, investment and remuneration. The Grantor and the Developer should agree on a time-frame, within which the indicators would be reviewed periodically and the targets revised. An adequate time-frame for review would be between 2 and 5 years, depending on the particular parameters of the contract and of the service, at which time more accurate and reliable operational data will be available. This timeframe should be stipulated clearly in the contract[6].

Periodic revision of targets, service parameters and operator remuneration

The time period for adjustments should be clearly set out and should mesh with other periodic review periods and transition periods as set out in other areas of the contract:

- The contract should detail the terms of reference of the party responsible for adjusting the targets and how that party is to be appointed, operate and be funded.
- To avoid the need to rebase, in situations where the original estimate values are significantly different from the new data, both parties should agree to extend the time-frame for achieving the targets. 'Trigger limits' between the original and new data values can be detailed in the contract to allow for that extension.

Both parties must agree upon a financial model that connects the baseline data with targets related to service levels, investment, tariffs and cash-flow profiles. Each party must also agree to accept and follow adjustments to the model resulting from the revised baseline data. The financial model should not be overly complicated so as each party understands and can easily interpret it. The model will detail the cash-flow curve and a flexible remuneration framework which interfaces with additional factors to consider such as investment amounts, service levels and debt service.

After the transition period has ended, the relevant adjustments have been made to the baseline data and performance targets have been adjusted, the arrangement will continue to operate with the penalty and incentive mechanisms fully operational. Data will continue to evolve through the life of the contract and it is recommended that further baseline data review periods be incorporated into the contract but with more stringent parameters for acceptable errors. An alternative suggestion is to set 'margins of tolerance' for key data indicators. If applied, the penalties should entail significant increases or decreases in contractual obligations for each party. These margins should be agreed upon for each indicator in the contract.

Also when the transition period has ended, the Grantor and the Developer should agree (negotiate) an investment plan and associated performance targets for the remainder of the PPP Agreement based on an adjustment mechanism agreed at the bid stage.

13.5 CONTRACTUAL PROVISIONS TO ADJUST FOR THE IMPLICATIONS OF IMPERFECT DATA

In coordination with the provisions detailed in Chapter 9, dealing with imperfect data will require a different approach to the PPP Agreement's contractual structure. There are two main recommendations related to contractual provisions:

- Data improvement clause.
- Mutual Convenience Termination/Break clause.

Data improvement clause

As can be seen from Figure 13.2, the basic process for dealing with imperfect data is that the Transaction Advisor should determine which data will be important for bidders in estimating the cost of service. Explicit assumptions are to be provided on these key variables in the bidding documents. Bidders should bid using those assumptions and after a predetermined initial period – say 12 to 24 months or so after contract award – the actual levels of those variables would be determined. Both the Developer and the Grantor would have the ability to terminate the Agreement using a Mutual Termination Clause (as described below). The tariff or other aspects of the PPP Agreement are adjusted through a well-specified process so that the Developer remains in the same situation as it would have been had the original assumptions held.

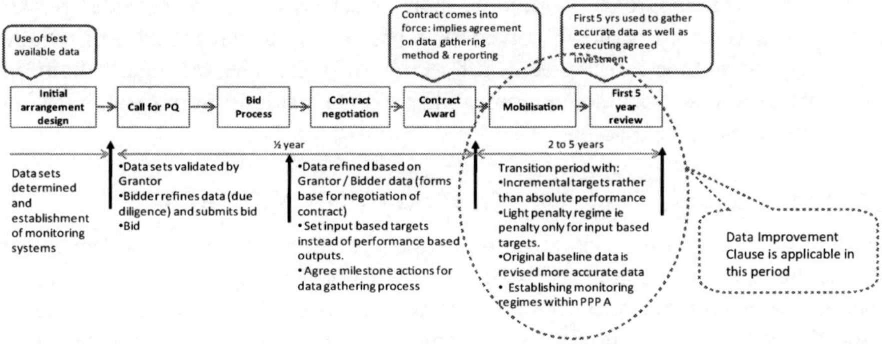

Figure 13.2 Representation of when to apply the Data Improvement Clause. *Source*: Based on Babtie (2005).

In order to deal with the data validation process a process similar to the one used for Extraordinary Adjustments is suggested. This is described in detail in Chapter 9[7](Section 12.7) and is a good model because it ensures that the Developer is in no worse situation than if good data had been available at the time of bidding[8]. This suggested approach would include:

- Once the PPP Agreement has been awarded, the Developer reports the actual values of the specified variables, and the affect this has on cost of service and the achievability of other contractual commitments. The Expert Panel assesses and verifies that the Developer's report is reasonable and undertakes a preliminary assessment of the affect on the Developer's financial position as a result of actual values being different from the assumptions in the bidding documents.
- The Expert Panel proposes changes in the tariff or service standards to take into account the revised information (initially presented by the Developer and validated by the Expert Panel). The purpose of such a process is to ensure that the revised data would put the Developer in the same financial position that he would have been if the true position had been as assumed in the bidding documents.
- The Expert Panel should issue the recommendation of the tariff or service standards. The procedure by which they are introduced into the PPP Agreement will be up to the Developer and the Grantor. They may, for example, simply reduce service standards or increase tariffs. If the parties cannot agree an Arbitrator should decide.
- Alternatively, the Grantor may provide the Developer with a financial model of the business prior to contract award. This was discussed in Chapter 8. This would constitute the basis of any future negotiations between the parties. Thus any negotiations resulting from improved data collection are reflected in the financial impact of deviations from the original assumptions. In this way the process would involve replacing the assumed data in the model with the actual data. This approach, although more mechanistic, may provide greater certainty to all parties involved. The use of a financial model will require the bid and the PPP Agreement to establish in the rules the legal validity of the financial model, the enforceability of data resulting from the application of the model and so on.

Mutual convenience termination or break clause

Prior to the end of the initial transition period, either the Grantor of the Developer would have the possibility of ending participation in the arrangement without cause. Under regular contracts, the right to terminate without cause is normally only granted to the Grantor[9] (if it is granted at all). Research done for this book suggests that normally under an early termination or in a default situation, the initiating party pays a penalty for breaching the agreement (Figure 13.3).

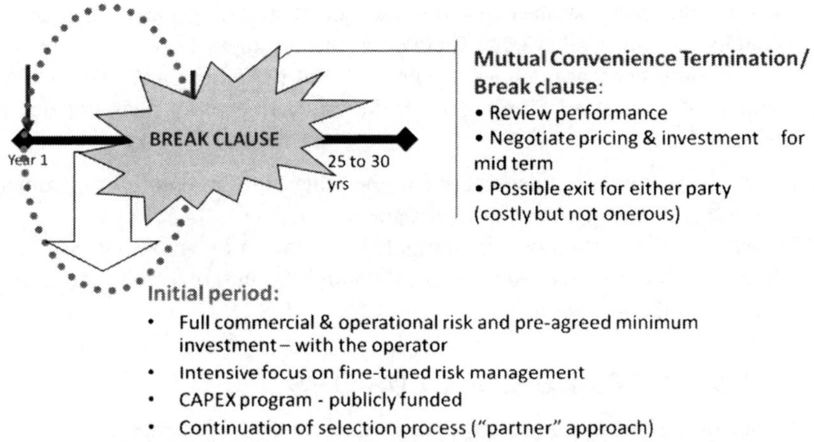

Figure 13.3 Mutual Convenience Termination or Break Clause.

Developing from Shugart's (2004) basic idea, each party should have the right to invoke a termination of convenience. The basic premise of this idea is that a party may not wish to continue with the contract. However, terminating the relationship would not be without consequences, that is, there would be a significant monetary cost to the party invoking such termination right. If both parties were to face a significant financial penalty for terminating the contract, it is unlikely that short-term issues will lead to termination[10] but rather, a greater focus will be placed on resolving the dispute. It should be noted that such a provision would include the following conditions:

- The right to terminate could only be invoked at a specific time at the end of the initial transition period.
- Clause could only be permissible prior to the tariff review in order to avoid messy negotiations for the regulator or expert panel.
- Invoking such a clause would incur a charge to be set at such a level to create an incentive for each party to remain in the relationship but should not be so great so as to place too great a financial risk on the arrangement in the initial stages of the relationship. Establishing the fee should be commensurate with the cost of covering the unwinding the PPP Agreement (for the Developer) and the costs of a transition and rebidding for the Grantor[11]. Three possible options could be used to handle the mutual convenience termination.
- The Developer and the government hand over a sealed statement to a third party by a specified date in which their decision to continue or cancel the arrangement is stated.
- In the event that both parties decide to terminate, neither party pays any penalties and losses are absorbed.

- If only one party wishes to terminate, the terminating party must pay the penalty fee to the other party dictated in the contract.
- If one party does not deliver a statement and the other party does deliver a statement requesting a termination, the requesting party does not pay any penalty fees.

One party can create a mild disincentive to the other party initiating a convenience termination by submitting a negative statement.

The risk to the Developer is mitigated by the provision that all capital expenditures whose cost recovery extends beyond the initial period would be the responsibility of the government.

13.6 SUMMARY AND CONCLUSIONS

The recommendations on the suggested approach to deal with imperfect data are:

(1) Establish monitoring systems at the bid stage and provide refined data package at the beginning of the PPP Agreement/negotiation stage.

(2) There should be an 'input-based' transition period focused on the Developer meeting specific early year targets, that is:
  - o Input targets specified in physical terms – for example, implementation of capital works that are immediate in nature eg investments to ensure security or reliability of supply.
  - o Gathering data in the early years to become part of operator's contractual objectives – targets.

(3) Performance monitoring should be based on:
  - o Measuring incremental improvement.
  - o In the early years there should be a basket of indicators, based on 'easy-to-measure' sets of data, reflecting overall utility performance.

(4) Both parties must agree upon a financial model that connects the baseline data with specific service targets. Each party must also agree to accept and follow the adjustments to the model as a result of the revised baseline data.

(5) Consider other types of adjustments other than tariff adjustments – cascade of options to avoid conflict during rebasing.

(6) After an initial 2 to 5 year period (the comprehensive review period), the Developer can negotiate adjusted performance targets and investment plans based on the new and more accurate data[12]. This adjustment system will be clearly detailed during the bid negotiation stage.

Finally, it must be noted that the data monitoring aspects become particularly important under the situation described herein. If such an approach is used under a PPP Arrangement the normal monitoring and control system (see Chapter 10, provisions related to service standards, as well as monitoring and enforcement) will have to be specifically adapted to permit accepting and incorporating the changing and improved data standards used. This needs to be adequately captured in the PPP Agreement and reflected in the institutional regulatory and monitoring arrangements (see Chapter 5 and some of the specific recommendations made in the next chapter).

ENDNOTES

[1]In this context, service is based on the level of unaccounted for water (UFW) in a given section. And is defined as the level of degradation in that given section of the network showing the difference between the level of water produced by a system and the level of water actually billed (and collected).

[2]The bid process, where appropriate, may benefit from the use of a peer review challenge process, whereby the contractual objectives set by the Grantor during bid preparation are subject to challenge by a peer team, independent of the transaction advisor, who are knowledgeable 'experts' in long term PPP schemes. This process would have the advantage of ensuring the requested objectives of the sector are plausible and reasonable and would be similar to the type of peer review that Developers themselves would do at due diligence.

[3]An adequate time-frame for a transition period is likely to be between 2 and 5 years, largely as a function of the status and level of preparedness of the specific utility in question. Discussion with technical experts (Adrian Harrisson of Babtie, and Scott McIntosh of MottMacdonald) suggests that it is not unusual that a period over 2 years is required to gather robust information on parameters like Unaccounted for Water, and different periods needed for different situations.

[4]It is recommended to use 'best value' or 'non-price' bidding criteria instead of just tariff criteria (e.g. lowest required tariff) and to focus more on the level of service as a target instead of the amount of investment required.

[5]Other examples of milestone actions aimed at gathering reliable data for which the developer would be responsible as input based targets, might include for example:

- Procure mapping or aerial photographs for all urban areas over an estimated population of 10,000.
- Complete house count survey: identify and map number of connections, number of households within 250 m of a shared connection, house count without direct access to piped supply, number and location of standpipes and water tanker points.
- Demarcation of complete supply zone.
- Establish metering monitoring regime of zone: procure, install and replace meters.
- Complete production monitoring programme and establish key production records reporting; install bulk meters at each water production site and service reservoir.
- Complete consumer numbers for each system.
- Establish consumer complaint database.
- Establish level of service monitoring regime.
- Establish demand/loss management regime.
- Implement programme of sanitation education measures for new connections.
- Commission integrated water resources development plan.

[6]Past experience has demonstrated that this is often the most contentious element of defining an arrangement and as such care should be taken in the contract design stage to ensure that contract adjustment mechanisms are clear, precise and understood by all parties.

[7]Emergency adjustments are an emerging option as demonstrated by their use in the UK Water Licenses (the 'Shipwreck Clause') and are implicit in the US rate of return regulatory approach as well as French civil jurisprudence. Such clauses are designed to avoid the need for renegotiation by providing a structured way of dealing with events which fundamentally alter the financial conditions under which the services are to be performed. In the context of Imperfect Data, the suggestion is that the Imperfect Data Clause that is being proposed here parts from the premise that the Developer must be protected against imperfect data but creates a strictly contractual procedure whereby as accurate data becomes available it can readily be integrated into the PPP agreement without the need for a renegotiation.

[8]The actual values for key variables may be different from the assumptions used by the Developer or as originally contained in the bidding documents.

[9]See discussion in Chapter 9.

[10]Many PPP Agreements using the concession model in Argentina were rescinded in the years after the 2001 economic crisis. However, in the case of the province of Salta, the PPP Agreement continued until May 2009 Juan Manuel Urtubey, governor of the Province, terminated the concession contract because of non-compliance and created a transition unit to provide services until a state company could be created whose shares would be held by the Province (90%) and workers (10%). (Télam (the Argentine national news agency): Salta rescinde concesión del servicio de agua potable, May 25, 2009).

[11]It must be noted that the payment required by each side may differ from each other. It would be important to establish that payment amounts should be equitable, e.g. the payment for each side could be X% of the estimate of the expected net costs to the other side arising from termination, where X is the same for both parties.

[12]The PPP Agreement should target and measure incremental improvement for a period of time, instead of an absolute level of performance, especially during the transition period. Instead, as a pragmatic example, one could consider targeting an absolute level of 35% of unaccounted water in year 5, the target could be an incremental improvement of 8% over the first five years. If the assessed level of unaccounted water reveals the baseline to be inaccurate, it will then be easy to transpose the improvement made by the operator to the correct values and to assess his performance against these more efficient and accurate targets.

Chapter 14

Improving procurement proposals for PPP Agreement

14.1 INTRODUCTION

A key recommendation resulting from the initial period of the PPP Agreement and the inadequacies of data mainly related to underground asset conditions is the need for enhanced data acquisition once the Developer has been contracted. This has been discussed extensively in the previous two chapters. Accordingly, the main recommendation resulting from Chapter 13 is the need for specific contractual provisions in which a mechanism is formulated for the progressive and systematic acquisition and improvement of data. This obviates the need for a Data Improvement Clause and for a Mutual Termination or Break Clause. However, although this is part of the PPP Agreement, the fact that there is a data acquisition and validation period with an option for both parties in the agreement to terminate such agreement, could be argued to be an 'enhanced procurement' process. This is why this chapter has been developed to specifically deal with this enhanced procurement process. The reason for this is that if a period of data acquisition and refinement is being used in a PPP Agreement, the decision as to whether a longer term engagement is continued would remain the Developer's choice. Below are suggestions as to how to carry out the Data Improvement Period.

14.2 ENHANCED PROCUREMENT TO DEAL WITH THE DATA IMPROVEMENT PERIOD

Prior to the end of the initial period, a chosen regulator or expert panel will be called upon to perform an evaluation of the accuracy and reliability of the data available. The precise procedures regulating this review will be clearly defined in the PPP Agreement in order to reduce discretionary levels and potential disputes. As mentioned before, the decision of the regulator or expert panel should be binding. In addition, as this is adjudication by a third party, it is not necessary for rebidding to take place. What is crucial to the process is that the review should not consider the price or cost bids contained in the bidder's original proposal through which the winning bidder was selected). The tariff review process should be based on the *actual* costs of the Developer over the initial period with considerations of future investments and O&M requirements, projected costs based on the Developer's historical costs, adjusted for future efficiency changes, and expected costs of a hypothetically efficient company[1].

Responsibility for capital investments during the initial concession period

Typically, under a traditional long term PPP Agreement, the Developer finances capital expenditure but the Grantor has strong opinions on how the capex should be directed. Under this revised approach, the responsibility for funding capital expenditure is split between the Grantor and the Developer. It is to be recommended that the Developer is granted full discretion over the investment program but that the Grantor be allowed to insert provisions/constraints to assist it in achieving his goals or to prevent any possible unwelcome developments that might arise from total Developer discretion, or issues over which the government has particular opinions even though the Developer does not deem them critical to his operations. Therefore, particular care should be taken over the wording in PPP Agreements and phrases such as 'both parties to agree' and 'aspects are to be coordinated' are to be avoided through lack of specifics. Prior to signing the PPP Agreement the Grantor and the winning Developer should agree upon a three-year rolling investment plan. This approach will minimise disagreements in the first years of the contract and highlight any possible gaps or issues not yet considered. Furthermore, given that the relationship between the Grantor and the Developer is crucial to the long term success of the PPP Agreement, this initial period (of relatively low risk or at least at a lower risk than if there were no possibility for an early termination of convenience as described in the chapter above), discussions on the proposed budget will give each party experience of working alongside each other and indicate any institutional personality mismatches that might hinder the relationship.

Both parties should share information on the scope of their capex responsibilities and will decide a detailed capital investment program for the forthcoming year and a conceptual program for the following two years. If no agreement can be reached, a structured process of dispute resolution stipulated and detailed in the concession contract, will proceed. The independent adjudicator (detailed below) will be consulted if no agreement can be reached at this stage.

Building on experience using independent adjudicators in the UK, it is to be recommended that each party agree on appointing this independent party at the beginning of the PPP Agreement. The adjudicator would be retained throughout the entire initial period of the contract. The adjudicator will have an important role in resolving disputes, if the Grantor and the Developer cannot reach an agreement (after a set period of time). In which case, the independent adjudicator is authorized to make binding decisions. If the Grantor fails to comply with the adjudicator's decisions, the Grantor will be liable to face claims for damages. The structure for making damage claims should be detailed in the PPP Agreement and the adjudicator will be encouraged to make swift and punitive decisions with little room for disputes or appeals.

14.3 PREQUALIFICATION AND AWARD RECOMMENDATIONS

The proposed recommendations differ from the usual criteria of other PPP Agreements reviewed in that they seek to go beyond the usual contract-based criteria and develop a greater understanding and examination of potential Developers. The key point behind this approach is to view a PPP Agreement as a business *relationship* rather than merely a contractual relationship. Therefore, the suggestion is to go beyond the regular levels of due diligence performed by Grantors to obtain a greater understanding of prospective candidates in their performance, management and experience profiles.

Evaluation at prequalification stage

Conventional initial criteria are used to screen candidates – number of similar projects, specific experience, and basic financial data. Those that pass the minimum criteria proceed to further evaluation.

- *Reference Case Study*: The second stage involves a greater in-depth examination and assessment of the candidates than traditionally used. The level of examination would be akin to that used prior to engaging in a joint-venture with another business. The candidates need to submit much greater detail about their experience by presenting a case study to be evaluated by an examining committee. The committee would study the reference case study submitted and would conduct interviews with people involved in the project and obtain further details from the prospective candidate if required. Should it be necessary, a second reference case study could be requested if the first case proves insufficient.
- *Business Qualification Statement*: This would be similar to a review conducted by a credit rating agency, requesting more detailed information than in traditional procurement, for example:
  - Profile of the candidate.
  - Organizational structure and management.
  - Corporate strategy.

- o Market standing.
- o Specific markets and experience.
- o Financial strengths.
- o Future challenges and risks.

The Business Qualification Statement would seek to go beyond the usual questions by requiring potential bidders to provide firm evidence of past experience, strategies and the thought-process behind some of the decisions made in past projects. The proposed pre-qualification strategy places more emphasis on past deeds rather than what Developers propose they will do in their bids. In addition, a scoring process that uses judgmental criteria differing from the usual merit-based scoring system or weighted average formulas could be used. Well reasoned arguments explaining the decisions need to be in writing[2].

Pre-Contract Signature assessment

It is recommended that after bids have been evaluated and substantially responsive bids are selected, a Pre-Contract Signature assessment be conducted on the bidder/Developer with the best bid price. This is an intensive process and if, after the evaluation, the first-ranked bidder is deemed inappropriate, the process is conducted on the next highest price-bid. The rationale behind the implementation of the assessment is to establish whether:

- The bidder is technically and financially capable of executing the project to an acceptable standard.
- The technical, financial, scheduling proposals and assumptions are realistic, given the parameters of the proposed contract, the risk matrix and the operational conditions (and whether there are realistic justifications given for these assumptions).
- The bid price is realistic given the technical requirements.
- There are any identifiable risks in any aspect of the proposal.

Conducting such a thorough analysis of the bid at this stage might lead to exposing flaws in the bid unable to withstand scrutiny. At the same time, it may also reveal elements of the contract requiring greater deliberation and revision. If new risks are identified, changes to the contract can be made barring any material modifications that might affect the bid price. The most important effect of such a rigorous evaluation would be to induce the bidders to take even greater care with the preparation of their bids.

After the evaluation is conducted and the bidder is found to be capable of executing the contract adequately, final negotiations can commence. However, a potential flaw is to be recognized behind this proposed plan in the possibility of evaluators who are biased to reject the highest-price bid. Therefore, a more impartial assessment might take place if evaluators were not aware of the figures in the price evaluation.

14.4 THE DEVELOPER'S REMUNERATION AND THE BID PRICE

It is recommended to use a single criterion for the evaluation of bids. These could include:

- Tariff level – lowest is best.
- Subsidy level – lowest is best.
- Up-front transfer payment – highest is best.
- Lump sum payment linked to performance standards – highest for both is best.

For the tariff and subsidy level, the Grantor should define the tariff profile over the initial period (3–5 years) and the Developer' remuneration package would be set to create an incentive for improved performance. As stated earlier, during the initial period capex financing would be provided by the Grantor, therefore the incentive payments would only need to cover incremental O&M costs. The remuneration should be set to cover the optimal O&M and other costs for the initial period (e.g. financing costs for working capital). Furthermore, the remuneration should be set at a generous level so that should adverse conditions ensue, the total expected remuneration is enough to cover the Developer's essential costs and allow him to perform basic functions without need of Grantor support.

Incentive payments should be based on:

- Key performance indicators that encourage Developers to increase water availability.
- Discrete variables and outputs (e.g. network model, leakage detection) that earn incentive payments when delivered but are decreased every month if not delivered.
- Service requirements of such import that if not met, significant damages may be claimed – for example wastewater quality.

For a lump sum payment linked to performance standards, the Grantor should decide upon the performance standards and judge how the Developers' remuneration could be calculated in view of these performance standards – base remuneration and incentive payments – along with the capex budget. Prospective Developers will then bid in the following manner:

- Bidders will quote the lump sum payment to be made to the Grantor should they be awarded the contract. The higher the lump sum, the higher the likelihood of winning.
- If a bidder is unable to pay a lump sum, he can bid on the annual payment he requires from the Grantor, in addition to other remuneration. The lower the payment required, the higher the likelihood of winning.

Using a transfer payment tackles the issue of low-balling that occurs when tariff levels are used as the price criterion for decisions. In this way regardless of the lump sum quoted, the Developer would receive sufficient remuneration from the tariff structure to enable him to focus on operating the concession. This would lessen the requirement for the Grantor to renegotiate the remuneration terms should the Developer suffer losses in his equity, return or any O&M deficits required. With this risk mitigated, Grantors can expect more serious bids from potential Developers.

14.5 SUMMARY AND CONCLUSIONS

This chapter covers the need for enhanced data acquisition once the Developer has been contracted. Building on the conclusions of Chapter 13 and introducing the need for a Data Improvement Clause and for a Mutual Termination or Break Clause, it is argued that the introduction of these clauses in a PPP Agreement is in effect an 'enhanced procurement' process. Therefore the process of continued selection of the developer is recommended even though a PPP Agreement has been signed and how it should be done. The principal recommendations are:

- Prior to the end of the initial period, a chosen regulator or expert panel should be called upon to perform an evaluation of the accuracy and reliability of the data available. There should be adjudication by a third party if the parties cannot agree.
- The tariff review process should be based on the *actual* costs of the Developer over the initial period with considerations for future investments and O&M requirements, projected costs based on the Developer's historical costs, adjusted for future efficiency changes, and expected costs incurred by a hypothetically efficient company.
- Selection of a potential Developer should go beyond the usual contract-based criteria and develop a deeper understanding and perception of potential Developers. The key essence behind this approach is to try to build up a business relationship in a concession agreement rather than a mere contractual relationship. A scoring process that uses judgmental criteria differing from the usual merit-based scoring system or weighted average formulas was also proposed. Decisions would have to be explained using well reasoned arguments and set out in writing.
- After bids have been evaluated and substantially responsive bids have been selected, it is recommended that a Pre-Contract Signature assessment be conducted for the bidder/Developer with the best bid price, ultimately allowing for any final scrutiny and exposure of flaws in the bid, identification of new risks and minor modifications prior to PPP Agreement going into effect.

This chapter also recommends using a single criterion for the evaluation of bids for example lowest tariff, lowest subsidy and so on. Given that one of the key recommendations in this chapter is that during the initial PPP Agreement period capex financing would be provided by the Grantor, incentive payments to the Developer would only need to cover incremental O&M costs. The remuneration should cover the optimal O&M and other costs (e.g. financing costs for working capital) in the initial period. Furthermore, the remuneration should be generous to cover the Developer's essential costs in adverse conditions and sufficient to allow him to perform basic functions without the need of Grantor support. If a lump sum payment linked to performance standards is the bidding criteria, the Grantor should decide on the performance standards and calculate the Developers' remuneration on the basis of attaining these performance standards.

ENDNOTES

[1]Given that this is a new way of doing things, it is to be highly recommended that the rules and procedures governing this process are developed thoroughly to avoid demotivating potential Developers. However, it is to be noted that this process would reduce the overall risk to the Developer as he will not be held responsible for capital investments that extend beyond the review period.

[2]The use of judgmental criteria requires a greater amount of discretion on the part of the evaluators and less reliance on complex quantitative formula. Therefore it would not necessarily be as transparently evident as a simply quantitative process. However, a list of the core qualities desired in the Developer is to be recommended. These qualities should then be applied to further-level criteria (i.e. whether the Grantor would value more regional understanding, local partners etc.) which would provide guidance to the evaluation committee. Sound decision making using judgmental criteria requires compartmentalizing the issues and applying discretion only to the areas that require it in which evaluators must provide sound justification for assessments made. Ultimately, the aim is to use the same rigour as in the evaluation of bids by the Grantor for short listing bidders. Using discretional evaluation may face critics who propose only mechanical evaluation. However, as discussed in the section on procurement, the European experience of PPPs suggests that even a modest amount of discretionary judgment may eliminate some of the mechanical evaluation system issues.

Chapter 15

Summary conclusions and recommendations

Throughout this book we have sought to determine, and make recommendations on, measures to optimize the use of Public Private Partnership (PPP) in developing infrastructure whilst ensuring sustainable long term provision of water and waste water services. To achieve this objective the focus has been on providing detailed recommendations on contractual issues and structures as mentioned below.

The provision of a public service requires a careful balance between the potentially divergent interests of the public and private sectors. On the one hand, the public sector has a responsibility and an obligation to provide services to its inhabitants and ratepayers. At the same time the private sector is entitled to receive adequate compensation for the risk of managing the provision of an infrastructure service, as well as compensation for putting its Private Capital at risk. Since, 1989 we have seen a surge in the number of 'Public Private Partnerships' (or PPPs) in many sectors but particularly in the water and waste water sectors. The considerable case histories amassed show that these PPP arrangements can be an effective tool for developing Water and Waste Water schemes and long term management and operation. However, it is also clear that there is a considerable range of effectiveness in these PPP arrangements, including some outright failures.

The following questions related to the effective design, development and implementation of PPP arrangements for the Water and Wastewater Services Sector have been discussed here:

(1) How do we best capture the benefits of the private sector and blend it with the strengths of the public sector?
(2) Which aspects of PPP arrangements work and which do not?
(3) Are there any common lessons to be derived?
(4) What special legal and financial instruments are needed to deal with PPPs in infrastructure, particularly when applied to the socially sensitive sector of water and sanitation services?

The methodology was to develop practical approaches for optimising PPPs, based on research in international case histories, literature and contractual document reviews, consultation with experts and fact finding missions. The approach was based on:

- Review of water and waste water sector issues, and general review of use of PPPs taking into account levels of service and cost implications.
- Establishing key issues in effective PPP development.

From this it was assessed that while current literature and guidelines provided a variety of commercial and technical recommendations, there was a marked lack of comprehensive guidelines for a legal and contractual framework specifically adapted to PPP projects in the Water and Wastewater Sector. To cover this need an approach has been developed to incorporate a practical mechanism to ensure optimal implementation of PPP arrangements and provide a review and detailed recommendation of contractual issues for optimal and sustainable long term provision of Water and Wastewater Service. Furthermore, it became apparent that dealing with limited or unreliable data, was an element that has not been comprehensively dealt with in current PPP Arrangements. As it has a 'make or break' effect on project viability an approach has been suggested to ensure long term contractual sustainability. The proposition involves two possible hybrid structures focusing on particular features which could enhance the overall bankability and sustainability of long term PPP Arrangements in the Water and Wastewater Sector.

The conclusions drawn show that for the PPP arrangement to be effective, the design of the structure for private sector involvement needs a clear definition of various aspects which include the roles played by the government and the private sector, the type of structure to be adopted for private sector participation, the applicable regulatory regime and its effects and the influence of specific current economic and political conditions. The type of PPP structure chosen will have a direct influence on the provision of services. Therefore, it is essential that an adequate legislative and regulatory framework be in place to ensure the long term sustainability of the chosen PPP Arrangement. Nevertheless, it is essential to realize that however much governments put the onus of risk onto the private sector, if things go wrong with the PPP Arrangement, it is unlikely, for social and political reasons, that the government will be able to walk away from resuming these obligations, which, in turn may be very costly. This is a key reason for ensuring PPP effectiveness and sustainability. The arrangement not only has to consider the skills, resources and responsibilities expected from the Developer but for project viability and effectiveness and the manner to harness the profit motive of the private Developer through the structure, the contract and associated incentives of the Arrangement.

Furthermore, since the service level provision is a key tenet of the Water and Waste Water sector, experience shows that an iterative process should be used to establish effective scheme service standards, sizing, costs, funding and financing. If service levels are taken as the principal objective PPP schemes, this iterative process would define an affordable scheme for which effective operational, financial and contractual arrangements have to be developed. At the same time, determination of affordable tariff levels is key to ensure customer ability to pay for services, and hence the financial sustainability of the services. For schemes in developing countries some form of subsidy is generally necessary, as affordable tariff levels are not high enough to provide full cost recovery. As levels of service improve details of any subsidy arrangements and any progressive program for change in tariff levels, need to be clearly defined in the contractual arrangements.

Under the PPP Arrangement, there is a need to monitor and control performance parameters. For many activities of a utility, a variety of performance measures can be defined. Additionally, following best current practice, measures should be output based and directly related to a customer focused approach. There is also a need to ensure that there is full stakeholder consultation at all stages of development and operation. Case history shows the positive effect of stakeholder involvement in design, development and implementation of PPP schemes (and also the potential negative effects when they are not involved). This leads to the premise that for long term effectiveness and sustainability customer and stakeholder consultation throughout the life of the scheme needs to be formalized. To reflect the importance of this issue the Grantor should establish in the contract terms specific provisions for stakeholder consultation.

In order for the Agreement to operate the design of PPP schemes require legal and regulatory frameworks. Having determined the performance measures to demonstrate the level of scheme effectiveness, the contractual and regulatory framework needs to incorporate mechanisms to monitor and control these performance measures. These mechanisms have to operate effectively throughout the life of the PPP Arrangement, and institutions should be established to be responsible for maintaining and operating the monitoring and control activities, and also regulatory activities. It is important to ensure that these institutions are have long term funding and resources, as well as having any necessary enforcement and remedial powers.

An essential step in the successful planning and delivery of major infrastructure projects once the appropriate service delivery scheme has been chosen, is the identification and allocation of responsibilities and risks between the private and public sector parties. The importance of analyzing and allocating these risks is in light of the need to balance the private sector risks so as to offer optimal benefits to the Grantor, but no longer constitute a significant impediment to private financing of infrastructure projects.

Principal types of PPP Arrangements analyzed show how the risk level to the developer increases according to type of arrangement chosen. Within these types of PPP schemes the Concession (or sometimes referred to as Build Operate Own and Transfer or Design Build Operate and Finance models) and Divestiture type arrangements generally involve the most comprehensive array of risks and responsibilities to be taken by a private Developer. The main reason these particular contract forms have been used as the starting point to develop comprehensive recommendations for contractual issues is because they represent the most complex contractual arrangements which have the greatest level of risk transfer to the private sector. A variety of ways are used to allocate responsibilities and risks and given the importance of this issue, a systematic approach is recommended for registering this allocation, such as a Risk Matrix.

The importance of funding and finance cannot be overstressed as they are two of the key drivers for choosing PPPs for Water and Waste Water Sector schemes. Policymakers should understand the implications that funding and finance has on project structure and underline this critical component for effective planning. The concept of 'bankability' has been reviewed in reference to investor/lender sentiments and their willingness to commit debt or equity capital toward a project. Emphasis has been placed on the importance of realizing that when considering bankability it goes beyond financial analysis alone and should include much wider project considerations. How to structure finance for water PPP projects has been explored and the merits and limitations of the different financing techniques available discussed. In addition, the role of public support and bilateral and multilateral financial organisations and the possible forms this may take has been examined, depending on the type of PPP structure chosen. The consideration and valuation of contingent liabilities should also be understood when undertaking PPPs in the water sector. These important issues should be carefully reflected in appropriate contractual arrangements.

Generally, existing public procurement methods have to be adapted to ensure most effective design, implementation and regulation of PPP projects. Issues that may arise during the whole procurement process have been highlighted and best practice options for dealing with them have been suggested. Key recommendations have been made in terms of ensuring transparent competitive bidding procedures and some key issues of other procurement approaches noted. It has also been highlighted that there is a benefit in having continuing stakeholder consultation throughout the procurement process, in order to develop the most optimal scheme. Finally, attention has been drawn as to how best manage the procurement process and the use of early consultation with the market is recommended to assess suitability and 'market appetite'. Moreover, a consultative process to contract development at the bidding stage is also recommended, provided the right framework has been established to engage with bidders directly and inclusions made are fair and transparent. At the end there is a review of the possible ways in which negotiations can be conducted and a recommendation that in whatever process chosen the grantor retain the elements of competitive tension as a means of achieving best value for money.

As a conclusion, Part III provides specific recommendations for effective procurement and implementation of the chosen PPP scheme, so as to achieve optimal results. The contractual arrangements for the PPP scheme need to embody ways of dealing with all the main factors influencing effective development and long term operation as determined in the previous section. The long term success of the PPP Arrangement will in a large measure be determined by the way that these factors are dealt with contractually. Furthermore, in this Part III findings are based on an outline contract form developed for the most complex types of PPP Arrangements (i.e. the Concession contracts). This has been developed as a practical checklist covering all the typical contract elements normally expected. Ways have been suggested of dealing with all clauses considered potentially influential in contractual and scheme performance. The core of the recommendations lies in the premise that the contractual obligations need to focus on achievement levels of service for the customer and based on output monitoring and not necessarily inputs. One of the main recommendations is to contract out the performance monitoring to an independent, reputable, international firm.

Furthermore, it is to be noticed that the sustainability of the PPP Agreement can be enhanced by avoiding on demand performance bonds, making performance bonds reciprocal (i.e. both the Grantor and the Developer are obliged to post a bond) and by allowing 'rights of set-off' in the contract. Other specific recommendations include an automatic compensatory mechanism permitting the Developer to adjust cash payments unilaterally, service obligations and any investment roll-out if the Grantor breaches the PPP Agreement in ways which increase the Developer's cost or reduce its revenue. The importance of allowing variations in the PPP Agreement have also been argued, that take into account pass through and other cost items (including financing costs) not under the Developer's control.

It has also been demonstrated that the unpredictable nature of long term PPP Agreements such as concession contracts may mean frequent re-negotiations in order to facilitate adjustments in specifications and performance requirements. Recognizing the fact that significant numbers of Water and Waste Water PPP Arrangements had to be renegotiated within a few years of contract award, practical suggestions have been made for contractual provisions in establishing legal and contractual mechanisms for potential future renegotiation of new contracts, together with ways of dealing existing PPP bids including the contentious issue of renegotiation. This leads to specific suggestions on development of contractual structures to be used to contend with the imperfections identified in a typical PPP design. These include the Output Based PPP Agreements and long term PPP Agreements incorporating the use of Trust funds. Under an Output Based long term PPP Agreement, the Developer commits to performance targets specified in the PPP Agreement and receives financing for capital expenditure from commercial capital markets. Tariff revenues may be supplemented by funds payable by a Subsidy Fund which pays (usually in grants) against prescribed results and on an output basis. The main advantage of introducing OBA subsidies is in reducing the burden on the Developer to recover all his costs. On the other hand, the Trust structure recognises that equity is expensive and as a result, implies an overall increased risk profile. Equity requirements on behalf of the Developer may be minimised with other financial resources that can be leveraged effectively through revenue security provided through pledges against public sector flows guaranteed by a third party.

In the supporting conclusions there are also recommendations on the optimization of specific areas of PPP viability and sustainability. Many PPP schemes are designed on information of existing and future water and waste water system conditions using less than perfect available data and estimates. The extent of the potential sub-optimal effect this had on the PPP design, implementation and on general project viability and sustainability was noteworthy. This is particularly obvious in the underground assets (e.g. piping, water mains and sewers) associated with provision of water and waste water services. Accordingly, effective ways have been discussed of transferring the risks of underground asset conditions and other imponderables to the Developer in a systematic and sequential manner. This can have a major impact on the original agreement, and specific ways have been developed to improve the bidding and contractual mechanisms to handle this and any future improvements in data. Arising from the particular issue of Imperfect Data and the specific characteristics of Water and Wastewater PPP Schemes, there is a particular need to improve public bidding and procurement procedures to deal with this. Given that there will be gaps in the information on which Developers make their proposals; it is recommended that the procedures be strengthened by using an approach which recognizes the need for structured data improvement and validation. This structured approach should start at the pre-qualification process, run through the development of the bidding process which would incorporate revisions and improvements in data and assumptions and would finally deal with improved data in the initial period of the PPP contract. Regarding the last point a proposal has been included for dealing adjustments in the Developers' risks and payments, as improved information becomes available during the initial years of the contract.

Due to poor initial information and a lack of knowledge of how the system really operates, the early years of water service PPP Agreements are historically challenging. As a result the initial bid price submitted by the Developer may be completely irrelevant within a short period of time. This may mean that tariff adjustments may be required to compensate for the reality encountered by the Developer as he starts operating. It is therefore recommended that PPP Agreements be designed to adapt to the initial period of a contract and provide more intelligent and flexible mechanisms to deal with the higher likelihood of problems within the initial few years of the contract. It would seem that there needs to be a change in emphasis: a shift in focus from purely technical and financial assessment of the Developer's proposal, to a thorough evaluation of the potential Developer's profile and experience levels. In parallel with this, recommendations have been made for prequalification and bidding processes including means of entering a long term PPP Agreement that considers the possibility of early termination by both parties in certain well defined circumstances.

Future considerations

While gathering information for the suggestions and recommendations set forth here, it became evident that there are no right or wrong PPP structures. The structure chosen should reflect, as best possible, the actual realities on the ground and the conditions of the water utility. There is no way of analysing the ways in which all limitations could be addressed. One might like to consider how the public sector in the future could restructure to create the same advantages as one of the PPP structures examined. One might like to examine how the private sector's profit motivation can be balanced with the government's obligation to provide services to the poor. It would also be interesting to investigate how subsidies are managed. Moreover it may be useful to assess how government water and wastewater entities have performed against some form of private sector participation. In this same vein, it would be interesting to assess whether specific national policies can better promote the use of PPPs in all its forms and if indeed specific issues of PPPs may be enshrined in law.

In the European context in particular, in the last 15 years the number of countries actively seeking accession rights to the EU increased dramatically. Major investment in water and waste water infrastructure was necessary to meet the accession requirement standards in water and waste water and environmental services. Many of these schemes have been developed under PPP arrangements, yet some of the member counties have not embraced the private sector as a strategy for more effective management, operation and financial resource. Although a few of the recent schemes in Bulgaria, Romania and Estonia have been included in the current publication, it might be beneficial to investigate the overall effectiveness of the PPP contractual approaches used in countries seeking accession to the European Union. The current recommendations could be used for comparison.

Although there are some specific recommendations here on how to improve the contractual structures they have not been tested under different legal regimes. One such regime is the Competition Directive of the European Union framework for PPPs. This is particularly important, as in some instances it has been argued that some PPP structures may require Government support (in the form of grants and or soft loans) to allow the Developer to include strategies for dealing with the very poor who may be unable to pay tariffs reflecting the full cost recovery of providing such services. These suggestions, whilst pragmatic, may fall foul of some of the EU directives on competition, government support and so on. This could be explored in more detail and the specific implications assessed. Furthermore, some of the suggestions included regarding procurement of the PPP arrangements should also be tested.

Future considerations might delve into the ex ante and ex post conditions of PPPs not only in the water and wastewater sector but other infrastructure and services sectors as well. Comparisons could be made between different PPPs in different infrastructure sectors and qualitatively and quantitatively assessed in their effectiveness in achieving the Government objectives. Similarly, it would be interesting to explore whether PPPs are being used as a way to 'mask' Government interventions. One example would be the recent clarification from the European Union stating that even if a member country calls a contractual arrangement a PPP, it can only fit the definition of such arrangement if there is more than 50% ownership (and control) by the private sector partner. In this same vein, it may be worth exploring what constitutes effective use of PPP, what studies have been done to show value for money and whether public cost comparisons have been used effectively in the decision making process.

Regulation is a hotly debated issue and one which also merits future consideration. The form of regulation, whether it is a price cap, rate of return, benchmark and so on will depend on the actual situation on the ground and the specific project's idiosyncrasies. Investigation could be carried out on the different ways in which regulation can help promote and resolve specific issues under the PPP Agreement. For example, in a developing country where the major challenges to the water sector is high leakage and insufficient water resources (and therefore major initial investment is required to help address this challenge), the regulatory regime may specifically incorporate a 'rate of return' type system which creates quick investment incentives. In more developed economies such a structure may not be applicable as the focus may be on operational efficiency requiring a combined regulatory structure following a 'price cap' style plus an efficiency indicator which may be more relevant.

Similarly, it may be interesting to undertake an assessment of different regulatory regimes and compare and contrast how regulation by contract vs. discretionary (or commission) regulation deals with the challenges of the Water and Wastewater Sector. Issues such as parameters of discretion, setting of tariffs, implications on decision making systems, transparency, accountability, avoidance of regulatory capture and political interference could be considered and compared and lessons learnt. In this respect, some of the hybrid structures that have been developed should be assessed to see how they have managed against traditional concession or discretionary systems.

Annex 1

European legal framework and its implications on PPP

The need for financing infrastructure services and in particular the water and wastewater sector is a challenge being faced by a number of governments seeking accession to the European Community. Moreover, Public authorities at all levels, including Member Countries, are increasingly interested in developing specific arrangements for cooperation with the private sector. Public Private Partnerships are one of the methods available, and as discussed under the previous chapter, there are various contractual forms available to undertake PPP. The choice of one of these forms is dependent on what a public authority seeks to achieve. Notably, within Community law there are no prescriptions at law for such relationships nor, in fact, whether a public authority should enter into a PPP. The decision to provide an economic activity is squarely in the hands of the public authority. If it wishes to delegate a part or the whole of this activity to a third party it is the prerogative of the public authority.

However, any delegation of responsibility may be the subject of Community law on procurement. Accordingly, the form of delegation through a PPP instrument needs to be examined against the basic principles of EU law and the framework it establishes.

In this annex we review the impact Community law may have on developing PPP structures. In the absence of a uniform legal framework for PPPs at European level and in view of the heterogeneity of national laws the question arises, whether uniform provisions are needed to safeguard transparency and effective competition between market participants. The EU has decided against the creation of a separate legal regime for PPPs. The EU also considers that, as a matter of principle, the law on public contracts needs to be applied whenever a private partner is to be selected.

The main purpose of Community law on public procurement is to create an 'internal market' in which free movement of goods and services can take place.

The intent is to ensure the right of establishment and the right to provide services (Article 43 and Article 49 of the EC Treaty) are respected. The implication of this is that, under Article 48, the principles of equal treatment, transparency and mutual recognition must be assured. Value for money must also be obtained when a public authority buys products or mandates third parties with performing services and/or works.

Given the above issues, this chapter starts with a general examination of the general principles of EU law in terms of competition and procurement. It follows that the impact of the EU Utilities Directive, the implications of public involvement and ownership, and state aid must be examined. General principles of EU Competition law must also be understood and a simple test structure is presented to exemplify the legal concepts behind EU thinking in this regard.

Having understood the basic principles of Community law, it follows that the specific rules applicable to PPP in the EU context must be understood. The selection of a private partner under a given PPP structure derives firstly from the definition of the contractual relationship which the party enters into with a public contracting authority (see previous chapter for contract types). Under Community secondary legislation, any contract for pecuniary interest concluded between a contracting body and a private partner in the form of an operator, which has as its objective the execution of works, or the execution of works and services, is designated as a 'public works or public services contract'. The concept of 'concession' is defined as a contract of the same type as a public contract except that as consideration for undertaking these works, a right to exploit has been granted or a combination of a right and a payment is made. Thus, under this section of the chapter we detail the manner in which PPP structures can be developed.

Given that our objective is to develop PPPs in the water and wastewater sector, it is contingent upon us to ensure that the PPP structure devised is consistent with relevant sector practice and legislation. The EU's framework for the sector is enshrined in the Water Framework Directive (WFD). The WFD is an inclusive approach to managing water as it flows through catchments from lakes, rivers and ground waters, to estuaries and the sea. The WFD creates a framework for all future actions by the EU on water policies.

So, in this annex we also examine the Water Framework Directive, its evolution as well as its main principles and objectives. We delve into the key principles behind the application of the WFD and its managerial and environmental objectives. Under each we identify the most salient issues as they apply to Accession Countries. These key principles, together with the principles of competition as they apply to PPP, will be utilised later on as the basis for the development of an analytical framework for examining more rigorously the provisions of the WFD against a real life situation in two Accession Countries: Bulgaria and Estonia.

Relevance of the EU law in the water sector

In this annex the relevant provisions of Community law that may apply to the water and wastewater sector will be highlighted. In so doing, the implications on procurement and competition need to be examined. Community law on public procurement seeks to create an 'internal market' in which free movement of goods and services can take place. The principles of equal treatment, transparency and mutual recognition must be assured. Value for money must also be assured. However, given the need to simplify and modernize procurement procedures of entities operating under the water, energy, transport and telecommunications sectors, a 'Utilities Directive' coordinating the procurement procedures of entities operating in these sectors has been developed. This Directive is specific to the sector and a distinction has thus been made between these sectors and other forms of public procurement – the 'Public Sector Directive'.

EU Utilities Directive

The EU Commission recognised the need to simplify and recast the procurement Council Directive 93/38/EEC of 14 June 1993 coordinating procurement procedures. The new Utilities directive of 31 March 2004, is based on Court of Justice case-law, in particular case-law on award criteria. One major reason for the introduction of rules coordinating procedures for the award of contracts in these sectors is the variety of ways in which national authorities can influence the behaviour of these entities. This influence may be exerted through a participation in their capital and/or through representation in the entities' managerial structure. Another main reason relates to the closed nature of the markets in which they operate, due to the existence of special or exclusive rights granted by the Member States concerning the supply, provision or operation of the service concerned. Nowhere is this more the case than in the water and wastewater sector where the incumbent has typically an exclusive and monopoly right to provide services in a given geographical area.

In this annex we detail, briefly, the main issues related to this Directive and the impact it has on the water and wastewater sector. Equal treatment and transparency: In order to guarantee the opening up to competition of public procurement contracts awarded by entities operating in the water sector, it is advisable to draw up provisions for Community coordination of contracts above a certain value. Such coordination is based on the requirements inferable from Articles 14, 28 and 49 of the EC Treaty and from Article 97 of the Euratom Treaty, namely the principle of equal treatment, of which the principle of non-discrimination is fundamental. Additionally, the principles of mutual recognition and proportionality, as well as transparency, must be respected. Accordingly, procurement needs to ensure that it follows these principles while providing sufficient flexibility. Similarly, the procedures developed should ensure equal treatment for contracting entities operating in the public and private sectors.

Single or multiple contracts: It must be noted that the Directive does not seek to prescribe whether the design and execution of works should be done separately or jointly. This decision should be determined by qualitative and economic criteria, which may be defined by national law.

Classification of contracts: the Directive provides for the classification of the different types of contracts for works as defined in Annex XII. Service contracts may under certain circumstances include works contracts.

Tendering and procurement processes: The Directive provides for the free provision of services without discrimination. Procurement techniques to be used by the Member Countries should comply with the principles of equal treatment, non-discrimination and transparency. It is appropriate for the contracting entities to apply common procurement procedures in respect of their activities relating to water. However, procurement rules of the type proposed for supplies of goods are inappropriate for purchases of water, given the need to procure water from sources near the area in which it will be used. Qualification systems developed as part of the tendering system should be set with objective criteria, which, at the contracting entities' choice, may concern the capacities of the economic operators and/or the characteristics of the works, supplies or services covered by the system.

Special or exclusive rights: The fact that an entity supplies water under a special or exclusive right granted by a competent authority of a Member State, does not mean that the provisions of this Directive do not apply to the entity concerned. However, it is possible – subject to not introducing distortions in competition – to exclude certain service, supply and works contracts awarded to an affiliate undertaking having its principal activity the provision of such services, supply or works to the group of which it is part, rather than offering them on the market. It is also appropriate to exclude certain service, supply and works contracts awarded by a contracting entity to a joint venture which is formed by a number of contracting entities for the purpose of carrying out activities covered by this Directive and of which that entity is part.

Technical Specifications: Technical specifications should allow public procurement to be opened up to competition and different technical solution to be presented should be possible. The Directive allows for the inclusion of specific performance conditions provided they are not directly or indirectly.

Mutual recognition: when evidence of a particular qualification is required as part of the tender process, the principles and rules of mutual recognition of evidence of qualification apply.

Award of contracts: Is to be made on the basis of objective criteria which ensure the principles of transparency, non discrimination and equal treatment, and which guarantee that all parties to the tender are treated in conditions of effective competition. The determination of the award criteria depends on the object of the contract and the award conditions should fulfil economic and qualitative criteria.

Implications of public involvement & ownership under EU

Although in this book we have looked at the introduction of the private sector, Community Law is not concerned with the mechanisms through which the private sector is involved, but rather the degree of control exercised by the public sector. This is subject to Article 85(1), which ensures that the rules contained in the Treaty must be maintained. In a practical sense, what this means in the context of this book is that even though a Government may have an 'interest' in a water utility, it must still abide by the basic principles of Community law and in particular the requirements of the Water Framework Directive.

At the same time, another important consideration in this book is the extent of public ownership in the sector. Actually, the Treaty does not prohibit public ownership, nor indeed does it limit it from enhancing its current area of service provision so long as it does not break any fundamental principles of Community law. Thus, Article 222 of the Treaty states that the free market ethic that characterises the economic system of the EU will in no way affect the system of property ownership. For us, the implication of such a position is that, in principle, public undertakings are subject to the same rules of Competition law as the private sector.

Implications of state aid

The amount of state aid in any particular situation may vary and its impact has to be carefully considered. The first thing to consider is if the country itself has abided by the rules regarding state subsidy whose range is limited in Article 87(1) of the Treaty. However, there are a number of cases deemed to be compatible with the common market but that should be carefully authorised by the Commission. These are set out in Article 87(2) and 87(3) of the Treaty.

Some of these are very important in the sector under consideration and those relevant issues should be noted, including: aid of a social nature granted to individual consumers provided there has been no discrimination, aid aimed at promoting the economic development of specific areas, & aid seeking to promote the implementation of a major project of common European interest, or rectifying a serious disruption in a Member State.

In these particular case studies, state aid was a definite condition under which both contracts with the private sector were obliged to operate. A government could not be seen to favour any one contractor through subsidies, either to businesses or consumers. Thus, the aid available had to be on a fair and transparent basis for all operators.

EU competition law

In the provision of public services such as water and wastewater it is essential to first ascertain whether or not EU Competition law is applicable and if so, in what context and to what extent. A number of principles for determining whether a firm has a dominant position have been stipulated by the European Court of Justice in the application of case law. These principles can be used as 'tests' to ascertain how a water utility, and hence a PPP option may be judged to be compliant with Competition Law. These tests are:

Test 1: The capacity of a firm to engage in actions that do not take into account their competitors, customers or suppliers

In case law, the European Court of Justice has defined dominant position to be 'a position of economic strength enjoyed by an undertaking which enables it to prevent competition being maintained in the relevant market by giving it the power to behave, to an appreciable extent, independently of its competitors, customers, and ultimately of its consumers'. This concept captures the potential imbalance of power between a water utility and its customer base, both consumers and government. As a natural monopoly in a geographical region, the water utility clearly has the power to exhibit this form of dominance and potentially abuse it. Hence an effective regulatory framework is required to mitigate the effect of this type of dominance through a series of monitors, controls and, as necessary, sanctions.

Test 2: The capacity to prevent other undertakings/firms from entering the market

The main barrier to other operators entering a given geographical market is the existence of a single water utility infrastructure system. It is likely to be economically unattractive to build a duplicate infrastructure system. If a water utility were able to control competitors from having access to that infrastructure, either through veto or excessive tariffs, then it would constitute a dominant position. Similarly if it were able to prevent a competitor building new infrastructure, it would be in a dominant position. Hence a regulatory framework is required to ensure fair competition. This could be managed through the contract letting process to avoid giving the private operator of an existing infrastructure an unfair commercial advantage. This could be through regulating mutual access to infrastructure, or a mechanism to maintain an equal playing field regarding the benefits of building and operating new infrastructure.

Test 3: The market share of the undertaking in question

The natural monopoly of the water utility in both its geographical and product market effectively gives customers little, or no choice of the supplier from whom they may buy water and wastewater services. Hence, the water utility's market share will be greater than 70–80%. The European Court of Justice generally regards a market share of greater than 70–80% as a dominant position. Usually a water utility is in this position and thus clearly dominant and consequently may be in the position to abuse it. Therefore, the regulatory regime must ensure that this type of dominance is mitigated through a series of monitors, controls and sanctions.

Test 4: The economies of scale realised by the undertaking/firm

Again, as a natural monopoly, the economy realised by a water utility with its almost total market share gives rise to economies of scale, particularly in the case of distribution. Undoubtedly this could prove a major economic barrier to competition. Hence, a regulatory regime should ensure that there is no economic advantage to be gained from being the first private sector operator in a geographical area, and that a mechanism is in place to ensure that competition is encouraged, if only in future bid contests.

Test 5: The capital strength of the firm and its access to capital markets

In the water sector, this type of dominance is only partially relevant to the argument around competitiveness. Indeed a large element of the benefit of using PPPs is that they exploit the private sector's ability to raise capital. There is however, a risk that with the growing importance of the 'Water Trans-nationals', the need for capital, with the subsequently dominant position of the 'Water Trans-national', may lead to a poor deal for the customer on an ongoing basis, and hence be anti-competitive.

Test 6: The ownership structure

After a water utility has introduced some form of private involvement, say through the selling of a strategic interest, the key issue is whether the public sector continues to exercise control over the water utility's activities.

Although this concept is not directly applicable in our case studies, Article 82 of the Treaty is concerned with competition that may 'affect trade between Member States'. However, the geographical market definition of a water utility and its historical constraint within national borders has meant that there is no cross-border trade in water services. Notwithstanding the current situation, it may be possible in the future for the same international operators in adjacent nations to trade water between them to gain economies of scale, or alternatively merge their commercial operations into a regional hub larger than a single national operation.

EU legal position on PPP

Member Countries, recent entrants to the EU as well as Accession Countries are relying more and more on different forms of PPP's to meet infrastructure investment needs. The form of PPP varies substantially between Member States and depending on the sector. National law also clearly influences the type of PPP structure chosen. In the water and wastewater sector, PPPs have increasingly been used by Accession Countries as a strategy to help meet EU accession criteria and more specifically the EU Water Framework Directive.

The EU's definition of 'PPP' is a blanket term for agreements between public authorities and private industry. Such agreements may concern public infrastructure projects or public services and are categorised as:

(a) Purely Contractual PPP which refers to a partnership based solely on contractual links between the different players. It covers a variety of set-ups where one or more tasks are assigned to the private partner, and which can include the design, funding, execution, renovation or exploitation of a work or service.
(b) Institutionalized PPPs refer to the establishment of an entity held jointly by the public partner and the private partner. The joint entity has the mandate to deliver work or a service for the benefit of the public.
(c) Cooperation agreements: these are usually between local authorities.

Recently significant steps have been taken by the Commission to develop an approach to PPP. To this end the Commission adopted the Green Paper on Public-Private Partnerships and Community Law on Public Contracts and Concessions as well as subsequent communications. This paper was the result of a consultative process and was one of the actions resulting from the European Initiative for Growth. The aim of this Green Paper has been to determine the suitability and applicability of Community law on public contracts and concessions involving PPP. It sets out the rules that should be applied when taking a decision to entrust a given activity a third party (usually the private sector) under a PPP. It must be noted however, that these rules are downstream of the economic and organisational choice made by a local or national authority. The Commission makes no value judgment regarding the decision to use PPP and thus externalise the management of public services or not. This decision remains squarely within the competence of public authorities.

Despite the basic guidelines or distinctions on PPPs, Community law on public contracts and concessions do not prescribe whether or not a PPP arrangement should be used. A Member Country, and thus by implication Accession Countries, are free to provide a public service themselves or to entrust it to a third party. Moreover, under PPP arrangements, public partners are primarily national, regional or local authorities. There may also be public law bodies which have been created for the specific purpose to fulfil certain specific interests under State control. For the purposes of EU interpretation, the term 'contracting body' or 'contracting authorities' is used to designate all of these entities. However, it must be noted that for selecting the partner in a PPP, the general principles of the EC Treaty apply, as do the current directives on public contracts. Under the EC Treaty all agreements in which a public authority awards a contract are subject to the general principles on freedom of establishment and freedom to provide services in Articles 43 to 49. It follows that PPP agreements must satisfy the requirements of transparency, equal treatment, proportionality and mutual recognition. In addition, PPP agreements that count as public contracts are subject to the application of the public procurement directives. Some special provisions apply to building concessions, while service concessions are not at present covered by secondary law. On the implementation of PPPs the national laws apply and they must be in agreement with the provisions of the EC Treaty, which take precedence. Additionally, the Commission has introduced some degree of flexibility under procurement law. Notably, it defined, on the basis of the rules and principles derived from the Treaty and applicable secondary legislation, the outlines of the concept of concession in Community law and the obligations incumbent on the public authorities when selecting the economic operators to whom the concessions are to be granted.

In the absence of a uniform legal framework for PPPs at European level and in view of the heterogeneity of national laws the question arises, whether uniform provisions are needed to safeguard transparency and effective competition between market participants. The EU has decided against the creation of a separate legal regime for PPPs, but considers that there is a need for legislative initiatives in the areas of concessions for the provision of services, and institutionalised public-private partnerships (IPPPs). EU also considers that as a matter of principle, the law on public contracts needs to be applied whenever a private partner is to be selected.

Community secondary legislation establishes that for any contract for the execution of works or services between a contracting body and an operator, these are to be designated as a 'public works or public services contracts'. The concept of 'concession' is defined as a contract of the same type as a public contract except for the fact that the consideration for the works to be carried out or the services to be provided consists either solely in the right to exploit the construction or service, or in this right together with payment.

Thus, in relation to all PPP contracts the following should be considered:

- PPP arrangements should emphasise fairness and transparency. Accordingly, transparency must be in evidence whenever public funds are involved, as well as ensuring that the PPP arrangement allows for the right of elected representatives to inspect agreements and documents. Furthermore, the EU takes the view that transparent rules should be applied for the award of public contracts to ensure effective competition and protection from corruption, in the citizen's interest.
- For those PPP arrangements or contracts which have already taken place, and in order to avoid legal uncertainty, transitional periods should be applied for existing contracts that have been concluded in good faith in accordance with national law.
- PPP arrangements should make adequate provisions to ensure protection of the citizen's rights to ensure secure safety, efficiency and quality standards.
- Member States are recommended to alleviate the task of the public sector by means of standard contracts and by stepping up the training of

the decision-makers who have the task of selecting the private partners for PPPs.

Furthermore the EU has decided not to establish a European agency for PPPs, but welcomes other ways of sharing experience, such as the networking of national and regional PPP authorities. PPPs are essentially a form of public contracting. Given the significant examples under case-law of the Court of Justice, the EU has decided against developing blanket regulations for PPPs. In any event, the decision has been taken by the Commission that the legislative process should as a matter of principle go carefully, because PPPs are often very complex agreements whose attractiveness can only survive in a climate of flexible regulation.

As stated earlier in this annex, in legal terms there needs to be a clear statement that, wherever a private partner is being selected, transparent procedures have been used for this selection. Transparency and fair competition between market participants should be used no matter what form of PPP is used.

Purely contractual PPP

As mentioned above, EU considers two types of PPP arrangements, Contractual PPPs and Institutional PPPs. In the case of PPPs that are characterised as public contracts the main issue is the question of the award procedure.

With PPP structures, the competitive procedure to be used tends to be slightly more complex. In this regard, a new award procedure has appeared on the scene to join the familiar public, restricted and negotiated procedures for procurement above the threshold values. This procedure combines elements of the tender procedure with those of the negotiated procedure and takes precedence over the negotiated procedure. It must be noted that difficulties do arise in defining precisely what is meant by ‘particularly complex’ contracts. In response to this, the Commission has clarified that a particularly complex project is that in which the public contractor is objectively unable to specify the technical means with which their needs and objectives can be fulfilled, and/or is objectively unable to specify the legal and/or financial conditions of a project.

Although PPP projects by their very nature need not be necessarily be more complex, each case must be assessed on its own merits. In any event, the Commission does allow for the competitive dialogue procedure to take preference for the award of contracts in the PPP sector as a principle because it combines the advantages of the restricted procedure with those of the negotiated procedure. Its flexible nature enables it to obtain optimum solutions under great competitive pressure. Although its disadvantages include its complexity and the extent of effort required from all involved, these are offset by its greater transparency compared with the negotiated procedure. This reduces the risk of anti-competitive behaviour.

PPPs as concessions

The concept of 'concession' is defined as a contract of the same type as a public contract except for the fact that the consideration for the works to be carried out or the services to be provided consists either solely in the right to exploit the construction or service, or in this right together with payment. The Commission takes the view that concessions should be of limited duration so that competitors are not excluded from competition for an unnecessarily long time. At the same time it recognises that under these types of PPPs, competitive dialogue is also the appropriate procurement procedure, as its flexibility is as a rule suited to the complexity of concessions, without jeopardising the principles of transparency, equal treatment and proportionality.

Institutionalised PPPs (IPPPs)

IPPPs are joint industrial creations combining private and public partners. The objective of such a PPP structure is to provide or supply a service on behalf of the public, typically in the field of public welfare. IPPPs may be created by the formation of a mixed-economy enterprise or company, but may also arise when a private company assumes control of what was a public concern, that is by changing the share ownership.

This raises the question whether the two processes, new formation or a change of ownership, are relevant from the point of view of procurement law. The Commission says that the private partner of such an enterprise must be selected for the tasks involved in a transparent way and without discrimination, and must be selected irrespective of the actual form of the contract under the procurement directives or the EC Treaty. The selection of a private partner in accordance with objective criteria will safeguard fair competition.

Case law under the judgments of the Court of Justice in the Teckal and Stadt Halle cases (C-107/98 and C-26/03) set the principle that the participation of the awarding authority in a mixed-economy company does not justify exemption from the principles of procurement law. What is crucial is to award the contract to a body with independent legal personality. Accordingly an exemption from procurement law is recognised when the awarding authority exercises a control over the mixed-economy enterprise similar to that control which it exercises over its own departments, and when the enterprise essentially acts for the public body. The two criteria must be cumulatively fulfilled to ensure that there is equivalence with internal entities of the contracting authority. The Stadt Halle case-law has further clarified the matter. The Court decided that public contracting authorities may award a contract to an undertaking that belongs to them without adhering to the procurement provisions only when they hold 100% of the undertaking's capital, in other words when there is not even a small private shareholding in the undertaking. Only then, says the Court, does the contracting authority exercise control as it would over its own departments.

In summary, the Commission considers it necessary, in view of the transparency requirement and the ban on discrimination, for procurement law to be applied when an IPPP is set up, in so far as the act of setting it up is combined with the assignment of a public contract, concession or other public tasks to a private undertaking. Additionally, this same principle would apply when an IPPP is established by the sale of shares in a public undertaking.

Water framework directive – background & development

It was not until the early 1970s that the EU Member States started developing legislation to reach a common approach to water policy. Initially this approach concentrated on deterring pollution and only for certain types of water or by certain pollutants. The legislation was not an all-encompassing solution aimed at developing a sustainable pollution policy.

Throughout the 1980s and 1990s the remit of pollution control has been gradually extended, culminating in the first draft of the Water Framework Directive (WFD) issued in February 1997. The main objective of the proposed WFD was to create a framework for all future actions by the EU on water policies, but at the same time be observant of regional variations. Thus, through the WFD the Commission sought to integrate the views of all stakeholders including local and national authorities, water users, water providers, industry, agriculture, consumers, environmentalists, non-Governmental Organisations (NGO's) and other bodies involved in the water and wastewater sectors. It also aimed at ensuring coherence of objectives and eradicating conflicting policies between the sectors for example water, wastewater and environment, as well as any conflicting policies and/or objectives between EU member countries. Fundamental to this approach was the need to rationalise water management policies by applying the concept of 'river basin' management system, which all governments would be required to adopt.

However, a great deal of debate between the differing bodies, that is, European Commission, European Parliament and the European Council, delayed the WFD implementation. It was not until September 2000 that finally the EU adopted the WFD that entered into force on the 22nd December 2000. In its current form, the WFD focuses on establishing the right conditions to encourage efficient and effective water protection at the local level, by providing a common approach and common objectives, principles, definitions and basic measures to achieve a sustainable water policy. Thus, 'sustainable use' ensures adequate supplies of water of good quality to be maintained for present and future generations. It must be noted however, that the WFD recognises that its objectives are ambitious and that Member States will not be able to comply with them immediately. It allows for extensions where improvements cannot reasonably be achieved.

The scope of the WFD covers all surface, ground transitional and coastal waters. Under Article 3, Member States must identify River Basin Districts and a competent authority to manage them. The WFD also recognises that the management of these districts may imply regional, national and international co-operation and may also involve co-operation with non-Member States. It also explicitly permits existing structures stemming from international agreements to be used for this purpose.

Thus, it can be appreciated from the previous section on the history and development of the WFD, that it has two equally important objectives. A managerial objective, aimed at creating integrated water management at the river basin level to ensure overall co-ordination of water policy in the EU, and an environmental objective, based on the premise of preventing further deterioration and achieving water quality of 'good status'. Good status means that it is not only about preventing deterioration but, if deterioration has already occurred, there is an obligation to restore water quality.

The parameters of the WFD, as specified in Article 1, are to protect, enhance and prevent further deterioration of the aquatic ecosystems, promote sustainable water use based on long-term protection of available water resources, enhance protection and improvement of the environment, ensure the progressive reduction and prevention of pollution of groundwater, and to mitigate the effects of floods and droughts. In order to do this, a sustainable environmental objective needs to be ensured, which will prevent deterioration and achieve 'good status' in all waters by 2015 that is 'good surface waters' meeting certain ecological and chemical criteria, and 'good ground waters' meeting certain quantitative and chemical criteria). Simply stated, 'status' means when the body of water in question deviates from its original or referential conditions. The objective of the WFD in this context is to set both quantitative and qualitative criteria to be achieved by the Member States. These detailed criteria stipulated in Annex 5 of the WFD are the basis for measuring sustainable use of waters.

The WFD is pragmatic in its approach to achieving its environmental objectives. It allows for extensions where improvements cannot reasonably be achieved. For example, exceptions may be granted when it would not be feasible, or would be disproportionately expensive (Article 4 (5)). Other exceptions might include a temporary deterioration due to unforeseen natural circumstances (Article 4 (6)). Needless to say, the WFD is very specific about water quality standards. Accordingly, under Article 7, Member States are required to identify water currently in use or to be used in the future for human consumption, and to take all necessary protective measures to ensure the quality of these waters. Member States are also required to comply with existing and future legislative requirements. This necessitates measures to combat pollution whilst protecting the environment, which is a key principle of the WFD. Article 10 establishes emission limits and environmental quality standards. The WFD also focuses on the need to co-ordinate existing and future legislative requirements. Each River Basin must have river basin management plans that include specific responsibilities to develop and set up appropriate administrative arrangements within each river basin, including the identification, or creation, of a competent authority to manage each River Basin District (RBD). In each plan, the characteristics of the river basin should be defined, a summary of the significant pressures on that particular basin identified, and a programme of appropriate measures to be developed. In the case of international RBD's, each Member Country has to set up the relevant administrative arrangements within the portion of international RBD lying within its territory. In addition, each River Basin should design and implement measures that will ensure that the objectives of the WFD are achieved, for example, pollution control measures, water abstraction control, implementation of existing EU legislation for the protection of water and so on. For international river basins that lie between Member States and Non-Member States, Member States 'shall endeavour to establish appropriate coordination with the relevant non-Member States, with the aim of achieving the objectives of this Directive throughout the river basin district'. However, it is up to the Member State to apply the rules of the WFD within its respective territory.

Although the WFD states that 'water is not a commercial product like any other', there are some basic economic principles that the WFD endorses. In particular it states that Member States should take into account the principle of full cost recovery and the 'polluter pays' principle as provided for in Article 9. It should also take note of the economic analysis, the long-term forecast of supply and demand, volume, price and cost of water services, as well as the relevant investments for the provision of these services.

The WFD is specific about water pricing policies to be attained by 2010 to allow for adequate incentives for the efficient use of water. In addition, the WFD makes a distinction between the contributions that industry, agriculture and households provide in order to recover the cost of water services.

In the compliance and implementation of these WFD criteria, Article 8 provides for a monitoring structure. In it, each Member State must establish a monitoring programme by the year 2006. Ecological and chemical status for surface waters will be monitored, as well as the chemical and quantitative status of groundwater. Such monitoring should be in accordance with Annex V. The aim of these provisions is so that each Member Country may perform different types of monitoring and inform the European Commission of these arrangements in accordance with certain reporting mechanisms. These reporting mechanisms provided for in Article 15 require each Member State to send copies of the river basin management plan to the Commission, the reports on the characteristics of river basins and the monitoring programmes that are being implemented, or proposed (Article 8). A further obligation is to submit an interim report describing progress to date and the analysis required under Article 5 within three years of implementing the river basin management plan.

Public participation is seen as an essential part of the WFD. Article 14 actively encourages each Member State to ensure that all interested parties are involved, especially in the production, review and updating of the river basin management plans which should be made available to the public for comment at least three years prior to adoption. All information used for the development of the plans should be made public.

Finally, it should be noted that compliance and implementation of the WFD cannot be fulfilled if there are no instruments in place to enforce adequate penalties for a breach in the WFD provisions. In this respect, the WFD allows penalties to be imposed by Member States under Article 23.

Annex 2

Sector specific risk matrix

Risk	Description	Proposed allocation
1. Design and construction		
Network expansion and rehabilitation	Risk of inadequate design and thus options need full assessment for example incorporation of badly corroded systems rather than construction of new network needs to be assessed carefully	To Grantor
Asset condition	Bid is dependent on the information provided about the assets. The actual state and value of the assets may differ from initial assumptions with implications for operating and maintenance costs.	Grantor/Developer
Third party/relevant authorities approval risk	Risk that water abstraction and other licences are not granted	To be discussed, this is a key risk that needs to be evaluated within a larger framework of bid submission and bid evaluation
Design life expectancy	Risk that the design proposed is not adequate for the duration of the PPP agreement	To Developer

(*Continued*)

Risk	Description	Proposed allocation
Consistency of design with urban surroundings	Risk that the major implications on architectural and urban design are not well understood	Shared
Permits and access	Risk that permits from city, municipality or borough are not granted	To Grantor (and a blanket permit to be provided as a condition of effectiveness to contract)
Redirection of utilities	Risk that utilities network are not located where there are supposed to be, or that their relocations is more costly or time consuming than what is required	To be discussed depending on the complexity of the tasks
Delay/increase costs due to change in design required by Grantor	Risk that Grantor require changes in design that are not warranted by the initial specification proposed and/or by the existing legislations/process	To Grantor
Change in design, quality and quantity standards due to legislative and or regulatory changes	A change in law/regulation leads to a requirement to change the design and increase costs and or delays	To Grantor, the Developer cannot protect itself against this type of risk
Increased design costs	Risk that the full design costs increased as the project is being developed, due to bad initial estimate of the design cost	To Developer
Change in design required by Developer	Risk that a change in design will lead to delay/increase costs due to a change required by the Developer	To Developer
Change in law/regulation during construction	Risk that a change in law/regulation during construction lead to delay in construction or additional costs	To Grantor

(Continued)

Risk	Description	Proposed allocation
Land acquisition	Acquisition of right to use the land	To Grantor
Access to land and site	Access to land with rights to build and operate project	To Grantor – benefit from expropriation rights at law and asks Grantor to manage the complexity of rights to site under the river
Site/ground conditions (except sub surface conditions and man made artefact)	Site ground conditions might lead to increase in costs or delays in construction time. Different from subsurface risk and risk of discovery of man made artefact (see below) artefacts)	To Developer- as long as the site ground conditions are properly documented, the Developer should be able to bear the risk that the site ground conditions are likely to lead to increased costs or delay in construction
Sub-surface risk	Risk that subsurface conditions are different than those anticipated. This may be important if Developer needs to build a dam or other major civil works	To Grantor – it would be very expensive for each bidder and very inefficient to have each bidder to conduct its own survey, and therefore to assess what are the sub surface conditions. So the risk should be borne by the Grantor, to the extent subsurface conditions are different than those indicated in the geological surveys performed by the Grantor
Man-made artefacts	The risk that the discovery of man-made objects during construction requires special treatment to remove them, introducing delay in time and increased costs	To Grantor – to the extent discovery was not foreseeable. This is a typical risk allocation, each bidder cannot know in advance whether (say) archaeological artefact of importance could be discovered, and if so how the Grantor/government would want to deal with them

(*Continued*)

Risk	Description	Proposed allocation
Increase in construction costs due to contractor	Price of labour, materials and other elements of construction cost	To Developer – able to manage in this risk through construction contract, subject to below
Time overruns due to contractor	Risk of delay in construction related to contractor error or external influences	To Developer – General risk of delay, to be transferred to construction contractor, subject to below.
Latent defect in existing work	If there are existing works that have been already undertaken, then Developer will need to take over such works. Such works might contain latent defects that might affect the quality and extent of the work need to be undertaken by the Risk that existing works on the project site	To Grantor
Site safety	Risk that accident happen on the site due to poor safety procedures	To Developer
Industrial action	Risk of strike by staff, construction companies or subcontractors (not general strike)	To Developer
Performance of sub contractor	Risk that one of the subcontractor to the Developer does not perform leading to increase in costs or delay	To Developer
Misinterpretation of specifications	Risks that the project specifications are misinterpreted and this leads to increase in costs and/or delays	To Developer
Force majeure time delay		Shared
Force majeure cost implications		Shared

(Continued)

Risk	Description	Proposed allocation
Development costs	Increase in development costs for the Developer	To Developer
Delays/costs due to Grantor changes		To Grantor
Cost increase due to increase in inflation, increase in interest rate or movement in Foreign exchange		Shared. Typically the risk that inflation costs, interest rate costs and Foreign exchange rate movement create higher than planned costs between bid submission and financial close are shared between the Grantor and the Developer. Following financial close, this tends to Developer risk
2. Operations and maintenance		
Obtaining and maintaining licenses to comply with regulatory requirements		Developer
Labour	Depending on the type of PPP agreement staff may be transferred to the Developer. This may require special agreement in terms of (a) a forced retrenchment programme (and who pays for it), and (b) the associated pension and other liabilities	Shared between Grantor and Developer
Interface with subcontractors		Developer
Monitoring of performance	The penalties to be imposed by the promoter and the method for withholding money from the Developer – or recovering money – must be fair and easily understood	Shared

(*Continued*)

Risk	Description	Proposed allocation
Standard of performance	Whether the service provided to users and to the benefit provided to the Grantor corresponds to the obligations set out any PPP agreement and at law	To Developer – as a general point, but subject to other aspects of risk allocation discussed here, for example change in law
Costs of compliance with general change in legislation and statutory requirements		Developer
Changes in scope of services due to Grantor requirements		Grantor
Change in scope of services due to Developer requirements		Developer
Danger to users through health and safety		Developer
Public liability		Grantor
Damage to infrastructure other than for Grantor default		Developer, shared
Cost of operation	Risk of increases in the cost of labour, utilities and the replacement of equipment (incl. inflation)	To Developer
Cost of maintenance	Risk of increases in the cost of maintenance and replacement of the works, and major maintenance (incl. inflation)	To Developer
3. Environmental		
i. Existing	Damage to the environment existing before commencement of work	To Grantor
ii. New		To Developer

(*Continued*)

Risk	Description	Proposed allocation
	Damage to the environment caused during or after commencement of work	
Noise	Excess noise causing noise pollution	To Developer
Stakeholder consultation	Risk that stakeholders have not been consulted and inadequate support is mustered to deal with urban disruptions during construction. Also Stakeholder interface risks such as adjacent property and land owners with differing objectives for the project	To Developer & Grantor
4. Financial		
Access to finance/financial close	Risk that the Developer is unable to attract the financing required for the project, at a sufficiently low cost	To Developer, before completion of the bidding process. The Grantor will use best efforts to facilitate financing and accelerate the due diligence process by lenders
Interest rate	Risk that the interest charged on financing increases	To Developer, though a Grantor will bear part of risk that interest rate changes between project award and financial close
Inflation	Risk that higher than forecast inflation increase project costs	To Developer, though a Grantor will bear part of risk that inflation rate changes between project award and financial close
Change in VAT and taxes		To Developer if applicable indiscriminately, to Grantor if change is limited in application
Forex	Risk that financing is obtained in foreign	To Developer, though this will need to be reviewed once clarity is achieved on

(*Continued*)

Risk	Description	Proposed allocation
	currency, but revenue stream is in local currency	currency of financing and availability of hedging
Tenor	Length of debt insufficient to achieve financial viability in the context of a long-term concession	To Developer, subject to the above review of the source of financing, for example if the Grantor decides that it would rather source debt in local currency, it may need to provide some support of debt tenor
5. Social/political/legal		
Protestor action	Disruption of construction or operation by individuals or groups against the project	To Grantor, unless caused by Developer act or omission
Legal challenge	Legal actions against the Grantor or the Developer that restrict the Developer's ability to build or operate the project	To Grantor, unless caused by Developer act or omission
Political interference	Governmental or political act interfering with the project or the Developer	To Grantor to the extent contrary to the concession agreement or law
Change in Law	Change in legislation, regulation or tax	To Grantor
Force majeure		Shared between Grantor and Developer
Nationalization/ expropriation	Governmental acts seizing the property of the Developer or the project	To Grantor

References & Bibliography

Abdala M. A. (1996). Welfare Effects of Buenos Aires' Water and Sewerage Services Privatization. Expectativa—Economic Consultants, Cordóba, Argentina. Available on line at http://www.aaep.org.ar/espa/anales/pdf/abdala.pdf (accessed 23 April 2012)

Alcázar M., Abdala L. and Shirley M. The Buenos Aires Water Concession (2001). http://www.econ.worldbank.org/view.php?topic=20&type=5&id=1065

Alcázar L., Xu L. C. and Zuluaga A. M. (2000). Institutions, Politics, and Contracts: The Attempt to Privatize the Water and Sanitation Utility of Lima, Peru. World Bank, Washington, DC. Available on line at http://wbln0018.worldbank.org/Research/workpapers.nsf/0/c75b0b2a6601d2a685256989006d7430/$FILE/WPS2478.pdf

Alexander I. (2000). Measuring the cost of capital: dealing with data problems. Processed. Circulated at December 2000 South Asia Forum for Infrastructure Regulation SAFIR training course.

Alexander I. (2002). Paying the Price: The True Cost of Public Provision of Services: Conference Background Paper: "Infrastructure Development—Private Solutions for the Poor: The Asian Perspective" October 2002. Available on line at http://www.ppiaf.org/conference/docs/Papers/paying%20the%20price.pdf

Al Farawati O. (2001). LEMA cites list of improvements, challenges after two years of managing Amman's water. *Jordan Times*, 16 July. Available on line at http://www.jordanembassyus.org/07162001002.htm

Alfaro R. (1997). Institutional Development of the Water and Sanitation Sector in Chile. Working paper UNDP—World Bank. Kennedy School of Government Case Study, Cambridge, MA.

Andres F., Guasch J. L. and Haven T. (2009). The Impact of Private Sector Participation in Infrastructure: Lights, Shadows, and the Road Ahead, World Bank 2009.

Anthony I. (1994). Regulation: Legal Form and Economic Theory. Clarendon Press, Oxford, UK.

Anwandter L. and Ozuna T., Jr. (2002). Can public sector reforms improve the efficiency of public water utilities? *Environment and Development Economics*, **7**, 687–700.

Araujo A. R. (1998). Procurement in Privately Provided Intrastructure (PPI) Projects Financed by the World Bank. Discussion Paper. Operational Core Services, Procurement Policy and Services Group. March. World Bank. http://rru.worldbank.org/Documents/Toolkits/Highways/pdf/85.pdf (accessed 30 November 2004)

Armstrong M., Cowan S. and Vickers J. (1995). Non linear pricing and price cap regulation. *Journal of Public Economics*, **58**(1), 33–55.

Armstrong Lawyers. (2009). Legal Terms, retrieved on 11 August 2009 http://www.armstronglawyers.com.au/index.php

Arrowsmith S. (1998). National and international perspectives on the regulation of public procurement: harmony or conflict? In: Public Procurement: Global Revolution, S. Arrowsmith and A. Davies (eds), Kluwer Law International, London.

Artana D., Nanajas F. and Urbiztondo S. (1999). Governance and regulation: a tale of two concessions in Argentina. In: Spilled Water: Institutional Commitment in the Provision of Water Services, W. Savedoff and P. Spiller (eds), Inter-American Development Bank, Washington, DC, pp. 197–248.

Asian Development Bank. (1993). Water Utilities Data Book. Asian Development Bank, Manila, Philippines.

Asian Development Bank. (1999). Handbook for the Economic Analysis of Water Supply Projects. Asian Development Bank, Manila, Philippines.

Asian Development Bank. (2000). Developing Best Practices for Promoting Private Sector Investment in Infrastructure: Water Supply. Asian Development Bank, Manila, Philippines. Available on line at http://www.adb.org/Documents/Books/Developing_Best_Practices/Water_Supply/default.asp#contents

Asian Development Bank. (2002). Beyond Boundaries: Extending Services to the Urban Poor. Asian Development Bank, Manila, Philippines.

Asian Development Bank. (2002d). Stakeholders and partnerships. In: Asian Development Bank. 2002 [Beyond Boundaries: Extending Services to the Urban Poor. Asian Development Bank, Manila, Philippines, Chapter Stakeholders and partnerships].

Asian Development Bank. (2003b). Chapter 16 – Civil Society Involvement. In: Asian Development Bank. 2003. [Asian Water Supplies]

Asian Development Bank. (2003c). Chapter 7 – Small-Scale Water Providers. In: Asian Development Bank. 2003. [Asian Water Supplies]

Bacon R. (1994). Restructuring the Power Sector: The Case of Small Systems. Public Policy for the Private Sector. Viewpoint Note no. 10. World Bank, Washington, DC. Available on line at http://rru.worldbank.org/viewpoint/HTMLNotes/10/10bacon.pdf

Babtie. (2005). Operator Roundtable Study on Impact of Imperfect Data–Final Summary Report. World Bank, Washington, DC.

Baietti A., Kingdom W. and van Ginneken M. (2006). Characteristics of Well-Performing Public Water Utilities. Water Supply & Sanitation Working Notes. Note No. 9. World Bank, Washington, DC.

Bajari P., MacMillan R. and Tadelis S. (2002). Auctions versus Negotiations in Procurement: An Empirical Analysis. Department of Economics. Stanford University, Palo Alto, CA.

Baker B. and Tremolet S. (2000). Utility Reform: Regulating Quality Standards to Improve Access for the Poor. Public Policy for the Private Sector. Viewpoint Note 219. World Bank, Washington, DC. Available on line at http://rru.worldbank.org/viewpoint/HTMLNotes/219/219Baker-10-24.pdf

Bakovic T., Tenenbaum B. and Woolf F. (2003). Regulation by Contract: A New Way to Privatize Electricity Distribution? World Bank, Washington, DC.

Ballance T. and Taylor A. (2001). Competition and Economic Regulation in Water: The Future of the European Water Industry. A report for Water UK by Stone and Webster Management Consultants.

Bank-Netherlands Water Partnership and World Bank. (2002). Private Sector Participation in Water Supply and Sanitation Services in Sub-Saharan Africa. Decision Makers Workshop. Dakar, Senegal. 13–15 February 2002. Available on line at http://wbln0018.worldbank.org/water/bnwp.nsf/files/Volume+1+-07-19-2002.pdf/$FILE/Volume+ 1+-07-19-2002.pdf

Barlow M. and Clarke T. (2002). Blue Gold. McClelland and Stuart, Toronto.

Basañes F., Uribe E. and Willig R. (eds) (1999). Can privatization deliver? Infrastructure for Latin America. Inter-American Development Bank, Washington, DC.

Bajari P., MacMillan R. and Tadelis S. (2002). Auctions versus Negotiations in Procurement: An Empirical Analysis. Department of Economics. Stanford University, Palo Alto, CA.

Beato P. (1997). *Participación del Sector Privado en los Sistemas de Agua Potable y Saneamiento: Ventajas, Riesgos y Obstáculos* (Private Sector Participation in Water and Wastewater systems: Advantages, Risks and Obstacles), Inter-American Development Bank, Washington, DC. Available on line at http://www.iadb.org/sds/IFM/publication/ publication_4876_157_e.htm

Beato P. and Vives A. (1996). Risk, fiscal, and efficiency issues in public–private arrangements for the provision of services. *Infrastructure*, **I**(3), 3–14.

Bechtel. (2002). *Cochabamba and the Aguas del Tunari Consortium.* Bechtel. Available on line at http://www.bechtel.com/PDF/Cochabambafacts.pdf

Benninga S. (2000). Financial Modeling, 2nd edn. MIT Press, Cambridge, MA.

Berger K. (2003). Renegotiation and adaptation of international investment contracts: the role of contract drafters and arbitrators. *Vanderbilt Journal of Transnational Law*, **36**(4), 1347.

Bertolini L. (2004). Regulating Utilities: Contracting Out Regulatory Functions. Public Policy for the Private Sector. Viewpoint Note 269. World Bank, Washington, DC. Available on line at http://rru.worldbank.org/PublicPolicyJournal/Summary.aspx?id=269

Bestani R. and Sagar A. (2004). *The Local Currency Financing Revolution.* Asian Development Bank, Mass, USA.

Bitrán G. A. and Valenzuela E. O. (2003). Water Services in Chile: Comparing Private and Public Performance. Public Policy for the Private Sector. Viewpoint Note 255. World Bank, Washington, DC. Available on line at http://rru.worldbank.org/Viewpoint/HTMLNotes/255/255Bitra-031103.pdf

Blöch H. (1999). The European Union Water Framework Directive: taking European water policy into the next millennium. *Water Science and Technology*, **40**(10), 67–71.

Blokland M., Braadbaart O. and Schwartz K. (eds.) (1999). Private Business, Public Owners: Government Shareholdings in Water Enterprises. Published for the Ministry of Housing, Spatial Planning, and the Environment of the Netherlands and Water Supply and Sanitation Collaborative Council, Netherlands.

Boardman A., Greenberg D., Vining A. and Weimer D. (2001). Cost–Benefit Analysis: Concepts and Practice, 2nd edn., Prentice-Hall, New Jersey.

Boland J. and Whittington D. (2000). The political economy of water tariff design in developing countries. In: Beyond Resource Wars: Scarcity, Environmental Degradation, and International Cooperation (Global Environmental Accord: Strategies for Sustainability and Institutional Innovation) (2001), D. Ariel (ed.), Massachusetts Institute of Technology, USA.

Bradbury R. (1992). Privatization of Natural Monopoly Public Enterprises. The Regulation Issue. Policy Research Working Paper 862. World Bank, Public Sector Management and Private Sector Development, Washington, DC.

Broches A. (1990). Arbitration Under the ICSID Convention. International Centre for Settlement for Investment Disputes, Washington, DC.

Brocklehurst C. and Janssens J. G. (2004). Innovative Contracts, Sound Relationships: Urban Water Sector Reform in Senegal. World Bank—Water Supply and Sanitation Sector Board Discussion Paper Series, Paper No. 1. World Bank, Washington, DC. Available on line at http://www.worldbank.org/watsan/pdf/WSS_Senegal.pdf

Brook P. and Cowen T. (2002). Deregulated private water supply. In: Reinventing Water and Wastewater Systems: Global Lessons for Improving Water Management, P. Seidenstat, D. Haarmeyer, S. Hakim (eds), Wiley Publishing.

Brook P. J. and Locussol A. (2001). Easing tariff increases: financing the transition to cost-covering water tariffs in Guinea. In: Contracting for Public Services: Output-based Aid and Its Applications [online], P. J. Brook and S. M. Smith (eds), World Bank, Washington DC.

Brook P. and Irwin T. (eds) (2003). Infrastructure for Poor People: Public Policy for Private Provision. World Bank, Washington, DC.

Brook P. and Smith S. (eds) (2001). Contracting for Public Services: Output-Based Aid and its Applications. World Bank, Washington, DC.

Brown A. C., Jon S. and Bernard T. with Gencer D. (2006). Handbook for Evaluating Infrastructure Regulatory Systems. The International Bank for Reconstruction and Development/The World Bank, Washington, DC.

Buckland R. and Fraser P. (2001). Political and regulatory risk in water utilities: beta sensitivity in the United Kingdom. *Journal of Business Finance and Accounting*, **28**(7–8), 877–904.

Budds J. and McGranahan G. (2003). Privatization and the provision of urban water and sanitation in Africa, Asia and Latin America. Human Settlements Discussion Paper Series (Water and Sanitation), International Institute of Environment and Development, London. Available at http://www.iied.org

Building Partnerships for Development in Water and Sanitation (BPD). (2001). Flexibility by Design: Lessons from Multi-Sector Partnerships in Water and Sanitation Projects. BPD, see other BPD reports on "Partnerships: Reports and Documents: Available on line at http://www.bpd-waterandsanitation.org/english/lessons/pdocs.asp, and "Partnerships: Practitioner Reports", available on line at http://www.bpd-waterandsanitation.org/english/lessons/ppractice.asp

Cabanero-Verzosa C. and Mitchell P. (2003). Communicating Economic Reform. World Bank, Washington, DC. Available on line at http://www.worldbank.org/developmentcommunications/Publications/carnegie3.pdf

Calabrese D. (2003). Public Communication Programs for Privatization: A Toolkit for World Bank Task Team Leaders and Clients. World Bank, Washington, DC: Available on line at http://www.worldbank.org/developmentcommunications/Publications/wb%20toolkit%20book%203.6.02.pdf

Calderón C. and Servén L. (2004). The Effects of Infrastructure Development on Growth and Income Distribution, Policy Research Working Paper, World Bank.

Camdessus M. (2003). Financing Water for All. Report of the World Panel on Financing Water Infrastructure.

Camdessus M. and Winpenny J. (2003). Financing Water For All. Report of the World Panel on Financing Water Infrastructure. World Water Council and Global Water Partnership. Available on line at http://www.gwpforum.org/gwp/library/FinPanRep.MainRep.pdf

Campbell-White O. and Bhatia A. (1997). Privatization in Africa. World Bank, Washington, DC. Available on line at http://www.worldbank.org/html/dec/Publications/Abstracts 97/09psd/psd23.html

Caplan K. and Jones D. (2001a). Partnership Analysis Report Series: La Paz/El Alto, Bolivia. Business Partners for Development, London.

Caplan K. and Jones D. (2001b). Partnership Analysis Report Series: Cartagena, Colombia. London: BPD Water & Sanitation.

Center for Public Integrity. (2003). The Water Barons: How a Few Powerful Companies are Privatizing Your Water. The International Consortium of Investigative Journalists, Washington, DC. A Project of the Center for Public Integrity.

Continental can Case OJ. (1972). L17 and Case 27/76United Brands v Commission (1978) ECR 207.

Chapman R. and Cuthbertson S. (1999). Sydney's Water: A Suitable Case for Private Treatment? Public Policy for the Private Sector. Viewpoint Note 80. World Bank, Washington, DC. Available on line at http://rru.worldbank.org/viewpoint/HTML Notes/80/80cuthbe.pdf

Chong A. and Rama M. (2000). Do Separation Packages Need to be so Generous? Simulations for Government Employees in Guinea Bissau. World Bank, Washington, DC. Available on line at http://www.worldbank.org/research/projects/downsize/pdfs/chongrama.prn.pdf

Clarke G. and Wallsten S. J. (2003). Universal service: empirical evidence on the provision on infrastructure services to rural and poor urban consumers. In: Infrastructure for Poor People: Public Policy for Private Provision, P. J. Brook and T. Irwin (eds), World Bank Publications.

Clarke G., Kosec K. and Walsten S. (2004). Has Private Participation in Water and Sewerage Improved Coverage? Empirical Evidence from Latin America. AEI-Brookings Joint Center for Regulatory Studies, Washington, DC.

Clarke G., Ménard C. and Zuluaga A. M. (2002). Measuring the welfare effects of reform: urban water supply in guinea. *World Development*, **30**(9), 1517–1537.

Coelli T., Estache A., Perelman S. and Trujillo L. (2003). A Primer for Efficiency Measurement for Utilities and Transport Regulators. World Bank Institute, Washington, DC.

Colin J. and Lockwood H. (2002). *Making Innovation Work Through Partnerships in Water and Sanitation Projects*. Business Partners for Development, London, UK. Available on line at http://www.bpd-waterandsanitation.org/english/docs/innovationsumm.pdf

Comisión de Regulación de Agua Potable Y Saneamiento. (2004). Website. Comisión de Regulación de Agua Potable Y Saneamiento: Cartagena. Available on line at http://www.cra.gov.co/htm/sec_quienes.htm

Competition Commission. (2000a). Competition Act 1998: Application to the Water and Sewerage Sectors. Competition Commission, UK. Available on line at http://www.ofwat.gov.uk/aptrix/ofwat/publish.nsf/AttachmentsByTitle/ca98application.pdf/$FILE/ca98application.pdf

Competition Commission. (2000b). ESMEE Fairbairn Lecture by Dr. Derek Morris, Chairman, Competition Commission: Competition Policy and Regulation in the UK: A New Era. Lancaster University, 29 November. Available on line at http://www.competition-commission.org.uk/our_peop/members/chair_speeches/html/chair.htm

Concession contract between Interagua and the City of Guyaquil, Ecuador.
Concession Contract between The City of Bucharest and APA Nova Bucuresti S.A, Privatization of Bucharest Water and Sanitation Services, Corporate Finance Services, IFC, Amy.
Cook P. and Kirkpatrick C. (1998). Privatisation, Enterprise Development and Economic Reform. Edward Elgar Publishing, Cheltenham.
Cook P. D. and Stevens J. (2004). Operator Round Table—Performance Reporting by Water and Sanitation Utilities on Consumer-Oriented Issues. World Bank, Washington, DC. Forthcoming.
Cook P. and Stevens J. (2005). Draft Final Report on Consumer-Oriented Reporting of Service Performance. Prepared for World Bank and Operator Roundtable. World Bank, Washington, DC.
Council Directive 75/440/EEC on Surface Water for Drinking Directive. (1975). O.J. (L 194) 26.
Council Directive 76/160/EEC on Bathing Water. (1976). O.J. (L 31) 1.
Council Directive 76/464/EEC on Dangerous Substances. (1976). O.J. (L 129) 23.
Council Directive 79/923/EEC on the Quality of Shellfish Water. (1979). O.J. (L 281) 47.
Council Directive 80/68/EEC on Groundwater. (1980). O.J. (L 20) 43.
Council Directive 91/271/EEC concerning Urban Waste Water Treatment. (1991). O.J. (L135) 40.
Council Directive 91/676/EEC concerning Nitrates. (1991). O.J. (L375) 1.
Cowan S. (1997a). Competition in the water industry. *Oxford Review of Economic Policy*, **13**(1), 83–92.
Cowan S. (1997b). Tight average revenue regulation can be worse than no regulation. *Journal of Industrial Economics*, **45**, 75–88.
Cowan S. (1998a). The use of economic instruments to regulate water pollution and abstraction. *Oxford Review of Economic Policy*, Winter, **14**(4), 40–49.
Cowan S. (1998b). Welfare consequences of tight price-cap regulation. *Bulletin of Economic Research*, **50**(2), 105–16.
Crampes C. and Estache A. (1996). Regulating Water Concessions: Lessons from the Buenos Aires Concession. Public Policy for the Private Sector. Viewpoint Note 91. World Bank, Washington, D.C. Available on line at http://rru.worldbank.org/viewpoint/HTML Notes/91/crampes91.pdf
Dailami M., Lipkovich I. and Van Dyck J. (1999). INFRISK: A Computer Simulation Approach to Risk Management in Infrastructure Project Finance Transactions. Policy Research Working Paper No. 2083. World Bank, Washington, DC.
Daly H. E. (ed.) (1973). Toward a Steady State Economy. Freeman, San Francisco, CA.
Daly H. E. (1974). The economics of the steady state. *American Economic Review*, March, 15–21.
Daly H. E. and Cobb J. B. (1989). For the Common Good. Beacon Press, Boston, MA.
Daly H. E. (1992). Allocation, distribution and scale: towards an economics which is efficient, just and sustainable. *Ecological Economics*, **6**(3), 185–193.
Daly H. E. (1996). Beyond Growth: the Economics of Sustainable Development. Beacon Press, Boston, MA.
Daly H. E. (1999). Ecological Economics and the Ecology of Economics. E Elgar Publications, Cheltenham.

David C. (2000). MWSS Privatization: Implications on the Price of Water, the Poor, and the Environment. Philippine Institute for Development Studies, Makati City, Philippines. Available on line at http://dirp4.pids.gov.ph/ris/pdf/pidsdps0014.pdf

Delmon J. (2001). Water Projects: A Commercial and Contractual Guide. Kluwer Law International, Boston, MA.

Delmon J. (2009). A Practical Guide to Private Participation in Infrastructure: Project Development and Project Finance. Kluwer Law International, World Bank Publications.

Department of the Environment. (1999). Instrument of Appointment by the Secretary for the Environment of Thames Water Utilities Limited as a Water and Sewerage Undertaker Under the Water Act 1989. Department of the Environment, London. Available on line at http://www.ofwat.gov.uk/aptrix/ofwat/publish.nsf/AttachmentsByTitle/lic_thames.pdf/$FILE/lic_thames.pdf

Department for International Development. (DFID), United Nations Development Programme (UNDP), and World Bank 1998. Workshop on Willingness to Pay for Drinking Water Supply and Sanitation Report. World Bank, Washington, DC.

Department of Treasury and Finance. (2001). *Overview*. Partnerships Victoria. Department of Treasury and Finance, State of Victoria, Australia. Available on line at http://www.partnerships.vic.gov.au

Department of Treasury and Finance. (2001). Risk Allocation and Contractual Issues: A Guide. Partnerships Victoria. Department of Treasury and Finance. State of Victoria, Australia. Available on line at http://www.partnerships.vic.gov.au

Department of Water Affairs and Forestry. (1997). South Africa Water Services Act. Available on line at http://www.dwaf.gov.za/default.asp

Diaz J. (2003). La Participacion del Sector Privado En Los Servicios De Agua Y Saneamiento En San Pedro Sula, Honduras. Inter-American Development Bank, Washington, DC.

Dinar A. and Subramanian A. (eds) (1997). Water Pricing Experiences, an International Perspective. World Bank, Washington, DC.

Dinar A. (ed.) (2000). The Political Economy of Water Pricing Reforms. World Bank, Washington, DC.

Directive 2000/60/EC of the European Parliament and of the Council of 23 October 2000 establishing a framework for Community action in the field of water policy. (2000) O.J. (L327) 1.

Dumol M. (2000). The Manila Water Concession: A Key Government Official's Diary of the World's Largest Water Privatization. World Bank—Directions in Development, Washington, DC.

EconOne Research. (2003). Output-Based Aid Arrangements for Private Sector Participation in Peri-Urban and Provincial Water Supply Systems. EconOne, Los Angeles, available at http://wbln0018.worldbank.org/ppiaf/activity.nsf/files/C101901-S-WTS-RF-KH-FR.pdf/$FILE/C101901-S-WTS-RF-KH-FR.pdf

EconOne Research, Environmental Resources Management and ESG International. (2003). Emerging Lessons in the Private Provision of Rural Infrastructure Services. A report for Public Private Infrastructure Advisory Facility. World Bank, Washington, DC. Available on line at http://www.ppiaf.org/Final%20Report%20Activity%20Pages/Reports/A022001-L-MS-BP-IW-Final%20Report%204-29-03.pdf

Ecuador, Ciudad de Guayaquil, Reglamento de Servicios.

Ehrhardt D. (2003). Impact of market structure on service options for the poor. In: Infrastructure for Poor People: Public Policy for Private Provision. P. J. Brook and T. C. Irwin (eds), World Bank, Washington DC.

Ehrhardt D. (2004). Operator Round Table—Final Report on Key Contract Provisions for Long Term PPP in the Water and Sanitation Sector. Washington, DC.: World Bank. Forthcoming.

Ehrhardt D. and Burdon R. (1999). Free Entry in Infrastructure. Policy Research Working Paper 2093. World Bank, Washington, DC. Available on line at http://rru.worldbank.org/Documents/95.pdf

Ehrhardt G., Halpern J. and O'Connor S. (2007). Economic Regulation of Urban Water and Sanitation Services: Some Practical Lessons. Water Sector Board Discussion Paper Series, Paper No. 9. World Bank, Washington, DC.

Ehrhardt D. and Irwin T. (2004). Avoiding Customer and Taxpayer Bailouts in Private Infrastructure Projects: Policy Toward Leverage, Risk Allocation, and Bankruptcy. World Bank Policy Research Working Paper 3274, April 2004. World Bank, Washington, DC.

Ehrhardt D. and Irwin T. (2004). Avoiding Customer and Taxpayer Bailouts in Private Infrastructure Projects: Policy Toward Leverage, Risk Allocation, and Bankruptcy. Processed.

El Serafy S. (1989). The proper calculation of income from depletable natural resources. In: Environmental Accounting for Sustainable Development, Y. J. Ahmad, S. El Serafy and E. Lutz (eds), The World Bank, Washington, DC.

Engel E., Fischer R. and Galetovic A. (2000). The Chilean Infrastructure Concessions Program: Evaluation, Lessons and Prospects for the Future. La Transformacion Economica de Chile Centro de Economia Aplicada, Santiago.

Environmental Resources Management, Stephen Myers Associates and Hydroconseil. (2003). Aggregation in Small Towns Water Supply and Sanitation. A report for the World Bank Town Water Initiative. Processed. World Bank, Washington, DC.

Environmental Resources Management. (2002). Emerging Lessons in Private Provision of Infrastructure Services in Rural Areas: Water and Electricity Services in Gabon. A report for Public Private Infrastructure Advisory Facility and World Bank. World Bank, Washington, DC. Available on line at http://rru.worldbank.org/ Documents/1506.pdf

Environmental Resources Management. (2003a). Cost Recovery in Water and Sanitation Projects. A report for Department for International Development: Knowledge and Research (KzR) Projects, London.

Environmental Resources Management. (2003b). European Union Water Initiative—Financing Strategy. A paper for the Department for International Department and the European Commission. Processed.

Environmental Resources Management. (2004). Contracting Out Utility Regulatory Functions. A report for the World Bank, London.

Esguerra J. (2003). The Corporate Muddle of Manila's Water Concessions. A report for WaterAid and Tear Fund. WaterAid, London. Available on line at www.wateraid.org.uk/other/startdownload.asp?openType=forced&documentID=379

Estache A., de Azevedo J. S. and Sydensticker E. (2000). Labor Redundancy, Retraining and Outplacement During Privatization: The Experience of Brazil's Federal Railway. Policy Research Working Paper No. 2460. World Bank, Washington, DC. Available on line at http://wbln0018.worldbank.org/Research/workpapers.nsf/0/96549e4f3809732a8525 69690050853e?OpenDocument

Estache A., Foster V. and Wodon Q. (2002). Accounting for Poverty in Infrastructure Reform: Learning from Latin America's Experience. World Bank Institute, Washington, DC.

Estache A. and Rossi M. A. (2002). How different is the efficiency of public and private water companies in Asia? World Bank Economic Review, **16**(1), 139–148. Washington, DC: World Bank. Available on line at http://wber.oupjournals.org/cgi/reprint/16/1/139.pdf

EEB Open Seminar on EU Water Policy under the WFD, 'Making the WFD Work', Brussels, Nov. 3–4, 2000 (visited Mar. 27, 2008). http://www.eeb.org/activities/water/EEB%20Seminar%20WFD

EU Cases and European Court of Justice decisions

Case 209-215, 218/78 Heintz van Landewijck v Commission. (1980). ECR 3125 at3250.

Case 62/86, AKZO v Commission. (1991) ECR I-3359.

Case 27/76, United Brands v Commission. (1978) ECR 273.

ECJ Decision 82/861, OJ. (1982) L 360/36.

Case 155/73, Sacchi. (1974) ECR 409.

Cases 188-199/80, France v Commission. (1982) ECR 2545.

EU Green Paper on Public-Private Partnerships and Community Law on Public Contracts and Concessions, COM (2004) 327 final Brussels, 30.4.2004.

EU Green Paper on Public-Private Partnerships and Community Law on Public Contracts and Concessions, COM (2005) 569 final, Brussels, 15.11.2005.

EU Communication from the Commission 'A European initiative for growth: Investing in networks and knowledge for growth and jobs', COM (2003) 690 final, 11 November 2003. This report was approved by the Brussels European Council on 12 December 2003.

EU – Commission Interpretative Communication on the application of Community law on Public Procurement and Concessions to Institutionalised Public-Private Partnerships (IPPP), Brussels, 05.02.2008 C(2007)6661.

European Freshwater Programme, 'Actions for Internalising Sustainability' (visited Mar. 26, 2008) http://www.wwffreshwater.org/initiatives/wfd

European Freshwater Programme, 'What role for water pricing' (visited Apr. 3, 2008) http://www.wwffreshwater.org/initiatives/wfd/integration

Environmental Issue Report No.19, Sustainable water use in Europe (visited Apr. 5, 2008) http://www.reports.eea.eu.int/Environmental_Issues/No_19/en

Environmental Assessment Report No.2, Integration of the Economy and the Environment. Finnerty, John D 1996. Project Finance: Asset-Based Financial Engineering. John Wiley and Sons, New York.

Florence B. B. and Llorente M. (2003). Negotiating and Letting a Contract: What Are the Consequences on Performance. French Ministry of Environment.

Flyvbjerg B. (2001). Making Social Science Matter: Why Social Inquiry Fails and How it can Succeed Again. Cambridge University Press, Cambridge, UK.

Flyvbjerg B., Mette Holm S. and Søren B. (2002). Underestimating costs in public works projects. Error or lie? *APA Journal*, Summer 2002, **68**(3), 279–295.

Flyvbjerg B., Nils B. and Werner R. (2003). Megaprojects and Risk, an Anatomy of Ambition. Cambridge University Press, United Kingdom, ISBN 0521 00946 4.

Foster V. (2001). Condominial Water and Sewerage Systems: Costs of Implementation of the Model, El Alto, Bolivia/Pilot Project. Water and Sanitation Program: Andean Region. Available on line at http://www.wsp.org/publications/and_elaltostudy.pdf.

Foster V., Gomez-Lobo A. and Halpern J. (2000). Designing Direct Subsidies for the Poor: A Water and Sanitation Case Study. Public Policy for the Private Sector. Viewpoint Note 211. World Bank, Washington, DC. Available on line at http://rru.worldbank.org/viewpoint/HTMLNotes/211/211foste.pdf

Fouchard G. (1999). Goldman on International Commercial Arbitration 22–29. In: E. Gaillard and J. Savage (eds), Kluwer Law International edition.

Franceys R. (1997). Private Sector Participation in the Water and Sanitation Sector: Private Waters? A Bias Towards the Poor, Occasional Paper No. 3. Department for International Development, London, UK. http://www.dfid.gov.uk/../Pubs/files/iudd_water3.pdf

Frangano F., Linares C., Lockwood H., Rivera D., Trevett A. and Yepes G. (2001). Strategic Paper No. 1: Case Studies on Decentralization of Water Supply and Sanitation Services in Latin America. Prepared for the USAID Bureau for Latin America and the Caribbean. Environmental Health Project, Washington, DC. Available on line at http://www.phishare.org/files/890_whole%20document.pdf

Galal A., Jones L., Tandon P. and Vogelsgang O. (1994). Welfare Consequences of Selling Public Enterprises: An Empirical Analysis. Oxford University Press, New York.

Gassner P. and Pushak N. Does the Private Sector Deliver on its Promises? Evidence from a Global Study in Water and Electricity Distribution, World Bank, December 2007. www.ppiafdev.org

Gentry B. and Abuyuan A. (1999). Global Trends in Urban Water and Wastewater Financing and Management: Changing Roles for the Public and Private Sectors. Yale School of Forestry and Environmental Studies for the Organization on Economic Cooperation and Development (OECD). April.

Global Expert Team on Public Private Partnerships. (2009), Report on Mexico.

Global Partnership on Output-Based Aid website Available on line at http://www.gpoba.org/

Goodland R. (2002). Sustainability: Human, Social, Economic and Environmental. This article is a sample from the Encyclopedia of Global Environmental Change. John Wiley & Sons, Ltd., Oxford. Found at http://eu.wiley.com/legacy/wileychi/egec/pdf/GA811-W.PDF

Gómez-Ibáñez J. (1996). Santiago's Sewerage. Kennedy School of Government, Cambridge, MA.

Gómez-Ibáñez J. (2003). Regulating Infrastructure: Monopoly, Contracts, and Discretion. Harvard University Press, Cambridge, MA.

Gómez-Lobo A., Foster V. and Halpern J. (2000). Better Household Surveys for Better Design of Infrastructure Subsidies. Public Policy for the Private Sector. Viewpoint Note 213. World Bank, Washington, DC. Available on line at http://rru.worldbank.org/viewpoint/HTMLNotes/213/213gomez.pdf

Gómez-Lobo A. and Sergio H. (2000). Broad Roads in a Thin Country: Infrastructure Concessions in Chile. Policy Research Working Paper, no. 2279. World Bank, Washington, DC.

Gómez-Lobo A. (2001a). Incentive-Based Subsidies: Designing Output-based Subsidies for Water Consumption. Public Policy for the Private Sector. Viewpoint Note 232. World

Bank, Washington, DC. Available on line at http://rru.worldbank.org/viewpoint/HTMLNotes/232/232Gomez-531.pdf

Gomez-Lobo A. (2001b). Making water affordable: output-based consumption subsidies in Chile. In: Contracting for Public Services: Output-based Aid and Its Applications. [online], P. J. Brook and S. M. Smith (eds), Washington, DC, USA.

Gotanda J. Y. (2003). Renegotiation and Adaptation Clauses in International Investment Contracts, Revisited, 36 Vanderbilt Journal of Transnational Law 1461.

Gray P. (1997). Colombia's Gradualist Approach to Private Participation in Infrastructure. Public Policy for the Private Sector. Viewpoint Note 113. World Bank, Washington, DC. Available on line at http://rru.worldbank.org/viewpoint/HTMLNotes/113/113gray.pdf

Gray P. (2001). Private Participation in Infrastructure: A Review of the Evidence. World Bank, Washington, DC. Available on line at http://rru.worldbank.org/strategy/discussions.asp

Gray P. and Irwin T. (2003a). Exchange Rate Risk. Public Policy for the Private Sector. Viewpoint Note 262. World Bank, Washington, DC. Available on line at http://rru.worldbank.org/viewpoint/HTMLNotes/262/262Gray-062703.pdf

Gray P. and Irwin T. (2003b). Exchange Rate Risk: Allocating Exchange Rate Risk in Private Infrastructure Projects. Public Policy for the Private Sector. Viewpoint Note 266. World Bank, Washington, DC. Available on line at http://rru.worldbank.org/viewpoint/HTMLNotes/266/266Gray-121203.pdf

Green J. (2003). Regulation and the Balancing of Competing Interests in England in Wales. A report for WaterAid and Tearfund. WaterAid, London. Available on line at www.wateraid.org.uk/other/startdownload.asp?openType=forced&documentID=378.

Green R. J. and Pardina M. R. (1999). Resetting Price Controls for Privatized Utilities: A Manual for Regulators. Economic Development Institute of the World Bank, Washington, DC.

Groom E., Halpern J. and Erhardt D. (2006). Explanatory Notes on Key Topics in the Regulation of Water and Sanitation Services. Water Supply and Sanitation Sector Board Discussion Paper Series. Paper No. 6. World Bank, Washington, DC.

Guasch J. L. and Spiller P. (1999). Managing the Regulatory Process: Designs, Concepts, Issues, and the Latin America and Caribbean Story. World Bank, Washington, DC.

Guasch J. L., Jean-Jacques L. and Stéphane S. (2005). Infrastructure Concessions in Latin America Government-led Renegotiation. World Bank Policy Research Working Paper 3749. World Bank, Washington, DC.

Guislain P. (1997). The Privatization Challenge: A Strategic, Legal, and Institutional Analysis of International Experience. World Bank, Washington, DC.

Gupta P., Lamech R., Mazhar F. and Wright J. (2002). Mitigating Regulatory Risk for Distribution Privatization: The World Bank Partial Risk Guarantee. World Bank, Washington, DC.

Gutierrez E., Calaguas B., Green J. and Roaf V. (2003). New Rules, New Roles: Does PSP Benefit the Poor? WaterAid and Tearfund, London.

Haarmeyer D. (1992). *Privatizing Infrastructure; Options for Municipal Water-Supply Systems*. Paper 151 on Public Policy Reason Institute Website: www.rppi.org/ps151.html (accessed 15 May 2011)

Hall D. and Lobina E. (2002). Water Privatisation in Latin America. Public Services International Research Unit, Greenwich, UK.

Hammond J. S., Keeney R. L. and Raiffa H. (1998). Even Swaps: A Rational Method for Making Trade-Offs. Harvard Business Review (March–April), Boston, MA.

Hanley N., Mourato S. and Wright R. (2001). Choice modeling approaches: a superior alternative for environmental evaluation? *Journal of Economic Surveys*, **15**(3), 453–557.

Harris C. (2003). Private Participation in Infrastructure in Developing Countries: Trends, Impacts, and Policy Lessons. World Bank Working Paper No. 5. World Bank, Washington, DC.

Harris C. (2004). Output Based Aid. World Bank Infrastructure Economics and Finance Presentation. April 16, 2004.

Harris C., Hodges J., Schur M. and Shukla P. (2003). Infrastructure Projects: A Review of Cancelled Projects. Public Policy for the Private Sector. Viewpoint Note 252. World Bank, Washington, DC. Available on line at http://rru.worldbank.org/Documents/252Harri-010303.pdf

Harrison A. n.d. (2009). Operator Round Table—Study on Impact of Imperfect Data. World Bank, Washington, DC. Forthcoming.

Helm D. (1994). British Utility Regulation: Theory, Practice and Reform. Oxford Review of Economic Policy, 10:17–39, Oxford, UK.

Hemming R. and Mansoor A. M. (1988). Privatization and Public Enterprises. Internal Paper No. 56, International Monetary Fund, Washington, DC.

Hodges J. (2003). Unsolicited Proposals: Competitive Solutions for Private Infrastructure. Public Policy for the Private Sector. Viewpoint Note 258. World Bank, Washington, DC. Available on line at http://rru.worldbank.org/Viewpoint/HTMLNotes/258/258Hodge-031103.pdf

Hodges J. T. and Dellacha G. (2007). Unsolicited Infrastructure Proposals: How Some Countries Introduce Competition and Transparency. PPIAF Working Paper No. 1. World Bank, Washington, DC.

Hoffman S. L. (1998). The Law and Business of International Project Finance. Kluwer Law International, Boston, MA.

Hubbard D. (2007). How to Measure Anything: Finding the Value of Intangibles in Business. John Wiley & Sons, 6, Canada.

Hubbard D. (2009). The Failure of Risk Management: Why It's Broken and How to Fix It. John Wiley & Sons, Canada.

Hudson Institute. (1999). The NAWC Privatization Study: A Survey of the Use of Public-Private Partnerships in the Drinking Water Utility Sector. The Hudson Institute, Indianapolis, IN.

Hunt S. and Shuttleworth G. (1996). Competition and Choice in Electricity. John Wiley & Sons Ltd., London.

Idelovitch E. and Ringskog K. (1995). Private Sector Participation in Water Supply and Sanitation in Latin America. The World Bank, Washington, DC.

Indepen Consulting. (2002). Water Merger Policy: Time for Review. A paper by Indepen Consulting in association with Accenture, London.

Inter-American Development Bank. (1998). Aguas del Illimani Water and Sanitation Project: Environmental and Social Impact Report. IADB, Manila, Philippines. Available on line at http://www.iadb.org/pri/projDocs/BO0172_R_E.pdf

International Institute for the Unification of Private Law (UNIDROIT). (1995). Principles of International Contracts, art **6.2**(2) 1994. 34 I.LM. 1067.

International Water Association. (2001). Water quality management in the coastal city in the period of considerable water consumption decrease. *IWA Water Science and Technology*, **44**(2–3), 337–342. Available online at http://www.iwaponline.com/wst/04402/ws044020337.htm

Irwin T., Klein M., Perry G. E. and Thobani M. (1997). Dealing with Public Risk in Private Infrastructure. World Bank, Washington, DC.

Irwin T. (2003). Public Money for Private Infrastructure: Deciding When to Offer Guarantees, Output-Based Subsidies, and Other Fiscal Support. World Bank Working Paper No. 10, World Bank, Washington, DC.

Johnson R. A., McCormally J. and Moore A. T. (2002). Long-term Contracting for Water and Wastewater Services. Reason Foundation, Los Angeles, CA.

Johnstone N. and Wood L. (ed.). (2001). Water and Sanitation in Low Income Neighborhoods: The Scope for Service Differentiation and Decentralized Management in Private Firms and Public Water: Realising Social and Environmental Objectives in Developing Countries. Edward Elgar, Cheltenham, UK.

Johnstone N., Wood L. and Hearne R. (2007). Private Sector Participation in Urban Water and Sanitation, 23 NRF 287.

Joskow P. (1974). Inflation and environmental concern: structural change in the process of public utility price regulation. *Journal of Law and Economics*, **17**, 291–327.

Kariuki M. and Alocor G. (2000). Delivery of Water Supply to Low-Income Urban Communities through the Teshie Tanker Owners Association: A Case Study of Public-Private Initiatives in Ghana. Infrastructure for Development: Private Solutions and the Poor, 31 May–2 June 2000, London, UK.

Katalyst Solutions in association with Castalia. (2003). Private Sector Participation in Infrastructure: How Can Poor Households in Developing Countries Benefit? Draft Report. A report for KfW.

Kaufman D. (1997). Listening to Stakeholders' Views about Their Development Challenges and World Bank Instruments. Harvard University and the World Bank, Washington, DC.

Kelman S. (1990). Procurement and Public Management. AEI Press, Washington, DC.

Kennedy School of Government. (1999). Aguas de Cartagena: The Privatization of Water in Cartagena, Colombia. Kennedy School of Government, Cambridge, MA.

Kerf M., Gray D., Irwin T., Levesque T. and Taylor R. (1998). Concessions for Infrastructure: A Guide to Their Design and Award. World Bank Technical Paper no. 399. World Bank, Washington, DC. Available on line at http://rru.worldbank.org/Toolkits/Documents/Concessions/cover.pdf

Kerf M. (2000). Do State Holding Companies Facilitate Private Participation in the Water Sector? Evidence from Cote d'Ivoire, the Gambia, Guinea and Senegal. Policy Research Working Paper No. 2513. World Bank, Washington, DC. Available on line at http://rru.worldbank.org/Documents/68.pdf

Kikeri S. (1998). Privatization and Labor: What Happens to Workers When Governments Divest? World Bank Technical Paper No. 396. World Bank, Washington, DC. Available on line at http://rru.worldbank.org/Documents/336.pdf

Kingdom B. (2002). Alternative Bidding and Regulatory Design of Concessions: Concept Note. Operator Roundtable Working Paper 28. World Bank, Washington, DC.

Kingdom W. (1999). Benchmarking Water and Sanitation Utilities: A Start-Up Kit. World Bank, Washington, DC. Available on line at: http://www.worldbank.org/watsan/pdf/benchmarking.pdf

Klein M. and Hadjimichael B. (2003). The Private Sector in Development: Entrepreneurship, Regulation, and Competitive Disciplines. World Bank, Washington, DC.

Klein M. (1996). Economic Regulation of Water Companies. Policy Research Working Paper 1649. Private Sector Development Department. World Bank, Washington, DC. Available on line at http://wbln0018.worldbank.org/Research/workpapers.nsf/0/5256f12f0ddd430f852567e00055ade2?OpenDocument

Klein M. and Irwin T. (1996). Regulating Water Companies. Public Policy for the Private Sector. Viewpoint Note 77. World Bank, Washington, DC. Available on line at http://rru.worldbank.org/PublicPolicyJournal/Summary.aspx?id=77

Klein M. (1998a). Bidding for Concessions—The Impact of Contract Design. Public Policy for the Private Sector. Viewpoint Note 158. World Bank, Washington, DC. Available on line at http://rru.worldbank.org/viewpoint/HTMLNotes/159/159klein.pdf

Klein M. (1998b). Bidding for Concessions. World Bank Policy Research Working Paper No. 1957. World Bank, Washington, DC. Available on line at http://www.worldbank.org/html/dec/Publications/Workpapers/WPS1900series/wps1957/wps1957.pdf

Klein M. (1998c). The Case for Periodic Rebidding of Concessions. World Bank: Public Policy for the Private Sector. Viewpoint Note 161. World Bank, Washington, DC. Available on line at http://rru.worldbank.org/viewpoint/HTMLNotes/161/161klein.pdf/

Komives K. and Cowen P. B. (1998). Expanding Water and Sanitation Services to Low-Income Households. Public Policy for the Private Sector. Viewpoint Note 178. World Bank, Washington, DC. Available on line at http://rru.worldbank.org/viewpoint/HTMLNotes/178/178komiv.pdf

Komives K. (1998). Designing Pro-Poor Water and Sewer Concessions: Early Lessons from Bolivia. World Bank, Washington, DC. Available on line at http://wbln0018.worldbank.org/Research/workpapers.nsf/0/35aa5d266231 5b9b8525682d006c611e/$FILE/wps2243.prn.pdf

Laffont J.-J. and Tirole J. (1993). A Theory of Incentives in Procurement and Regulation. MIT Press, Cambridge, MA.

Lam P.-L. (2002). Hong Kong's Electricity Industry beyond 2008. The Hong Kong Polytechnic University, Hong Kong.

Landsburg S. (2003-03-03). Is your life worth $10 million?. Everyday Economics (Slate). http://www.slate.com/id/2079475/. Retrieved 2008-03-17.

Lee T. R. (1996). Alternatives for Private Participation in the Provision of Water Services, 20 NRF 333.

Levy B. and Spiller P. (1996). The Institutional Foundations of Regulatory Commitment: A Comparative Analysis of Telecommunications Regulation. Cambridge University Press, Cambridge.

Levy S. (1996). Build, Operate, Transfer: Paving the Way for Tomorrow's Infrastructure. J. Wiley and Sons, New York, NY.

Lewis C. M. and Mody A. (1997). The Management of Contingent Liabilities: A Risk Management Framework for National Governments. Chapter 6 in Dealing with Public Risk in Private Infrastructure. World Bank, Washington, DC.

Linares C. A. (2003). Institutions and the Urban Environment in Developing Countries: Challenges, Trends, and Transitions. Yale School of Forestry and Environmental Studies, United States of America.

Littlechild S. (2002). Competitive Bidding for a Long-Term Electricity Distribution Contract. Review of Network Economics 1 (No. 1, March).

Llorens F. and Soler-Couteaux P. (2001). Guide de l'Affermage du Service de Distribution d'Eau Potable. Association des Maires de France, Paris, France.

Lobina E. and Hall D. (2003). Problems with Water Concessions: A Review of Experience. Public Service International Research Unit, Greenwich, UK.

London Economics. (1997). Water Pricing: The Importance of Long Run Marginal Costs. A report for Ofwat. London, UK.

Lorrain D. (ed.) (1997). Urban Water Management: French Experience around the World. Levallois-Perret, Paris.

Lyonnaise des Eaux. (2010). Working Together for a Sustainable Future: How Public-Private Partnerships are Finding New Solutions to Water Needs. Lyonnaise des Eaux, Paris, France.

Mackie P., Nellthrop J. and Laird J. (2005). Notes on the Economic Evaluation of Transport Projects. Transport Economics, Policy and Poverty Thematic Group Transport Note No. TRN-2.Wa. World Bank, Washington, DC.

Mankiw N. G. (2006). Principles of Economics. 4th ed., Thomson/South-Western, London.

Mandri-Perrott C. (2003). Financing Water Issues in Recent International Privatizations. *Update Water Resources*, p. 125.

Mandri-Perrott C. (2005). Successful Divestiture of a Water Utility: The Case of Tallinn, Estonia, January 2005, Institute for Public Private Partnerships, http://www.ip3.org/pub/2005_publication_001.htm

Mandri-Perrott C. (2008b). Output-Based Aid in India: Community Water Project in Andhra Pradesh, http://www.gpoba.org/docs/OBApproaches21_IndiaWater.pdf

Mandri-Perrott C. (2009). Public Private Participation in the Water and Wastewater Sector: Developing Sustainable Legal Mechanisms. Water and Law Policy Series, IWA Publishing, United Kingdom.

Mandri-Perrott C. and Dominic P. (2007). Output-based aid in Colombia: Connecting poor households to natural gas service, http://www.gpoba.org/docs/OBApproaches17_ColombiaNatGas.pdf

Mann P. C. (1993). Water Utility Regulation: Rates and Cost Recovery. Reason Foundation Policy Study No. 155. Reason Foundation, Los Angeles. Available on line at http://www.rppi.org/ps155.html

Marek Swinarski M. (1999). The development of waste water treatment systems in Gdanski in 1871–1998. *European Water Management*, **2**(4), 69–76. Available on line at http://www.valt.helsinki.fi/projects/enviro/articles/Swina.pdf

Marin P. (2002). Output-Based Aid (OBA): Possible Applications for the Design of Water Concessions. International Finance Corporation: Private Sector Advisory Services, Washington, DC. Available on line at http://rru.worldbank.org/HotTopics/Hot_Topics_Marin.asp

Mariño M., Stein J. and Wulff F. (1998). Management Contracts and Water Utilities: The Case of Monagas State in Venezuela. Public Policy for the Private Sector. Viewpoint Note 166. World Bank, Washington, DC. Available on line at http://rru.worldbank.org/viewpoint/HTMLNotes/166%5C166marin.pdf

Markham B. (1993). Just Do It-Drafting the Arbitration Clause in an International Agreement, *10 Journal of International Arbitration* 4 (December 1993), 29–44.

Mas I. (1997). Managing exchange rate- and interest rate-related project exposure: Are guarantees worth the risk? In: Irwin, Klein, Perry, Thobani (eds), The World Bank Research Observer, Vol. 14, No. 2 (August, 1999), Oxford University Press, pp. 229–245.

Masons. (2003). Masons Water Yearbook: 2002–2003. Masons, London.

Matsukawa T., Sheppard R. and Wright J. (2003). Foreign Exchange Risk Mitigation for Power and Water Projects in Developing Countries. Energy and Mining Sector Board Discussion Paper No. 9. World Bank, Washington, DC. Available on line at http://www.worldbank.org/energy/pdfs/Energy_ExchangeCvr.pdf

McIntosh A. C. and Yñiguez C. E. (eds) (1997). Second Water Utilities Data Book: Asian and Pacific Region. Asian Development Bank, Manila, Philippines. Available on line at http://www.adb.org/Documents/Books/Second_Water_Utilities/default.asp

McPhail A. (1993). The five percent rule for improved water service: can households afford more? *World Development*, **21**(6), 963–973.

Megginson W. L. and Netter J. M. (2001). From State to Market: A Survey of Empirical Studies on Privatization. *Journal of Economic Literature*, **39**(2), 322–328.

Mehta M. (2003). Meeting the Financing Challenge for Water Supply and Sanitation: Incentives to Promote Reforms, Leverage Resources, and Improve Targeting. World Bank and Water and Sanitation Program. World Bank, Washington, DC. Available on line at http://www.wsp.org/FINANCE_REVIEW_PRESS.pdf

Mehta M., Virjee K. and Njoroge S. (2007). Helping a New Breed of Private Water Operators Access Infrastructure Finance. PPIAF Gridlines Note No. 25. World Bank, Washington, DC.

Ménard C. and Clarke G. (2002a). A transitory regime: Water supply in Conakry, Guinea. In: Public Economics and Regulation and Competition Policy, Shirley (ed.), World Bank, Development Research Group, Washington, DC.

Ménard C. and Clarke G. (2002b). Reforming water supply in Abidjan, Côte d'Ivoire: A mild reform in a turbulent environment. In Shirley (ed.), World Bank, Development Research Group, Trade, Washington, DC.

Ministre de l'Intérieur. (2001). Convention: Gestion Déléguée des Services d'Assainissement Liquids et de Distribution d'Eau Potable et d'Électricité de la Wilaya de Tanger. Ministère de l'Intérieur: Morocco.Ministry of Water and Irrigation. 2004. "Management Contract Directorate." Web site for the Ministry of Water and Irrigation. Ministry for Water and Irrigation, Jordan. Available on line at http://www.pmu.gov.jo/pmumc1.htm

Munasinghe M. (Lead Author); Munasinghe Institute for Sustainable Development (Content Partner); Cutler J. Cleveland (Topic Editor). (2008). Economic, social, and environmental elements of development. In: Encyclopedia of Earth, C. J. Cleveland (ed.), Environmental Information Coalition, National Council for Science and the Environment, Washington, DC. [First published in the Encyclopedia of Earth February 10, 2007; Last revised September 23, 2008; Retrieved August 22, 2009]. http://www.eoearth.org/article/Economic,_social,_and_environmental_elements_of_development

Mozammel M. and Zaktlokal B. (2002). Strategic Communication in PRSP. World Bank, Washington, DC. Available on line at http://www.worldbank.org/development communications/Publications/PRSP%20Strategic%20Comm.%20Chapter.pdf

National Economic Research Associates. (1997). A Guide to the Metropolitan Waterworks and Sewerage System Concession Agreements. A report to the Board of Trustees. National Economic Research Associates, London.

Nelson S. C. (1989). Alternatives to litigation of international disputes. *International Lawyer*, **23**(Spring), 187–206.

Neubauer D. W. and Meinhold S. S. (2007). Judicial Process: Law, Courts, and Politics in the United States. Thomson Wadsworth, Belmont.

Nevitt P. and Fabozzi F. (2001). Project Financing, 7th edn. American Educational Systems, New York.

NERA. (1997). A Guide to the Metropolitan Water Works and Sewerage System Concession Agreements, A Report to the Board of Trustees.

Nichols L. Monitoring Evaluation Plan Program Framework Indicators January 2004, Available at: http://www.ehproject.org/PDF/Activity_Reports/AR%20124%20WAWI%20M&E.pdf

Nickson A. (2001). Establishing and Implementing a Joint Venture: Water and Sanitation Services in Cartagena, Colombia. Working Paper # 442-03, GHK International, London. Available on line at http://www.ghkint.com/downloads/Cartagena.pdf

Nickson A. and Vargas C. (2002). The limitations of water regulation: the failure of the Cochabamba concession in Bolivia. *Bulletin of Latin American Research*, **21**(1), 99–120. Available on line at http://www.idd.bham.ac.uk/research/publications/ water_reg.pdf

Nickson A. and Franceys R. (2003). Tapping the Market—The Role of Government in Adjusting Economies. Palgrave Macmillan Press, New York.

Nouha H., Berradi M., Dinia M. and El Habati M. (2002). Gestion de la Demande en Eau: Partenariats Publics-Privé: Cas du Maroc: La Distribution de l'Eau Potable.

Office of Government Commerce. (2002). Standardisation of PFI Contracts: General (Revised 2002). HM Government, London, UK. Available on line at http://www.partnershipsuk.org.uk

Office of Government Commerce. (2005). Successful Delivery Toolkit (Version 5.0). HM Government, London, UK.

Office of Government Commerce. (2006). Competitive Dialogue Procedure. HM Government, London, UK.

Ofwtat. (1998). Levels of Service for the Water Industry in England and Wales 1997–1998 Report. Ofwat: Birmingham, UK. Available on line at http://www.ofwat.gov.uk/aptrix/ofwat/publish.nsf/AttachmentsByTitle/los98.pdf/$File/los98.pdf.Ogus

Ofwat. (2001). Leakage and Efficient Use of Water Report 2000-2001. Ofwat, Birmingham, UK. Available on line at http://www.ofwat.gov.uk/aptrix/ofwat/publish.nsf/AttachmentsByTitle/leakage.pdf/$FILE/leakage.pdf

Ofwat. (2003). Financial Performance and Expenditure of the Water Companies in England and Wales 2002-2003 Report. Ofwat: Birmingham, UK. Available on line at http://www.ofwat.gov.uk/aptrix/ofwat/publish.nsf/AttachmentsByTitle/fp_report2002-03.pdf/$FILE/fp_report2002-03.pdf

Ofwat. (2003b). Levels of Service for the Water Industry in England and Wales 2002–2003 Report. Ofwat: Birmingham, UK. Available on line at http://www.ofwat.gov.

uk/aptrix/ofwat/publish.nsf/AttachmentsByTitle/los2003_quest.pdf/$FILE/los2003_quest.pdf

Ofwat. (2003c). Security of Supply and the Efficient Use of Water 2002–2003 Report. Ofwat: Birmingham, UK. Available on line at http://www.ofwat.gov.uk/aptrix/ofwat/publish.nsf/AttachmentsByTitle/leakage_02-03.pdf/$FILE/leakage_02-03.pdf

Ofwat. (2004). Competition, Acquisitions, Mergers, and Company Structures. Web page for Ofwat. Available on line at http://www.ofwat.gov.uk/aptrix/ofwat/publish.nsf/Content/navigation-publications-competitionacquisitionsmergerscompanystructures

Organisation for Economic Co-operation and Development. (2000). Global Trends in Urban Water Supply and Waste Water Financing and Management: Changing Roles for the Public and Private Sectors. Organisation for Economic Co-operation and Development, Paris. Available on line at http://www.olis.oecd.org/olis/2000doc.nsf/c5ce8ffa41835d64c125685d005300b0/c125692700623b74c1256999005c9298/$FILE/00086209.PDF

Organisation for Economic Co-operation and Development. (2003). *Public-Private Partnerships in the Urban Water Sector*. Organisation for Economic Co-operation and Development Policy Brief, April 2003. Organisation for Economic Co-operation and Development, Paris, France. Available on line at http://www.oecd.org/dataoecd/31/50/2510696.pdf

Operators Roundtable. (2002). Discussion Paper on Key Contractual Provisions. Operator Roundtable Working Paper.

Palmer K., Cockburn M., Storer D. and Hulls D. (2003). Funding Johannesburg—Beyond the Rhetoric Delivering the Water and Sanitation Targets. Cambridge Economic Policy Associates, Cambridge.

Petrazzini B. A. (1996). Competition in Telecoms: Implications for Universal Service and Employment. Public Policy for the Private Sector. Viewpoint Note 96. World Bank, Washington, DC. Available on line at http://rru.worldbank.org/viewpoint/HTMLNotes/96/96ptrazzini.pdf

Phillips C. F. (1993). The Regulation of Public Utilities: Theory and Practice. Public Utilities Reports, Inc., Arlington, VA.

Picciotto R. and Wiesner E. (eds). (1998). Evaluation and Development. The Institutional Dimension. Transaction Publishers, New Brunswisk, NJ.

Pickering N. (1998a). Chile's Water System (A): The Privatization Debate. Kennedy School of Government, Cambridge, MA.

Pickering N. (1998b). Chile's Water System (B): Regulatory Reform. Kennedy School of Government, Cambridge, MA.

Pickrell D. H. (1992). A desire named streetcar. *Journal of North American Planning Association*, **58**(2), 158–176.

Platts. (1999). Infrastructure: French Win in Amman. Global Water Report. Issue 65.

Platts. (2001). Privatisation: No Way Back for Jordan. Global Water Report. Issue 126. Available on line at http://uknewsletters.platts.com

Plummer J. (2000). Favourable Policy and Forgotten Contracts: Private Sector Participation in Water and Sanitation Services in Stutterheim, South Africa. Working Paper 442 01. GHK International, London. Available on line at http://www.ghkint.com/pub_ wp1.htm

Plummer J. and Nhemachena G. (2001). Preparing a Concession—Working Towards Private Sector Participation in Water and Sanitation Services in Gweru, Zimbabwe. Working Paper 44204. GHK International, London.

Plummer J. (2002). Focusing Partnerships: A Sourcebook for Municipal Capacity Building in Public-Private Partnerships. Earthscan Publications, Ltd., London.

PNSO, Philippines National Statistics Office. Cencus of Population 2007. April 2008. http://www.census.gov.ph/data/census2007/index.html (accessed September 12, 2008).

Poirier J. P. (2002). *Developing a Fair and Transparent Bid Process*. Operator Roundtable Working Paper. World Bank, Washington, DC.

Poirier J. P. (2002b). Early Years of Concession Contracts: Elements of Design of a Baseline Data Revision Period. Operator Roundtable Working Paper. World Bank, Washington, DC.

President of the Philippines. (1994). Republic Act No. 7718: An Act Amending Certain Sections of Republic Act No. 6957, Entitled 'An Act Authorizing the Financing, Construction, Operation and Maintenance of Infrastructure Projects by the Private Sector, and for Other Purposes. Government of the Philippines, Manila.

Presidency Conclusions Helsinki European Council. (visited 10 and 11 December 2008). http://europa.eu.int/rapid/start/cgi/guesten.ksh?p_action.gettxt=gt&doc=DOC/99/16|0|RAPID&lg=EN

Pricewaterhouse Coopers. (2001a). Case Study: Sofia Water. Privatisation International. Water Sector Report 23. Pricewaterhouse Coopers. Available on line at http://www.pwcglobal.com/uk/eng/about/svcs/pfp/Sofia_Water_JGibbs_Feb_2001.pdf

Pricewaterhouse Coopers. (2001b). Forms of Private Sector Participation in the Water Sector. Pricewaterhouse Coopers. Available on line at http://chhattisgarh.nic.in/opportunities/Annexure%202.4.pdf

Pricewaterhouse Coopers. (2004). Operator Round Table—Improving Concession Bidding and Design in Water Supply and Sanitation: Risk Mitigation. World Bank, Washington, DC. Forthcoming.

Pricewaterhouse Coopers Securities LLC, IFI Risk Mitigation Instruments and Direct Sub-Sovereign Lending, Final Report, 2004.

Pritchard R. (1996). Economic Development, Foreign Investment and the Law: Issues of Private Sector Involvement, Foreign Investment and the Rule of Law in a New Era. International Bar Association Series. Kluwer Law International and International Bar Association, Boston.

Project Finance. (2003). PPP/Municipal Finance. Project Finance Magazine, April 1, 5.

Project Finance. (2005). Deal Analysis DLR Woolwich: Fab prefab. Project Finance Magazine, October 1, 12.

Public Citizen. (2003). Water Privatization Fiascos Website cases on Buenos Aires, Manila, Cochabamba, Atlanta, Jakarta, Nelspruit, and the United Kingdom. Available on line at http://www.citizen.org/cmep/Water/cmep_Water/fiascos/

Public Private Infrastructure Advisory Facility and Water and Sanitation Program. (2002a). Paper 1: Water Tariffs and Subsidies in South Asia: Understanding the Basics. Public Private Infrastructure Advisory Facility and Water and Sanitation Program. Washington, DC. Available on line at http://www.wsp.org/publications/Water%20Tariff%201_press_27th%20Feb.pdf

Public Private Infrastructure Advisory Facility and Water and Sanitation Program. (2002b). Paper 2: Water Tariffs and Subsidies in South Asia: A Scorecard for India. Public Private Infrastructure Advisory Facility and Water and Sanitation Program. World Bank, Washington, DC. Available on line at http://www.wsp.org/publications/Water%20Tariff%202_Press_27th%20Feb.pdf

Public Private Infrastructure Advisory Facility and Water and Sanitation Program. (2002c). Paper 3: Water Tariffs and Subsidies in South Asia: Tariffs in Six Cities in South Asia: Do They Target Subsidies Well?. Public Private Infrastructure Advisory Facility and Water and Sanitation Program. World Bank, Washington, DC. Available on line at http://www.wsp.org/publications/Water%20Tariff%203_press_27th%20Feb.pdf

Public Private Infrastructure Advisory Facility and Water and Sanitation Program. (2002d). New Designs for Water and Sanitation Transactions: Making Private Sector Participation Work for the Poor. Public Private Infrastructure Advisory Facility and World Bank. World Bank, Washington, DC. Available on line at http://www.wsp.org/publications/global_newdesigns.pdf

Public Private Infrastructure Advisory Facility and Water and Sanitation Program. (2003a). Paper 4: Water Tariffs and Subsidies in South Asia: Can Subsidies Be Better Targeted? Public Private Infrastructure Advisory Facility and Water and Sanitation Program. World Bank, Washington, DC. Available on line at http://www.wsp.org/publications/WaterTariff4_Press-27_05.pdf

Public Private Infrastructure Advisory Facility and Water and Sanitation Program. (2003b). Paper 5: Water Tariffs and Subsidies in South Asia: Do Current Water Subsidies Reach the Poor? Public Private Infrastructure Advisory Facility and Water and Sanitation Program. World Bank, Washington, DC. Available on line at http://www.wsp.org/publications/WaterTariff5_Press-27_05.pdf

Public Private Infrastructure Advisory Facility and World Bank Institute. (2002). Financial Modeling of Regulatory Policy: Theory and Practice CD-ROM. World Bank, Washington, DC.

Public Private Infrastructure Advisory Facility and World Bank. (2001). Toolkit: A Guide for Hiring and Managing Advisors for Private Participation in Infrastructure. World Bank, Washington, DC. Available on line from http://rru.worldbank.org/toolkits/hiring_advisors/index.htm

Public Private Infrastructure Advisory Facility and World Bank. (2002). Public Private Partnerships in Highways Toolkit. World Bank, Washington, DC. Available on line at http://rru.worldbank.org/toolkits/highways/

Public Private Infrastructure Advisory Facility and World Bank. (2004). *Labor Issues and Infrastructure Reform: A Toolkit*. World Bank, Washington, DC.

Public Services International Research Unit. (1997). Management of Joint Venture Water Companies. Public Services International Research Unit. Available on line at http://www.psiru.org/educ/resources/Res32.htm

Public Services International Research Unit. (2002). Water Privatisation and Restructuring in Central and Eastern Europe. Public Services International Research Unit, Greenwich. Available on line at http://www.psiru.org/reportsindex.asp

Rais J. C., Esquivel M. E. and Sour S. (2002). La Concesión De Los Servicios De Agua Potable Y Alcantarillado Sanitario En Tucumán Republica Argentina. Public Private Infrastructure Advisory Facility, Washington, DC. Available on line at http://wbln0018.worldbank.org/ppiaf/activity.nsf/files/A092099-M-WTS-RF-AR-FR.pdf/$FILE/A092099-M-WTS-RF-AR-FR.pdf

Rama M. (1999). Public sector downsizing: an introduction. *The World Bank Economic Review*, **13**(1), 1–22. Available on line at http://www.worldbank.org/research/journals/wber/revjan99/pdf/article1.pdf

Raymundie O. (1996). *La Gestion Deleguée des Services Publics en Europe (Le Cas de la Distribution d'Eau Potable): Un Modele Exportable?* Institut International de Droit du Développement. 18–19 April 1996, France.

Rees J. A. (1998). Regulation and Private Participation in the Water and Sanitation Sector, 22 NRF 95.

Research Triangle Institute. (2001). Willingness to Pay for Improved Water Supply in the Kathmandu Valley, Nepal: Final Report. Research Triangle Institute, North Carolina.

Ringskog K. (2000). Private Sector Participation in Water Supply and Wastewater: Opportunities and Risks. World Bank, Washington, DC.

Rivera D. (1996). Private Sector Participation in the Water Supply and Wastewater Sector: Lessons from Six Developing Country Case Studies. World Bank, Washington, DC.

Roche R., Revels C. and Amies M. (2001). Franchising in Small Town Water Supply. A report for the Bank-Netherlands Water Partnership. Washington, DC. Available on line at http://www.wsp.org/publications/smv_franchising_doc.pdf

Roth G. (1987). The Private Provision of Public Services in Developing Countries. World Bank, Washington, DC.

Ros A. J. (1999). Does ownership or competition matter? the effects of telecommunications reform on network expansion and efficiency. *Journal of Regulatory Economics*, **15**(1):65–92.

Saghir J., Sherwood E. and Macoun A. (1999). Management Contracts in Water and Sanitation: Gaza's Experience. Public Policy for the Private Sector. Viewpoint Note 177. World Bank, Washington, DC. Available on line at http://rru.worldbank.org/viewpoint/HTMLNotes/177/177saghi.pdf

Saghir J., Schiffler M. and Woldu M. (2000). Urban Water and Sanitation in the Middle East and North Africa Region: The Way Forward. The World Bank, Washington, DC. Available on line at http://wbln0018.worldbank.org/mna/mena.nsf/f34b224d37365b3f852567ee0068bd93/2421f467c2c0262685256951006660e9/$FILE/way-english.pdf

Salacuse J. W. (1991). Making Global Deals: Negotiating In the international Market Place. Houghton, Mifflin, Boston, p. 147.

Salacuse J. W. (2000). Direct foreign investment and the law in developing countries. *15 Foreign Invest. L.J.*, 382, 387.

Saltiel G. (2003). La Participación Del Sector Privado En Los Servicios De Agua Y Saneamiento En La Provincia De Salta En Argentina. Inter-American Development Bank, Washington, DC. Available on line at http://www.iadb.org/sds/doc/IFM%2DEstudio%5FAgua%5FArgentina%2DS.pdf

Saltiel G. and Cledan M.-P. (2008a). Output-Based Disbursements in Mexico: Transforming the Water Sector in Guanajuato (September 2008), http://www.gpoba.org/docs/OBApproaches20_Mexico_OBD.pdf

Santos E. C. (2003). Manila Water Supply Regulation. Presentation at the Third World Water Forum. 18 March 2003. Available on line at http://www.adb.org/Documents/Events/2003/3WWF/Cities_Santos.pdf

Savedoff W. and Spiller P. (1999). Spilled Water: Institutional Commitment in the Provision of Water Service. The Inter American Development Bank, Washington, DC.

Schwartz K. and Blokland M. (2002). Government-owned public limited companies in the dutch water sector. In: Reinventing Water and Wastewater Systems: Global Lessons for Improving Management, P. Seidenstat, D. Haarmeyer and S. Hakim (eds), John Wiley and Sons, Inc, New York.

Seidenstat P., Haarmeyer D. and Hakim S. (eds) (2002). Reinventing Water and Wastewater Systems: Global Lessons for Improving Water Management. John Wiley and Sons, New York.

Shirley M. (1998a). Book review: Armstrong, M., Cowan, S. and Vickers, J. regulatory reform: economic analysis and British experience. Information Economics and Policy, 10, September, 389–401.

Shirley M. (1998b). Performance contracts: A tool for improving public services? In: Evaluation and Development: The Institutional Dimension, Picciotto and Wiesner (eds), The World Bank, United States of America.

Shirley M. (1999). Bureaucrats in business: the roles of privatization versus corporatization in state-owned enterprise reform. *World Development*, **27**(1), 115–136.

Shirley M. (ed.) (2002). Thirsting for Efficiency: The Economics and Politics of Urban Water System Reform. Pergamon, Oxford, UK.

Shirley M. and Patrick W. (2000). Public versus Private Ownership: The Current State of the Debate. World Bank, Washington, DC. Available on line at http://econ.worldbank.org/files/1175_wps2420.pdf

Shirley M. and Xu L. C. (1996). Information, Incentives, and Commitment: An Empirical Analysis of Contracts between Governments and State Enterprises. Paper 1769. World Bank, Washington, DC. Available on line at http://www.worldbank.org/html/dec/Publications/Workpapers/WPS1700series/wps1769/wps1769.pdf

Shugart C. (1998). Regulation-by-Contract and Municipal Services: The Problem of Contractual Incompleteness. PhD dissertation, Harvard University.

Shugart C. (2002). Case Study: Sofia Water System Concession. A presentation for the 2nd Berlin Summer School on Private Participation in Infrastructure. Available on line at http://wip.tu-berlin.de/ppi/course_material/2002/presentations/p13-shugart-sofia_water_system_concession-case_study.pdf

Shugart C. n.d. Operator Round Table—Improving Concession Bidding and Design in Water Supply and Sanitation: Procedures for Selection of the Concessionaire. World Bank, Washington, DC, Forthcoming.

Singh B., Ramasubban R., Briscoe J., Griffin C. and Kim C. (1998). Willingness to Pay for Water in Kerala, India. Centre for Social and Technological Change, Bombay, India.

Smith W. (1997a). Covering political and regulatory risk: Issues and options for private infrastructure arrangements. In: Dealing with Public Risk in Private Infrastructure, Irwin, Klein, Perry and Thobani (eds), The World Bank, Washington, D.C.

Smith W. (1997b). Utility Regulators: Decisionmaking Structures, Resources, and Start-up Strategy. Public Policy for the Private Sector. Viewpoint Note 129. World Bank, Washington, DC. Available on line at http://rru.worldbank.org/viewpoint/HTMLNotes/129/129smith.pdf

Smith W. (1997c). Utility Regulators: Roles and Responsibilities. Public Policy for the Private Sector. Viewpoint Note 128. World Bank, Washington, DC. Available on line at http://rru.worldbank.org/viewpoint/HTMLNotes/128/128smith.pdf

Smith W. (1997d). Utility Regulators: The Independence Debate. Public Policy for the Private Sector. Viewpoint Note 127. World Bank, Washington, DC. Available on line at http://rru.worldbank.org/viewpoint/HTMLNotes/127/127smith.pdf

Smith W. (2003). Regulating infrastructure for the poor: Perspectives on regulatory system design. In: Infrastructure for Poor People: Public Policy for Private Provision, P. J. Brook and T. Irwin (eds), World Bank Publications.

Snijar F. and Syme H. (1998). Understanding the Cost Recovery Challenge: A Study of Ten Rural Communities North West Province and Northern Province. A DANIDA Report for the Department of Water Affairs and Forestry South Africa.

Sohail M. (2003). Putting Poor People First, Making Partnering Cork for All—A Guideline. Margraf Publishers, Weikersheim, Germany.

Sofia Concession Agreement. (23 December 1999). Municipality of Sofia and Sofijska Voda AD concession agreement in relation to the supply of water and wastewater services for the Municipality of Sofia.

Sofia Pre-qualification Information Memorandum in relation to the Private Sector Participation to operate the Sofia water supply and wastewater system, currently operated by the municipally owned company, ViK Sofia (January 1999), Bulgaria.

Sofia Concession Agreement. (23 December 1999). Municipality of Sofia and Sofijska Voda AD concession agreement in relation to the supply of water and wastewater services for the Municipality of Sofia.

Solis B., Cacnio F. and Inocencio A. (2003). Water Prices in Low Income Areas on the West Zone. Presentation to the Workshop on Water Services and the Urban Poor: the Power of Policies and Regulations. Manila, Philippines. 25–26 September 2003.

Solo T. M. (1998). Competition in Water and Sanitation: The Role of Small-Scale Entrepreneurs. Public Policy for the Private Sector. Viewpoint Note 165. World Bank, Washington, DC. Available on line at http://rru.worldbank.org/viewpoint/HTMLNotes/165/165solo.pdf

Sommer D. (2001a). Multi-Utilities: Policy: Promotion, Tolerance, or Control? Public Policy for the Private Sector. Viewpoint Note 228. World Bank, Washington, DC. Available on line at http://rru.worldbank.org/viewpoint/HTMLNotes/228%5C228Somme-327.pdf

Sommer D. (2001b). Multi-Utilities: Trends: Blurring Industry Boundaries. Public Policy for the Private Sector. Viewpoint Note 227. World Bank, Washington, DC. Available on line at http://rru.worldbank.org/viewpoint/HTMLNotes/227/227Somme-327.pdf

Standard and Poor. (2001). Water and Wastewater Utilities, Projects, and Concessions: Standard & Poor's Rating Criteria. Standard and Poor, New York. Available on line at http://www.waterinvestments.com/sponsors/SandP/WaterUtilitiesRatingCriteria.asp

Stiggers D. (1997). Trinidad & Tobago Private Participation in Water and Waste Water Services. Inter-American Development Bank Workshop, Washington, September 15–16, 1997. Washington, DC.

Stiggers D. (2004). Bankability: a framework to unlock the development potential of water and waste water utilities; PA Consulting Group, Viewpoint publication available at http://www.paconsulting.com/App/Templates/Publications

Stiggers D. (2007). Water Toolkit e-Learning Modules 1 to 8, From Approaches to Private Sector Participation in Water Services. A Toolkit. World Bank, Washington, DC, PPIAF, at http://www.worldbank/PPtoolkitwater

Stiggers D. and Cledan M.-P. (2007). Armenia Draft Report to the Asian Development Bank.

Stone & Webster Consultants. (2002). Depreciation Charges and Broad Equivalence at the 2004 Periodic Review: Setting Out a Way Forward. A report for Water UK.

Suleiman R. (2002). Privatisation of Jordan's Capital Water Utility: Assessment and Evaluation of Water Supply and Wastewater Services of Amman Governorate.

Master's thesis, Department of Land and Water Resources Engineering, Stockholm. Available on line at http://www.lwr.kth.se/Publikationer/PDF_Files/LWR_EX_2002_31.PDF

Superintendencia de Servicios Sanitarios. (2004). Website. Available on line at www.siss.cl.

Tallinn, Resolution no. 210, City of Tallinn (June 15 2000), Estonia.

Tallinn, Shareholders Agreement, AS Tallinna Vesi, (Draft 24 November 2000), Estonia.

Tallinn, City of Tallinn Tallinna Vesi. Services Agreement for (a) ensuring the functioning maintenance of the public water supply sewerage system of Tallinn (b) extraction of fire fighting water from fire hydrants in the public water supply system and extraction of water from public water extraction points (c) conduct of rain water drainage water other soil surface water from public roads streets, squares to the public sewerage system treatment thereof. (Draft 24 November 2000), Estonia.

Tallinna Linn AS Tallinna Vesi and [Investor], Share Sale and Subscription Agreement, Final Draft 24 November 2000, Estonia.

Tallinn Public Water Supply Sewerage Act February 10th 1999 (entered into force 5 March 1999).

Tallinn, Estonia Water Act May 11, 1994, Riigikogu Proceedings, TNN 1564 (1994).

The Economist. (2003). Priceless: A Survey of Water. 19 July. The Economist, London, UK.

Torres de Mästle C. and Izaguirre A. K. (2008). Recent Trends in Private Activity in Infrastructure: What the Shift Away from Risk Means for Policy. Gridlines series, No. 31, PPIAF, World Bank, Washington, DC.

Treasury H. M. (2003). The Green Book: Appraisal and Evaluation in Central Government. Treasury Guidance, London, United Kingdom.

Treasury H. M. (2004) Standardisation of PFI Contracts (version 3). H.M. Treasury. London, UK.

Trémolet S. (2002a). Can Private Multi Utilities Help Expand Service to Rural Areas?. Public Policy for the Private Sector. Viewpoint Note 248. World Bank, Washington, DC. Available on line at http://rru.worldbank.org/viewpoint/HTMLNotes/248/248Tremo-062802.pdf

Trémolet S. (2002b). Pro-Poor Regulation: Challenges and Implications for Regulatory Design. Conference Background Paper: Infrastructure Development: Private Solutions for the Poor: The Asian Perspective, October 2002. Available on line at http://www.ppiaf.org/conference/docs/Papers/Regulation.pdf

Trémolet S. (2002c). Rural Water Service: Is a Private National Operator a Viable Business Model?. Public Policy for the Private Sector. Viewpoint Note 249. World Bank, Washington, DC. Available on line at http://rru.worldbank.org/viewpoint/HTMLNotes/249/249Tremo-062802.pdf

Trémolet S. and Browning S. (2002). The Interface between Regulatory Frameworks and Tri-Sector Partnerships. A report for Business Partners for Development Research and Surveys Series. Available on line at http://www.bpd-waterandsanitation.org/english/docs/regulation1.pdf

Trémolet S., Browning S. and Howard C. (2002). Emerging Lessons in Private Provision of Infrastructure Service in Rural Areas: Water Services in Cote D'Ivoire and Senegal. A paper for Public Private Infrastructure Advisory Facility. World Bank, Washington, DC. Available on line at http://www.ppiaf.org/Final%20Report%20Activity%20Pages/ Emerging%20Lessons%20Page.htm

Tynan N. and Kingdom W. (2002). Effective Water Service Provision: Performance Targets for a Well-run Utility. World Bank, Washington, DC. Available on line at http://rru.worldbank.org/HotTopics/Hot_Topics_Tynan_Kingdom.asp

Tynan N. (2003). Returns to Scale in Water Systems in Developing Countries: Some Econometric Evidence. Processed, World Bank.

Tynemarch Systems Engineering and Stone & Webster Consultants. (2002). Capital Maintenance Planning: A Common Framework: Volume 1 Overview. UKWIR. Reference No. 02/RG/05/3, World Bank, Washington DC.

UNDP, World Bank Water, Sanitation Program. (1999). Willing to Pay but Unwilling to Charge: Do 'Willingness-to-Pay' Studies Make a Difference? UNDP, World Bank, and Water and Sanitation Program – South Asia, New Delhi, India. Available on line at http://www.wsp.org/publications/sa_willing.pdf

United Nations. (2003). United Nations Guidelines for Consumer Protection. Department of Economic and Social Affairs, United Nations, New York.

United Nations Conference on Trade and Development. (UNCTAD) www.unctad.org

Van den Berg C. (2000). Concessions: Who Wins, Who Loses, and What To Do About It. Viewpoint Note 217. World Bank, Washington, DC. Available on line at http://rru.worldbank.org/viewpoint/HTMLNotes/217/217Vande-10-23.pdf

Van Wynsberghe R. and Khan S. (2007). Redefining case study. *International Journal of Qualitative Methods*, **6**(2), Article 6. Retrieved from http://www.ualberta.ca/~iiqm/backissues/6_2/vanwynsberghe.htm (accessed 21 April 2012)

Van Ginneken M., Tyler R. and Tagg D. (2004). Can the Principles of Franchising Be Used to Improve Water Supply and Sanitation Services? A Preliminary Analysis. World Bank: Water Supply and Sanitation Sector Board Discussion Paper Series. Paper No. 2, January 2004. World Bank, Washington, DC. Available on line at http://wbln0018.worldbank.org/water/bnwp.nsf/files/WSS_Franchise_published.pdf/$FILE/WSS_Franchise_published.pdf

Van Ginneken M. and Kingdom W. (2008). Key Topics in Public Water Utility Reform. Water Working Notes. Note No. 17. World Bank, Washington, DC.

Veevers-Carter P. (2008). Output Based Aid, World Bank Infrastructure Economics and Finance Presentation, April 16, 2008.

Vickers J. and Yarrow G. (1988). Privatization: An Economic Analysis. The MIT Press, Cambridge, MA.

Vives A. (1997). Private infrastructure: ten commandments for sustainability. *The Journal of Project Finance*. Spring, 20–30.

Vives A. (2000). *Los fondos de Pensiones en el Financiamiento Privado de Proyectos de Infraestructura: Diseño de Regulaciones e Instrumentos* (Pension funds in the private financing of infrastructure projects: design rules and instruments). InterAmerican Development Bank, Washington, DC. Available on line at http://www.iadb.org/sds/doc/ IFM-119-S.pdf

Waelde T. W. and Kolo A. (2004). Renegotiation and Contract Adaptation in the International Investment Projects: Applicable Legal Principles and Industry Practices. Centre for Energy, Petroleum/Mineral Law & Policy, University Of Dundee, Scotland, UK.

Walker I., Ordonez F., Serrano P. and Halpern J. (2000). Pricing, Subsidies, and the Poor: Demand for Improved Water Services in Central America. Policy Research Working

Paper No. 2468. World Bank, Washington, DC. Available on line at http://econ.worldbank.org/files/1278_wps2468.pdf

Wallsten S. (2002). Does Sequencing Matter? Regulation and Privatization in Telecommunications Reform. Policy Research Working Paper # 2817. World Bank, Washington, DC. Available on line at http://econ.worldbank.org/files/13266_wps2817.pdf

Ward D. R. (2002). Water Wars: Drought, Flood, Folly and the Politics of Thirst. Riverhead Books, New York.

Water and Sanitation Division and the World Bank Institute (WBI). (1999). The Private Sector in Water: Competition and Regulation. World Bank, Washington, DC. Abstract available on line at http://www.worldbank.org/html/fpd/urban//mun_fin/toolkit/water_competition.html

Water and Sanitation Authority. (1995). Interim Operating Agreement. WASA: St. Joseph, Trinidad.

Water and Sanitation Authority. (1999). Interim Operating Agreement Improvements 1996–1999. WASA: St. Joseph, Trinidad.

Water and Sanitation Program. (2000). The Cancellation of the Pune Water Supply and Sewerage Project. Water and Sanitation Program: South Asia Region, New Delhi, India. Available on line at http://www.wsp.org/publications/sa_pune.pdf

Water and Sanitation Program. (2001). The Buenos Aires Concession: The Private Sector Serving the Poor. Water and Sanitation Program: South Asia, New Delhi, India. Available on line at http://www.wsp.org/publications/sa_buenos.pdf

Water and Sanitation Program. (2002a). Lao PDR: Water Supply and Sanitation: Development of Small Towns' Management Models. Water and Sanitation Program, Washington, DC. Available on line at http://www.wsp.org/publications/LAO_STS.pdf

Water and Sanitation Program. (2002b). Setting the Stage: The Process of Preparing for Pro-poor Water Sector Reforms in Kathmandu. Water and Sanitation Program, Washington, DC. Available on line at http://www.wsp.org/publications/psp_kathmandu_press.pdf

Water and Sanitation Program. (2002c). Vietnam: Evolving Management Models for Small Towns Water Supply in a Transitional Economy. Water and Sanitation Program, Washington, DC. Available on line at http://www.wsp.org/publications/eap_vietnam_models.pdf

Water and Sanitation Program. (2002d). Water Services in Small Towns in Africa: The Role of Small and Medium-Sized Organizations. Water and Sanitation Program, Washington, DC. Available on line at http://www.wsp.org/publications/af_bg_small.pdf

Water Utility Partnership for Capacity Building in Africa and Water and Sanitation Program. (2003). *Better Water and Sanitation for the Urban Poor: Good Practice from Sub-Saharan Africa*. Water Utility Partnership for Capacity Building and Water and Sanitation Program, Washington, DC.

Waterunc.com. (1999). Suez Lyonnaise des Eaux Wins Jordan's First Private Water and Wastewater Contract. Waterunc.com. Available on line at http://www.waternunc.com/gb/lyon24gb.htm

Webb M. and Ehrhardt D. (1998). Improving Water Services through Competition. Public Policy for the Private Sector. Viewpoint Note 164. World Bank, Washington, DC. Available on line at http://rru.worldbank.org/viewpoint/HTMLNotes/164/164 webb.pdf

Wenyon S. and Jenne C. (1999). Water and sewerage privatization and reform. In: Can Privatization Deliver? Infrastructure for Latin America, Basañes, Uribe, Willig, (eds), Inter-American Development Bank, 1993.

Whittington D., Lauria D. and Mu X. (1991). A study of water vending and willingness to pay for water in Onitsha, Nigeria. *World Development*, **19**(2/3), 179–198.

Whittington D. (1992). Possible adverse effects of increasing block water tariffs in developing countries. *Economic Development and Cultural Change*, **41**, 75–87.

Whittington D., Lauria D. and Kyeongae C. (1993). Household sanitation in Kumasi, Ghana: a description of current practices, attitudes and perceptions. *World Development*, **21**(5):733–748.

Whittington D. and Swarna V. (1994). *The Economic Benefits of Potable Water Supply Projects to Households in Developing Countries*. Asian Development Bank: Economics and Development Resource Center, Manila, Philippines.

Whittington D. (1998). Administering contingent valuation surveys in developing countries. *World Development*, **26**(1), 21–30.

Whittington D., Davis J., Miarsono H. and Pollard R. (2000). Designing a 'neighborhood deal' for urban sewers: a case study of semarang, Indonesia. *Journal of Planning Education and Research*, **19**(3):297–308.

Whittington D., Pattanayak S. K., Yang J.-C. and Kumar B. K. C. (2002). Do Households Want Privatized Municipal Water Services? Evidence from Kathmandu, Nepal. Economy and Environment Program for Southeast Asia (EEPSEA). Available on line at http://www.eepsea.org/ev.php?ID=8204_201&ID2=DO_TOPIC

Williamson B. and Mumssen Y. (1999). Economic Regulation of Network Industries. A Report for the New Zealand Treasury, Prepared by National Economic Research Associates: London. Available on line at http://www.treasury.govt.nz/workingpapers/2000/twp00-5.pdf

WHO and UNICEF. (2000). Global Water and Sanitation Assessment 2000 Report. WHO and UNICEF. Available on line at www.who.int/docstore/water_sanitation_health/Globassessment/Foreword1.htm

Willoughby, Infrastructure and the Millennium Development Goals, 2 October 2004.

Woodhouse E. (2003). The 'Guerra del Agua' and the Cochabamba concession: social risk and foreign direct investment in public infrastructure. 39 Stan. J. Int'l L. 295, Stanford University, Stanford, CA.

World Bank. (1994). World Development Report: Infrastructure for Development. Oxford University Press, Washington, DC. World Bank.

World Bank. (1997). Standard Bidding Documents—Supply and Installation of Plant and Equipment. World Bank, Washington, DC.

World Bank. (1999). Guidelines: Procurement under IBRD Loans and IDA Credits. January 1999. World Bank, Washington, DC.

World Bank. (1999a). *Guidelines: Procurement under IBRD Loans and IDA Credits.* , January 1999, World Bank, Washington, D.C.

World Bank. (2001). New Designs for Water and Sanitation Transactions: Making Private Sector Participation Work for the Poor. Water and Sanitation Program/PPIAF. World Bank, Washington, DC.

World Bank. (2002). Bolivia Water Management: A Tale of Three Cities. Précis. Number 222. World Bank, Washington, DC.

World Bank. (2004a). A Guide to Using Private Firms to Improve Water Services. World Bank, Washington, DC.

World Bank. (2004a). A guide to using private firms to improve water services" Draft as of July 2004.

World Bank. (2004b). Operational Guidance for World Bank Group Staff—Public and Private Sector Roes in Water Supply and Sanitation Services. World Bank Group and Water Supply and Sanitation Sector Board, World Bank, Washington, DC.

World Bank. (2004b). Operational Guidance for World Bank Group Staff [on] Public and Private Sector Roles in Water Supply and Sanitation Services. Water Supply and Sanitation Board, World Bank, Washington, DC.

World Bank. (2004c). PPIAF PPI Database, 2009. World Bank, Washington, DC.

World Bank International Finance Corporation Oxford University Press. (2004). Doing Business in 2004: Understanding Regulation. World Bank and Oxford University Press, Washington, DC.

World Bank. (2005). Infrastructure at the Crossroads: Lessons Learned from 20 Years of World Bank experience. World Bank, Washington, DC.

World Bank. (2006a). Integrated River Basin Management – From Concepts to Good Practice. Briefing Notes. Bank-Netherlands Water Partnership Program/World Bank Institute. World Bank, Washington, DC.

World Bank. (2006b). Approaches to Private Sector Participation in Water Services. A Toolkit. World Bank, PPIAF, Washington, DC.

World Bank. (2006c). Infrastructure and the World Bank: A Progress Report. World Bank, Washington, DC.

World Bank. (2008). Sustainable Infrastructure Action Plan FY 09–11. World Bank, Washington, DC.

World Bank. (2009). Country Classification. World Bank, Washington, DC.

Yates (1990). Drafting Force Majeure and Related Clauses, 3 J.C.L. pp. 186–187.

Yescombe E. R. (2002). Principles of Project Finance. Academic Press, Boston.

Yin R. (1994). Case Study Research: Design and Methods. 2nd edn., Sage, Thousand Oaks, CA.

Yin R. (2003). Case Study Research: Design and Methods. 3rd edn., Sage, Thousand Oaks, CA.

Yin R. (2004). The Case Study Anthology. Sage, Thousand Oaks, CA.

Lightning Source UK Ltd.
Milton Keynes UK
UKOW05n0059140913

217152UK00001B/63/P